SHELBY MUSTANG MUSCLE PORTFOLIO 1965-1970

Compiled By R.M. Clarke

ISBN 1 85520 1623

Brooklands Books Ltd.
PO Box 146, Cobham, KT11 1LG
Surrey, England.

Printed in Hong Kong

BROOKLANDS
BOOKS

BROOKLANDS ROAD TEST SERIES

AC Ace & Aceca 1953-1983
Alfa Romeo Alfasud 1972-1984
Alfa Romeo Alfetta Coupes GT. GTV. GTV6 1974-1987
Alfa Romeo Giulia Berlinas 1962-1976
Alfa Romeo Giulia Coupes Gold Portfolio 1963-1976
Alfa Romeo Giulia Coupes 1963-1976
Alfa Romeo Giulietta Gold Portfolio 1954-1965
Alfa Romeo Spider Gold Portfolio 1966-1991
Alfa Romeo Spider 1966-1990
Allard Gold Portfolio 1937-1959
Alvis Gold Portfolio 1919-1967
American Motors Muscle Cars 1966-1970
Armstrong Siddeley Gold Portfolio 1945-1960
Aston Martin Gold Portfolio 1972-1985
Austin Seven 1922-1982
Austin A30 & A35 1951-1962
Austin Healey 100 & 100/6 Gold Portfolio 1952-1959
Austin Healey 3000 Gold Portfolio 1959-1967
Austin Healey Sprite 1958-1971
Avanti 1962-1990
BMW Six Cylinder Coupes 1969-1975
BMW 1600 Col. 1 1966-1981
BMW 2002 1968-1976
BMW 316, 318, 320 Gold Portfolio 1975-1990
BMW 320, 323, 325 Gold Portfolio 1977-1990
Buick Automobiles 1947-1960
Buick Muscle Cars 1965-1970
Buick Riviera 1963-1978
Cadillac Automobiles 1949-1959
Cadillac Automobiles 1960-1969
Cadillac Eldorado 1967-1978
High Performance Capris Gold Portfolio 1969-1987
Chevrolet Camaro SS & Z28 1966-1973
Chevrolet Camaro & Z-28 1973-1981
High Performance Camaros 1982-1988
Camaro Muscle Portfolio 1967-1973
Chevrolet 1955-1957
Chevrolet Corvair 1959-1969
Chevrolet Impala & SS 1958-1971
Chevrolet Muscle Cars 1966-1971
Chevelle and SS 1964-1972
Chevy Blazer 1969-1981
Chevy EL Camino & SS 1959-1987
Chevy II Nova & SS 1962-1973
Chrysler 300 Gold Portfolio 1955-1970
Citroen Traction Avant Gold Portfolio 1934-1957
Citroen DS & ID 1955-1975
Citroen SM 1970-1975
Citroen 2CV 1949-1988
Shelby Cobra Gold Portfolio 1962-1969
Cobras and Cobra Replicas Gold Portfolio 1962-1989
Cobras & Replicas 1962-1983
Chevrolet Corvette Gold Portfolio 1953 1962
Corvette Stingray Gold Portfolio 1963-1967
Chevrolet Corvettee Gold Portfolio 1968-1977
High Performance Corvettes 1983-1989
Daimler SP250 Sport & V-8250 Saloon Gold Portfolio 1959-1969
Datsun 240Z 1970-1973
Datsun 280Z & ZX 1975-1983
De Tomaso Collection No.1 1962-1981
Dodge Charger 1966-1974
Dodge Muscle Cars 1967-1970
Excalibur Collection No.1 1952 1981
Facel Vega 1954-1964
Ferrari Cars 1946-1956
Ferrari Dino 1965-1974
Ferrari Dino 308 1974-1979
Ferrari 308 & Mondial 1980-1984
Ferrari Collection No.1 1960-1970
Fiat-Bertone X1/9 1973-1988
Fiat Pininfarina 124 + 2000 Spider 1968-1985
Ford Automobiles 1949-1959
Ford Bronco 1966-1977
Ford Bronco 1978-1988
Ford Consul. Zephyr Zodiac MkI & II 1950-1962
Ford Cortina 1600E & GT 1967-1970
Ford Fairlane 1955-1970
Ford Falcon 1960-1970
Ford GT40 Gold Portfolio 1964-1987
Ford RS Escorts 1968-1980
Ford Zephyr Zodiac Executive MkIII & MkIV 1962-1971
High Performance Capris Gold Portfolio 1969-1987
High Performance Escorts Mk1 1968-1974
High Performance Escorts Mk II 1975-1980
High Performance Escorts 1980-1985
High Performance Escorts 1985-1990
High Performance Fiestas 1979-1991
High Performance Mustangs 1982-1988
Holden 1948-1962
Honda CRX 1983-1987
Hudson & Railton 1936-1940
Jaguar and SS Gold Portfolio 1931-1951
Jaguar XK120 XK140 XK150 Gold Portfolio 1948-1960
Jaguar MkVII VIII IX X 420 Gold Portfolio 1950-1970
Jaguar Cars 1961-1964
Jaguar Mk2 1959-1969
Jaguar E-Type Gold Portfolio 1961-1971
Jaguar E-Type 1966-1971
Jaguar E-Type V-12 1971-1975
Jaguar XJ12 XJ5.3 V12 Glold Portfolio 1972-1990
Jaguar XJ6 Series II 1973-1979
Jaguar XJ6 Series III 1979-1986
Jaguar XJS Gold Portfolio 1975-1990
Jeep CJ5 & CJ6 1960-1976
Jeep CJ5 & CJ7 1976-1986
Jensen Cars 1946-1967
Jensen Cars 1967-1979
Jensen Interceptor Gold Portfolio 1966-1986
Jensen Healey 1972-1976
Lamborghini Cars 1964-1970
Lamborghini Countach Col No.1 1971-1982
Lamborghini Countach & Urraco 1974-1980
Lamborghini Countach & Jalpa 1980-1985
Lancia Stratos 1972-1985
Land Rover Series I 1948-1958
Land Rover Series II & IIa 1958-1971
Land Rover Series III 1971-1985
Land Rover 90 & 110 1983-1989
Lincoln Gold Portfolio 1949-1960
Lincoln Continental 1961-1969
Lincoln Continental 1969-1976
Lotus and Caterham Seven Gold Portfolio 1957-1989
Lotus Cortina Gold Portfolio 1963-1970
Lotus Elan Gold Portfolio 1962-1974
Lotus Elan Collection No.2 1963-1972
Lotus Elite 1957-1964
Lotus Elite & Eclat 1974-1982
Lotus Turbo Esprit 1980-1986
Lotus Europa Gold Portfolio 1966-1975
Marcos Cars 1960-1988
Maserati 1965-1970
Maserati 1970-1975
Mazda RX-7 Collection No.1 1978-1981
Mercedes 190 & 300SL 1954-1963
Mercedes 230/250/280SL 1963-1971
Mercedes Benz SLs & SLCs Gold Portfolio 1971-1989
Mercedes Benz Cars 1949-1954
Mercedes Benz Cars 1954-1957
Mercedes Benz Cars 1957-1961
Mercedes Benz Competion Cars 1950-1957
Mercury Muscle Cars 1966-1971
Metropolitan 1954-1962
MG TC 1945-1949
MG TD 1949-1953
MG TF 1953-1955
MG Cars 1959-1962
MGA & Twin Cam Gold Portfolio 1955-1962
MGB MGC & V8 Gold Portfolio 1962-1980
MGB Roadsters 1962-1980
MGB GT 1965-1980
MG Midget 1961-1980
Mini Cooper Gold Portfolio 1961-1971
Mini Moke 1964-1989
Mini Muscle Cars 1961-1979
Mopar Muscle Cars 1964-1967
Morgan Three-Wheeler Gold Portfolio 1910-1952
Morgan Cars 1960-1970
Morgan Cars Gold Portfolio 1968-1989
Morris Minor Collection No.1
Mustang Muscle Cars 1967-1971
Oldsmobile Automobiles 1955-1963
Old's Cutlass & 4-4-2 1964-1972
Oldsmobile Muscle Cars 1964-1971
Oldsmobile Toronado 1966-1978
Opel GT 1968-1973
Packard Gold Portfolio 1946-1958
Pantera Gold Portfolio 1970-1989
Panther Gold Portfolio 1972-1990
Plymouth Barracuda 1964-1974
Plymouth Muscle Cars 1966-1971
Pontiac Tempest & GTO 1961-1965
Pontiac Firebird and Trans-Am 1973-1981
High Performance Firebirds 1982-1988
Pontiac Fiero 1984-1988
Pontiac Muscle Cars 1966-1972
Porsche 356 1952-1965
Porsche Cars in the 60's
Porsche Cars 1960-1964
Porsche Cars 1964-1968
Porsche Cars 1968-1972
Porsche Cars 1972-1975
Porsche Turbo Collection No.1 1975-1980
Porsche 911 1965-1969
Porsche 911 1970-1972
Porsche 911 1973-1977
Porsche 911 Carrera 1973-1977
Porsche 911 Turbo 1975-1984
Porsche 911 SC 1978-1983
Porsche 914 Gold Portfolio 1969-1976
Porsche 914 Collection No.1 1969-1983
Porsche 924 Gold Portfolio 1975-1988
Porsche 928 1977-1989
Porsche 944 1981 1985
Range Rover Gold Portfolio 1970-1992
Reliant Scimitar 1964-1986
Riley 11/2 & 21/2 Litre Gold Portfolio 1945-1955
Rolls Royce Silver Cloud Gold Portfolio 1955-1965
Rolls Royce Silver Shadow 1965-1981
Rover P4 1949-1959
Rover P4 1955-1964
Rover 3 & 3.5 Litre Gold Portfolio 1958-1973
Rover 2000 + 2200 1963-1977
Rover 3500 1968-1977
Rover 3500 & Vitesse 1976-1986
Saab Sonett Collection No.1 1966-1974
Saab Turbo 1976-1983
Shelby Mustang Muscle Portfolio 1965-1970
Stubebaker Gold Portfolio 1947-1966
Stubebaker Hawks & Larks 1956-1963
Sunbeam Tiger & Alpine Gold Portfolio 1959-1967
Thunderbird 1955-1957
Thunderbird 1958-1963
Thunderbird 1964-1976
Toyota Land Cruiser 1956-1984
Toyota MR2 1984-1988
Triumph 2000. 2.5. 2500 1963-1977
Triumph GT6 1966-1974
Triumph Spitfire Gold Portfolio 1962-1980
Triumph Stag 1970-1980
Triumph Stag Collection No.1 1970-1984
Triumph TR2 & TR3 1952-60
Triumph TR4-TR5-TR250 1961-1968
Triumph TR6 Gold Portfolio 1969-1976
Triumph TR7 & TR8 1975-1982
Triumph Herald 1959-1971
Triumph Vitesse 1962-1971
TVR Gold Portfolio 1959-1990
Valiant 1960-1962
VW Beetle Collection No.1 1970-1982
VW Golf GTi 1976-1986
VW Karmann Ghia 1955-1982
VW Kubelwagen 1940-1975
VW Scirocco 1974-1981
VW Bus. Camper. Van 1954-1967
VW Bus. Camper. Van 1968-1979
VW Bus. Camper. Van 1979-1989
Volvo 120 1956-1970
Volvo 1800 Gold Portfolio 1960-1973

BROOKLANDS ROAD & TRACK SERIES

Road & Track on Alfa Romeo 1949-1963
Road & Track on Alfa Romeo 1964-1970
Road & Track on Alfa Romeo 1971-1976
Road & Track on Alfa Romeo 1977-1989
Road & Track on Aston Martin 1962-1990
Road & Track on Auburn Cord and Duesenburg 1952-1984
Road & Track on Audi & Auto Union 1952-1980
Road & Track on Audi 1980-1986
Road & Track on Austin Healey 1953-1970
Road & Track on BMW Cars 1966-1974
Road & Track on BMW Cars 1975-1978
Road & Track on BMW Cars 1979-1983
Road & Track on Cobra, Shelby & GT40 1962-1983
Road & Track on Corvette 1953-1967
Road & Track on Corvette 1968-1982
Road & Track on Corvette 1982-1986
Road & Track on Corvette 1986-1990
Road & Track on Datsun Z 1970-1983
Road & Track on Ferrari 1950-1968
Road & Track on Ferrari 1968-1974
Road & Track on Ferrari 1975-1981
Road & Track on Ferrari 1981-1984
Road & Track on Ferrari 1984-1988
Road & Track on Fiat Sports Cars 1968-1987
Road & Track on Jaguar 1950-1960
Road & Track on Jaguar 1961-1968
Road & Track on Jaguar 1968-1974
Road & Track on Jaguar 1974-1982
Road & Track on Jaguar 1983-1989
Road & Track on Lamborghini 1964-1985
Road & Track on Lotus 1972-1981
Road & Track on Maserati 1952-1974
Road & Track on Maserati 1975-1983
Road & Track on Mazda RX7 1978-1986
Road & Track on Mazda RX7 & MX5 Miata 1986-1991
Road & Track on Mercedes 1952-1962
Road & Track on Mercedes 1963-1970
Road & Track on Mercedes 1971-1979
Road & Track on Mercedes 1980-1987
Road & Track on MG Sports Cars 1949-1961
Road & Track on MG Sprots Cars 1962-1980
Road & Track on Mustang 1964-1977
Road & Track on Nissan 300-ZX & Turbo 1984-1989
Road & Track on Peugeot 1955-1986
Road & Track on Pontiac 1960-1983
Road & Track on Porsche 1961-1967
Road & Track on Porsche 1968-1971
Road & Track on Porsche 1972-1975
Road & Track on Porsche 1975-1978
Road & Track on Porsche 1979-1982
Road & Track on Porsche 1982-1985
Road & Track on Porsche 1985-1988
Road & Track on Rolls Royce & B'ley 1950-1965
Road & Track on Rolls Royce & B'ley 1966-1984
Road & Track on Saab 1955-1985
Road & Track on Toyota Sports & GT Cars 1966-1984
Road & Track on Triumph Sports Cars 1953-1967
Road & Track on Triumph Sports Cars 1967-1974
Road & Track on Triumph Sports Cars 1974-1982
Road & Track on Volkswagen 1951-1968
Road & Track on Volkswagen 1968-1978
Road & Track on Volkswagen 1978-1985
Road & Track on Volvo 1957-1974
Road & Track on Volvo 1975-1985
Road & Track - Henry Manney at Large and Abroad

BROOKLANDS CAR AND DRIVER SERIES

Car and Driver on BMW 1955-1977
Car and Driver on BMW 1977-1985
Car and Driver on Cobra, Shelby & Ford GT 40 1963-1984
Car and Driver on Corvette 1956-1967
Car and Driver on Corvette 1968-1977
Car and Driver on Corvette 1978-1982
Car and Driver on Corvette 1983-1988
Car and Driver on Datsun Z 1600 & 2000 1966-1984
Car and Driver on Ferrari 1955-1962
Car and Driver on Ferrari 1963-1975
Car and Driver on Ferrari 1976-1983
Car and Driver on Mopar 1956-1967
Car and Driver on Mopar 1968-1975
Car and Driver on Mustang 1964-1972
Car and Driver on Pontiac 1961-1975
Car and Driver on Porsche 1955-1962
Car and Driver on Porsche 1963-1970
Car and Driver on Porsche 1970-1976
Car and Driver on Porsche 1977-1981
Car and Driver on Porsche 1982-1986
Car and Driver on Saab 1956-1985
Car and Driver on Volvo 1955-1986

BROOKLANDS PRACTICAL CLASSICS SERIES

PC on Austin A40 Restoration
PC on Land Rover Restoration
PC on Metalworking in Restoration
PC on Midget/Sprite Restoration
PC on Mini Cooper Restoration
PC on MGB Restoration
PC on Morris Minor Restoration
PC on Sunbeam Rapier Restoration
PC on Triumph Herald/Vitesse
PC on Triumph Spitfire Restoration
PC on VW Beetle Restoration
PC on 1930s Car Restoration

BROOKLANDS HOT ROD 'MUSCLECAR & HI-PO ENGINE SERIES

Chevy 265 & 283
Chevy 302 & 327
Chevy 348 & 409
Chevy 350 & 400
Chevy 396 & 427
Chevy 454 thru 512
Chrysler Hemi
Chrysler 273, 318, 340 & 360
Chrysler 361, 383, 400, 413, 426, 440
Ford 289, 302, Boss 302 & 351W
Ford 351C & Boss 351
Ford Big Block

BROOKLANDS MILITARY VEHICLES SERIES

Allied Mil. Vehicles No.1 1942-1945
Allied Mil. Vehicles No.2 1941-1946
Dodge Mil. Vehicles Col. 1 1940-1945
Military Jeeps 1941-1945
Off Road Jeeps 1944-1971
Hail to the Jeep
Complete WW2 Military Jeep Manual
US Military Vehicles 1941-1945
US Army Military Vehicles WW2-TM9-2800

BROOKLANDS HOT ROD RESTORATION SERIES

Auto Restoration Tips & Techniques
Basic Bodywork Tips & Techniques
Basic Painting Tips & Techniques
Camaro Restoration Tips & Techniques
Chevrolet High Performance Tips & Techniques
Chevy-GMC Pickup Repair
Custom Painting Tips & Techniques
Engine Swapping Tips & Techniques
Ford Pickup Repair
How to Build a Street Rod
Mustang Restoration Tips & Techniques
Performance Tuning - Chevrolets of the '60s
Performance Tuning - Ford of the '60s
Performance Tuning - Mopars of the '60s
Performance Tuning - Pontiacs of the '60s

BROOKLANDS BOOKS

CONTENTS

ACKNOWLEDGEMENTS

Since we first put out a Brooklands Book on the Shelby Mustangs, several years ago, enthusiast interest in these cars has increased and a good deal of new material about them has come to light. When demand exhausted our stocks of this first volume, therefore, we decided to produce this much larger book on the cars, which contains some 40% more pages.

As with our earlier volume — and, indeed, all the Brooklands Books reprint series — we could not have published this one without the co-operation of those who own the original copyright to the material it contains. So we are pleased to acknowledge the debt we owe to the publishers of *Autocar, Auto Topics, Car and Driver, Car Collector, Car Craft, Car Life, Car Life Special, Cars, Classic and Sportscar, Complete Ford Book, High Performance Cars, Hot Rod, Motorcade, Motor Trend, Mustang, Road & Track, Road Test, Sports Car Graphic and Sports Car World.* Our thanks also go to motoring writer James Taylor for his brief introduction to this book.

R.M. Clarke

You can't keep a good man down. When Carroll Shelby was instructed by his doctors to give up motor racing in 1960, he was damned if he was going to keep away from the sport. If he couldn't drive competitively he could at least build cars for other people to drive. So it was that he established Shelby American Inc. in Los Angeles for exactly that purpose.

It wasn't long before Carroll Shelby was making his name all over again. In 1962, Shelby American took a British AC Ace roadster, crammed a Ford V8 under its shapely bonnet, and produced the first of the AC Cobras - a line of magnificent cars which would dominate SCCA A-Production racing for most of the 1960s.

Naturally, Shelby established close links with Ford management at Dearborn during this exercise, and it was only natural for Ford to turn to Shelby when it wanted to develop a hot version of its new Mustang. Shelby certainly came up with the goods. For 1965, he developed the legendary GT-350 and its lightweight racing derivative, the GT-350R, which took the National P-Production title in 1965, 1966 and 1967.

For 1966, the cars were little changed, with just the addition of an automatic transmission option and a rear seat for those who wanted it. For 1967, the GT-350 was supplemented by the GT-500 with its 428ci V8; but after that, things began to go downhill. Two things were happening. First, the stock Mustang was becoming more and more of a performance car (which it never had been when first introduced) and second, Ford was bowing to customer demand and making Shelby's roadburners more comfortable and heavier. New safety and exhaust emissions regulations for 1968 did the rest. Despite Cobra badging and the new King of the Road, designation for the GT-500KR, power and performance were coming down.

At the end of 1967, production was transferred from LA to Ionia, Michigan; and a year later, Ford took it in-house at its Southfield plant. For 1969, you could even get air conditioning as an option, which was a long way from Shelby's original concept. Production stopped in 1970, after only 10,825 Shelby Mustangs of all years had been built.

But you can't keep a good car down. In recent years, the Shelby Mustangs — particularly the early cars — have become collectable. More to the point, perhaps, they retain the fun value which they always had. This book helps to explain why they are regarded so highly today, more than a quarter of a century after the first one was built.

James Taylor

THE GT-350 MUSTANGS

The Shelby Touch Transforms the Car

BY JIM WRIGHT

There's a good chance future automotive historians and buffs alike will look back on the present day Cobras and Mustang GT-350s and equate them with the same names that we revere today: Mercer, Mercedes, Duesenberg, Alfa – or you name it. And for the same reasons: The cars look good, they go fast and they're constantly proving their mettle on the automotive battlefields of the world. In short, they are cars of character.

Shelby-American's latest offering – the Mustang GT-350 is one of the most exciting cars to hit the enthusiast's market in a long time. The modifications that have transformed it from production car to *gran turismo* embody much of the vast savvy and experience that Shelby's crew of designers, engineers, builders and racers have gained through long, hard years of building and preparing and racing high performance machinery in all the far corners of the world. The GT-350 appears to offer everything that one expects from a high performance machine and at a much more reasonable price than one is used to paying.

In a way, the GT-350 is more than just an exciting car. Since S-A functions in part as a kind of research and development arm of Ford's special vehicles department, the GT-350 is also an important step in the development of a car that's going to be around for a long time to come: the production Mustang. On the surface, it takes quite a few modifications to transform a stock Mustang into a thoroughbred GT-350. Closer inspection reveals that all the various component modifications and engineering could very easily be programed into the standard factory assembly line. So it's not at all unlikely that today's custom-built GT-350 might well be tomorrow's production-line high-performance Mustang.

The competition version of the GT-350 is showing great promise that it will make its name as familiar in the winner's circle as the Cobra has. Immediately after its February introduction a competition GT-350 was entered in the annual Southern Polar Prix at Green Valley Raceway in Texas. Running in B Production class, the car scored an impressive first-time-out victory. At Pomona, Calif., a month later, the car again scored a class win and ran third overall behind a Cobra and a Sting Ray in the big-bore production race. In the first Sports Car Club of America race of the year (Tucson, Ariz.) offering national points, the car decisively whipped a well-prepared E-Jag as it again won its class.

In the Texas race the car was driven by Ken Miles, Shelby's competition manager and the man largely responsible for the development of the GT-350's suspension. In the other two races the car was driven by Jerry Titus, popular West Coast automotive editor and race driver. Miles, who might be a little biased (understandably so), says that the GT-350 is one of the easiest handling race cars he's ever driven. Titus, who isn't at all biased, says just about the same thing. He feels that the car has all the power that it'll need to stay in front of SCCA Class B competition and that the suspension does a good job of transmitting the power to the ground. Both drivers agree that the GT-350's biggest assets are: (1) It's a completely predictable car that always lets the

PHOTOS: PAUL E. HANSEN, DEAN BATCHELOR, CARLYLE BLACKWELL, JIM WRIGHT

SPECIAL CAP on cast-alloy wheel hubs carries the Carroll Shelby insignia, identifying car as GT-350.

MORE SPECIAL touches are the pin-locks for the hood, the air scoop that scoops, and a small Mustang emblem on a blank grille.

driver know what's happening in advance; (2) It is a forgiving car that will make up for a lot of mistakes on the driver's part. Since both Miles and Titus are considered to be among the top sports car drivers in the country, those are pretty meaningful endorsements.

The biggest question concerning the GT-350 and the one that we most often hear from standard Mustang owners and potential GT-350 buyers is, "What makes the car so different from the standard Mustang?" The next most-asked question is, "Is it worth the added cost?"

Both questions can best be answered by casting an analytical eye at the product.

In appearance, the GT-350 still retains much of the original Mustang flair although there have been enough changes made to give it an identity of its own.

First off, all non-functional exterior trim is removed, leaving an extremely clean, uncluttered design. At the front the gunmetal anodized, stamped aluminum honeycomb grill insert is retained but the heavy crossbars, along with the large Mustang emblem, have been removed. A small Mustang tri-color medallion is carried, off-center to the left, on the grille insert.

One of the GT-350 design goals was to come up with a lighter car. By various means they've chopped about 200 lb. out of the car. The standard hood has been replaced by a fiberglass unit. A functional air scoop, molded into the center of the hood, supplies fresh, cool air to the carburetor. Standard hood hinges and springs are retained but the locking mechanism has been replaced by competition-type locking studs with safety pins. These are located at each forward corner of the hood. They allow a weight saving and also provide a positive method of keeping the hood down while traveling on rough roads and at high speeds.

Several interior modifications have been made to provide greater driver comfort and convenience. In keeping with the car's high-performance, dual-purpose character, there is room for only the driver and one passenger. The standard, folding rear seat has been replaced by a fiberglass shelf which mounts the spare tire and also provides luggage space (as well as saving a few pounds of weight). The spare is covered with a fitted vinyl cover which adds a finished touch.

Greater driver comfort is provided by the use of a wood-rimmed, competition steering wheel with aluminum spokes. This wheel has a smaller diameter than the standard Mustang wheel and also is formed with less "dish." This moves it about 2 in. closer to the dash and provides just that much more elbow room. This feature, combined with the fore and aft adjustment of the individual bucket seat, will give just about any size driver the leg and elbow room necessary to maintain precise and absolute control over the car. The bucket seats are well-shaped to provide a high degree of comfort and support. The padding is on the firm side and has a "just right" feel that will be appreciated on long trips.

Standard Mustang instruments are retained but are backed up by a tachometer and oil pressure gauge. The added instruments are carried in their own fiberglass pod which is

FINISHED IN WHITE with a pair of wide blue racing stripes and the triple-stripe trim on the rocker panel, the GT-350 looks fast and purposeful just parked along the roadside.

GT-350 Mustangs

mounted in the valley in the center of the top of the dashboard. They are angled slightly toward the driver to make them easier to read. A dash-mounted toggled switch operates the horn.

Heavy-duty, 3-in. wide nylon seat belts are installed. These are quick release competition units and are fully approved by all racing organizations. The belts are swivel mounted to heavy-duty eye-bolts.

In a later chapter we'll outline the engine modifications that are a part of this package. Other power-train modifications include the use of a Warner T-10, 4-speed, manual transmission, safety-strapped driveshaft and beefed-up rear axle with a self-locking differential.

The Warner gearbox was chosen because it has an aluminum case which makes it much lighter than the Ford-built 4-speed, and because of the availability of special ratios. The transmission is a close-ratio type with a 2.36:1 first gear, 1.62 second, 1.20 third and direct fourth.

A standard 2.75-in. diameter, high performance propeller shaft is used. S-A adds a strap-metal safety loop, located at the forward end of the shaft, to prevent the shaft from dropping to the ground in case of U-joint or shaft failure.

The rear axle assembly carries special axles and bearings and should be able to stand up to just about anything a GT-

INTERIOR TREATMENT of "street" GT-350 includes 3-in. wide safety belts, wood-rimmed steering wheel, electric tachometer and oil pressure gauge in pod atop dash, and back seat replaced by fiberglass cover.

NORMAL GT-350 engine has horsepower increased by bolt-on treatment, such as the high-riser intake manifold. Valve covers are for show. Competition oil pan, however, has added capacity for better cooling.

FRONT DISCS are standard Mustang items; upper arm mounting points are lowered; rear axle is fitted with Konis.

350 owner can hand it. The standard ratio is 3.89:1. Optional gear sets include ratios of 3.70, 4.11 and 4.33:1. The locking differential is standard equipment. Shelby's choice here is an excellent one: Detroit Automotive Products' "NoSPIN" unit. This one costs more than the average locking differential, but because of its positive engagement features and superior quality it will keep on doing its job after the average unit has failed.

Front wheel disc brakes are standard equipment on the GT-350 and work in conjunction with drums at the rear. The discs are identical to standard Mustang units except that they use a very special heavy-duty lining material that was developed for grand prix racing. At the rear, big 10-in. diameter by 2.5 in. wide cast-iron drums are fitted with metallic linings. The combination provides a high level of fade-

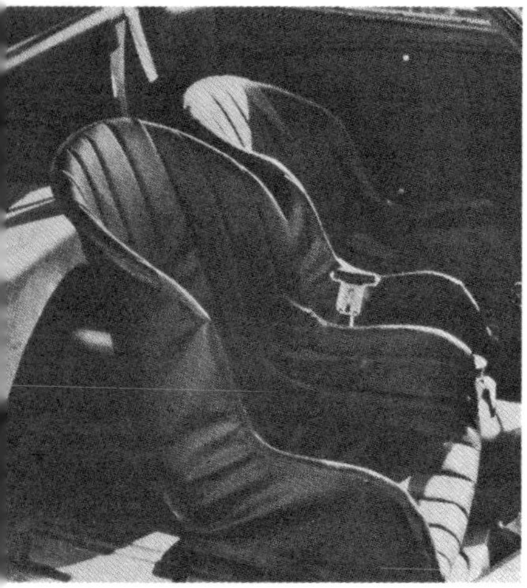

COMPETITION VERSION is much more stark; real bucket seats replace Detroit types; hefty roll bar is optional, but fully-instrumented dash is part of the race-ready package. All frills are removed.

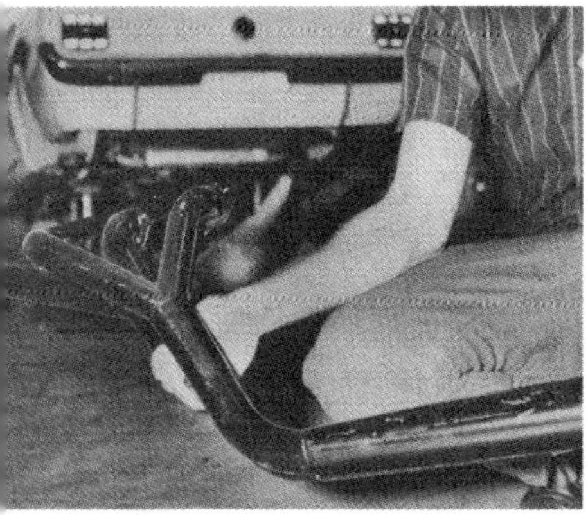

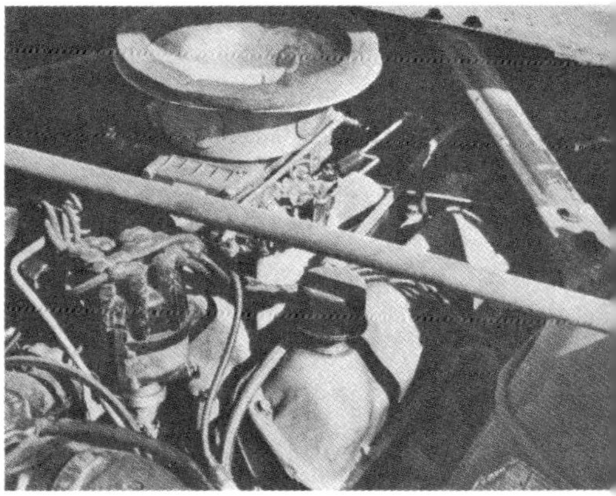

HUGE HEADERS are used with the racing-prepared engines. Hole in hood matches up with air horn atop the carburetor to give pressurized flow of cool air. Street engines have stock air cleaner; racing engines, air horns.

CONTROL ARMS are added to rear, extend forward into car body. Wider, segmented brakes are substituted for stock units.

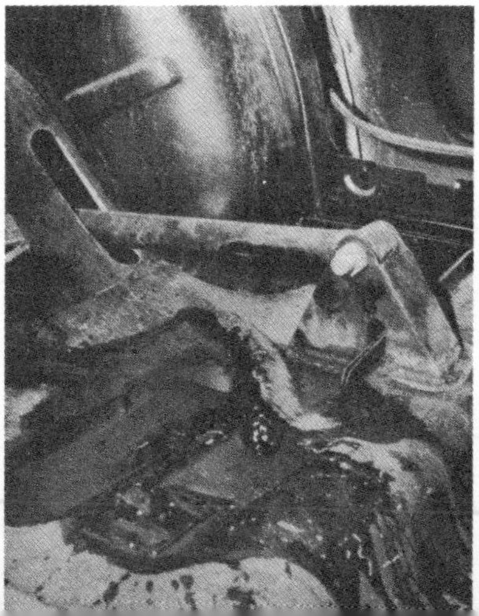

GT-350 Mustangs

proof braking power that is more than a match for the car's speed potential.

The most important and significant modifications are found in the chassis and suspension. Here Shelby's design goal was to obtain increased front and rear cornering power and stability to provide the GT-350 with a crisper handling feel along with quicker response characteristics. This obviously had to be accomplished without destroying the reputation that the original Mustang had gained as the car that had predictable handling characteristics.

On paper, the necessary modifications look pretty simple, but actually they represent untold hours of computation, experimentation and test driving. At the front, it was found that only one major physical alteration would be required to produce the desired results. This was the relocation of the inner pivots of the upper control arms. To do this they simply drilled two new mounting holes exactly 1 in. below the original ones. This resulted in changing two design parameters: (1) The camber change that takes place in the front wheels as they move through their travel limits from jounce to rebound has been increased from the standard 0° per in. of travel to 0.6° per in. of travel around design height; (2) The front roll center height has been raised from 2.6 in. to 4.9 in. above ground level.

By increasing the camber change they've decreased the amount that the front wheels will lean with body roll during cornering by 30% (0.7° per degree of body roll vs. the standard 1° per degree of body roll). This means the front wheels remain more nearly perpendicular to the ground during cornering, thereby improving their cornering power and bite while at the same time reducing the tendency for the front end to plow.

Raising the front roll center height re-established the basic understeer balance by slightly increasing the front end portion of the total weight shift under cornering, thereby automatically decreasing, by an equal amount, the rear axle weight shift. The improved front wheel attitude made this possible without re-introducing the tendency for the front end to plow. Raising the front roll center height also had the effect of reducing body roll by 8%. A further increase in roll stiffness was gained by increasing the diameter of the front anti-roll bar from 0.84 to 1.00 in.

Steering system modifications have resulted in quicker, more precise steering that is fairly light at all speeds. The standard Pitman and idler arms were replaced by specially forged, longer units that reduce the overall steering ratio from the 21:1 offered in the factory high-performance Mustang package down to 19:1. This reduces the between-lock steering wheel turns from 3.73 down to 3.5.

The added power of the GT-350 engine meant that a more positive way of locating the rear axle was needed, so a set of sturdy torque reaction arms is used to take the torque and braking loads and provide increased lateral stability. Extensive testing proved that the location of the torque arms was critical insofar as their ability to effectively eliminate spring wind-up was concerned. The best location was found to be directly over, and in line with, the spring and extending from the centerline of each axle forward to a point directly in line with and over the front pivot of the spring, thus forming a parallelogram. To accomplish this it was necessary to cut a hole in the frame section to allow the arms to extend into the rear quarter-panel area. Steel brackets are welded to the axle housing and inside the rear quarters. The torque arms are fabricated from 1 in. diameter, heavy-wall steel tubing with forged steel, rubber bushed ends. Rubber boots are fitted to the forward end of each arm to seal their entry holes and keep dirt and water from entering the quarter-panel areas.

As is the practice in current race car suspension design, spring rates are kept fairly low (95 lb./in. front and 110 lb./in. rear – same as the factory Mustang handling package) with a strong emphasis placed on good shock absorber control. Here again S-A has disregarded cost in favor of quality and performance and has chosen what most experts agree are the best shock absorbers available: Konis. These

LIGHTER, PLASTIC windows are used side and rear on competition GT-350. The rear window is modified to act as air vent. Battery of street GT-350 is installed in trunk; competition car has huge fuel tank and funnel to prevent sloshing.

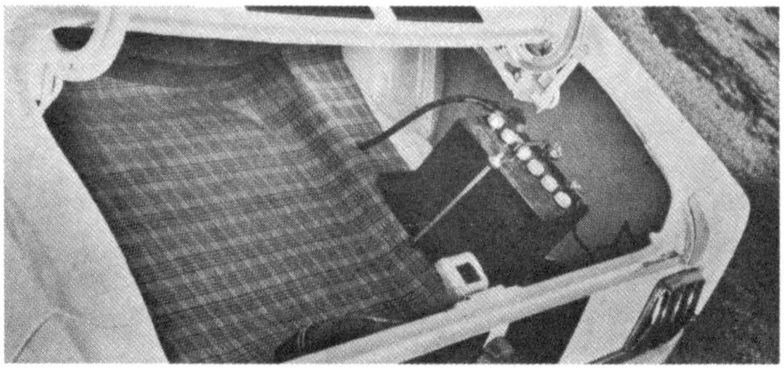

Dutch-built dampers enjoy a world-wide reputation for performance and reliability and are fully and easily adjustable for changing road conditions. Interestingly enough, it is only the choice of shock control settings that separates the street GT-350 suspension from that of the competition GT-350.

Another change that's done much for improved handling is the use of wide-base (6-in. rims) 15-in. diameter wheels along with a specially developed tire. These tires are made by Goodyear and called "Blue Dots." They are a low profile, nylon tire (7.75-15) featuring a rounded shoulder. In testing, they have been run 120 min. at 120 mph and 20 min. at 130 mph.

To provide a little better front to rear weight distribution, the battery had been relocated in the right rear corner of the luggage compartment.

At the present the only color combinations available are white exteriors and black interiors. The white exteriors are set off with blue stripes along each rocker panel and, as a dealer-added option, broad blue racing stripes running from hood to trunk.

That's about the extent of the modification it takes to convert a stock Mustang into a GT-350. To turn a stock GT-350 into an all-out competition car requires most of those already mentioned, plus just a few more. The really important difference is in the engine – which we've already described in preceding chapter.

The next biggest change is a further reduction in weight. Approximately 300 lb. more have been eliminated, bringing the competition version down to 2500 lb., ready-to-race. This was accomplished by using more fiberglass (bumpers, front under grille panel) and replacing side and rear window glass with plastic. The rear window is also designed to act as a vent to exhaust the interior of the car. This is necessary because the side "gill" vents have been eliminated and covered over by aluminum, and the doors have been gutted of window actuating mechanisms.

The interior has also been gutted as far as the rules will allow. No carpeting or sound deadening materials are used. The driver's seat is replaced with a lightweight fiberglass bucket that would almost provide a secure seat without the use of seat belts.

Other interior changes include a full set of instruments and a fire extinguisher. An approved roll-bar is available as an option, or the buyer can install his own.

A larger capacity radiator is used and an oil radiator, mounted low in front of the water radiator, is standard equipment. As a tribute to the standard GT-350 suspension, it is identical to the race version's. As we have said, the only difference is in the choice of shock absorber settings. Wheels and tires are another story. The competition car uses alloy wheel – 15 in. diameter, but with 7 in. wide rims. Tires are Goodyear's Sports Car Specials.

Brakes also remain stock, but with added cooling provisions. At the front the rubber dirt shields on the calipers are removed because they can melt under the heat of competition. The metal shields protecting the discs are also removed for increased cooling. Ducting of direct fresh, cool air to the discs is a part of the competition package. At the rear, the stock backing plates are liberally drilled and fitted with air cooling ducts. Standard GT-350 front and rear brake linings are retained.

A larger, 37-gal. gas tank is provided, and by a rather simple expedient at that. The stock Mustang gas tank is formed in two halves that are bolted together. The top half is flat and actually forms the floor of the luggage compartment while the bottom half is the part that has the volume. Shelby-American takes two bottom halves and bolts them together to get the increased capacity. A large diameter opening with a quick release cap and built-in funnel provide for fast refueling.

The street version GT-350 carries a $4500 price tag, while the race car goes for $6000. Both prices are quite a bit more than you have to pay for a factory Mustang, even one with the high-performance package. The buyer who wants to get something that's head and shoulders above anything else he can buy in this particular class will find the first price not at all high. And as for the second, where else can you buy a fully, and professionally, prepared race car for anywhere near that? If you didn't want to buy either you could probably get together enough of the parts to rework a stock Mustang, but you'd probably end up spending more than their price to equal either of them.

A STACK of raw, empty Mustangs awaits the S-A touch. The finished product comes out ready to tackle anything. The competition car has extra-wide radiator opening for better cooling, and an oil cooler for the engine.

G.T. 350

Precise control takes on a new meaning behind the wheel of Shelby American's new Mustang GT-350! The complete suspension system has been re-designed to comply with a computer plotted geometry that allows the GT-350 to stick like nothing you've ever driven. Reflex quick steering, Koni shocks, and Goodyear 130 mph "Blue Dots" let you pick your exact line... and stick to it! No use mentioning the GT-350's 306 hp Cobra hi-riser 289 or the competition proven disc brakes until you've actually sampled a few corners at speed. The most complex blind apex closing radius bend becomes the expert driver's challenge instead of an exercise in frustration. When you go down to test the GT-350, ask the salesman to bring your present car along for comparison... you won't believe it! Suggested list price $4547.00

for more information write Dept S, Shelby American, 6501 W. Imperial Hwy., Los Angeles, California 90009

SHELBY G.T.350 IS "SON OF COBRA"

Shelby-American presents America's fastest, race-bred stock car! You'll thrill to its performance, you'll hold your breath behind the wheel!! Every minute is something new never seen before!!!

SEE 306 Ford horses packed into the wild and wooly 289 engine... four-barrel carburation, high rise aluminum manifold and a hand-built tuned exhaust system. **SEE** distinctive Mustang styling to which has been added such "everybody will notice" features as rear quarter panel windows, rear brake scoops and "there goes a G.T. 350" striping. **SEE** computer designed suspension with front anti-roll bar, competition shocks, front disc brakes and torque controlled rear axle. **SEE** the G.T. 350 scream from zero to sixty in 5.7 seconds. **SEE** the G.T. 350... the car that's not for everybody. Just for you.

FIRST in Bp at Cumberland. **Bob Johnson**	FIRST overall at Kent, Washington. **Jerry Titus**	FIRST in Bp at Elkhart Lakes. **Tom Yeager**	FIRST overall at Mid-Ohio. **Bob Johnson**	FIRST second, third and fourth in Bp at Lime Rock. **Johnson, Donahue, Krinner and Owens**	FIRST in Bp at Willow Springs. **Chuck Cantwell**

WRITTEN AND DIRECTED BY **CARROLL SHELBY** PRODUCED BY **SHELBY-AMERICAN, INC.**

Carroll Shelby designed his COBRA GT to go like it looks

A brand-new Ford 302 cubic inch V-8 delivers for the GT 350. On the GT 500, a Ford 428 cubic inch V-8 is standard, with a new 427 V-8 powerhouse as a super-performance option. □ Four-speed transmissions are standard, close-coupled automatics are low-cost options. □ Great handling comes from competition-engineered suspension, 16-to-1 ratio power steering, adjustable shocks, heavy duty driveline and rear axle, and special high performance 130-MPH rated nylon tires. Front disc brakes, of course. □ And with this superb performance, Cobra GTs deliver head-turning styling *and* luxury, too. □ Interiors gleam with unique simulated wood grain trim on instrument panel, steering wheel, console and door panels. □ The exterior styling features *work* for you. □ Hood scoops supply extra carburetor air, fastback louvers are air extractors. □ Safety has not been overlooked—wide-rim wheels, integral overhead bar and shoulder harnesses are included. □ Carroll Shelby's unique fastbacks and new-for-'68 convertibles, are design-based on the Mustang, winner of two consecutive Trans-Am road racing championships. □ And that means real economy, a surprisingly low price for you. □ All four Cobra GTs say "Let's go!" □ See your Shelby Cobra dealer—and get going!

Shelby COBRA GT 350/500 POWER BY Ford

want more

F.P.M.?*

try a Shelby GT

*That's Fun-Per-Mile—yours aplenty in these *two* great new GT cars from Shelby American.

There's more F.P.M. when you have reserve performance. The GT 350 carries a 306 horsepower 289 cubic inch V-8. The GT 500 is powered by the 428 cubic inch V-8, descended from the 1966 LeMans winning Ford GT.

The F.P.M. is higher when steering is competition-quick, suspension is firm. Driving's *fun* with the safety of an integral roll bar, shoulder harnesses, disc front and drum rear brakes, wide-path 4-ply nylon tires.

Driving's *fun* with the Shelby brand of comfort and style. For more F.P.M., see your Shelby dealer P.D.Q.

SHELBY G.T. 350 and 500 The Road Cars Powered by Ford

Shelby American, Inc., 6501 W. Imperial Highway, Los Angeles 90009

Take a 2+2, inject some Cobra venom, tone up the leg muscles, add lightness, and you get a...

MUSTANG GT-350

by John Ethridge, *Technical Editor*

WHEN FORD DECIDED to go racing with the Mustang, they wisely first consulted their oracle for such matters, Carroll Shelby. Shelby's Los Angeles-based company, Shelby American, builds the all-conquering Cobras. Starting with the 1965 season, Shelby has taken over all Ford's road-racing operations. He knew that the first order of business was to get the SCCA to accept a raceable version of the Mustang in a competitive production sports car class. This was ultimately done (Class B), but with the stipulation that at least 100 more-or-less similar models of the proposed two-seater be built. Realizing that at first, anyway, they couldn't find 100 racing customers, they decided to make a street version as well.

Our test car was the street GT-350, which is the model most of Shelby's customers will order. Shelby receives the 2+2 semi-completed from the Ford assembly plant, and this forms the basis for either the street or competition car. Anyone who intends to drive on the street wouldn't want, need, or be willing to pay for the racing goodies and features of the competitive version. Gumball tires, magnesium wheels, oil cooler, plastic rear and side windows, gutted interior, and fiberglass seats are for the race course.

Although the most apparent difference between our test car and an ordinary Mustang is the absence of rear seats and the interior-mounted spare tire, the real difference is in suspension, engine, and weight. Between removal of the rear seats plus cutting and paring here and there, they've trimmed exactly 200 pounds off the weight of the Mustang 2+2 we tested in January. There are 35 more horses under the fiberglass hood, provided by a large Holley four-barrel carb on a special high-riser intake manifold, plus exhaust headers leading into twin pipes and straight-through mufflers.

Suspension has actually gotten the most attention of all. The words here are: control, limit, locate, stiffen, and snub. The Mustang performance handling kit springs are retained for use all around, but the shocks are special Konis. The rear axle got a pair of stabilizer bars, located above and parallel to the forward half of the leaf spring. Each chassis anchor for the bars appears to occupy what would've been part of any rear-seat passenger's anatomy with the standard 2+2 seats. This probably accounts for the absence of rear seats in the street GT-350. Rebound and jounce limiters complete rear suspension modifications.

Most noticeable change in the front suspension is a very substantial-looking anti-roll bar. The Shelby people tell us there are subtle changes in front suspension geometry, but we couldn't detect them by looking.

A weight-transfer program was carried out along with the weight-reduction program. Relocating the battery in the trunk and using the fiberglass hood were the major steps taken to reduce the proportion of weight carried by the front wheels. This permits use of the power-steering ratio in a manual setup. This improved steering transforms the car. It's wonderfully quick and light – even with the big tires.

When you start the engine, you're first impressed with a raucous note from the twin exhausts. These emerge just in front of each rear wheel. They're actually louder *inside* the car, because there's no insulation or undercoating. Not everyone will appreciate the cockpit noise and firm ride, but we're sure no enthusiast will ever complain. The engine's willing and responsive, and the carb/manifold setup hasn't made it temperamental in the least. Maybe it's just a bit cold-

PHOTOS BY PAT BROLLIER, DARRYL NORENBERG

AT NEAR MAXIMUM CORNERING CONDITIONS, THERE'S VERY LITTLE LEAN. BLUE SPOT TIRES STICK AND DON'T ROLL UNDER LIKE ORDINARY ONES.

STRONG LIMITED-SLIP DIFFERENTIAL LETS YOU LEAVE TWO LONG, BLACK STREAKS. BUT TOO MUCH WHEELSPIN WILL HURT ACCELERATION.

Big Holley carb's almost as long as engine block. Added cross-member triangulates engine bay, complicates rotor cap removal.

Custom wheel, designed especially for the GT-350, is sole accessory. Side exhaust's handy for issuing challenges at drags.

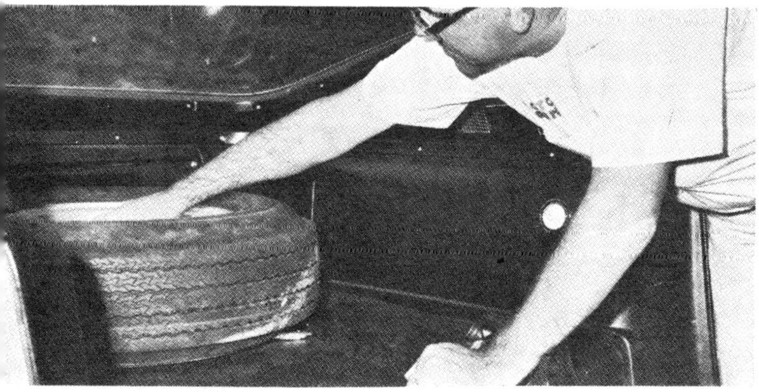

Spare perches atop aluminum panel where seats used to be. Tidy cover slips on for appearance. Panel won't support extra weight.

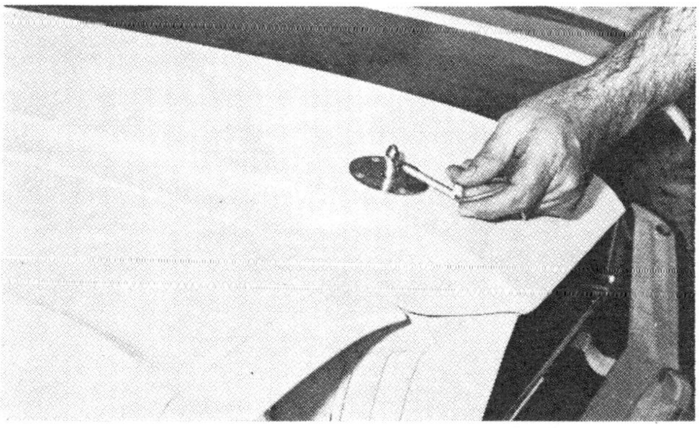

Fiberglass hood is hinged at rear and pin-latched same as racing stock cars. Airscoop actually admits air to the air cleaner!

BRAKES COULD HARDLY BE BETTER. THEY SIMPLY BRING THE CAR TO A STRAIGHT-LINE STOP IN A VERY SHORT DISTANCE, WITH NO WHEEL LOCK.

natured on cold mornings, but it warms up quickly and runs normally within a block or two.

Handling is much improved over even the handling-kit-equipped normal Mustang. The GT-350's best handling qualities won't show with half-hearted cornering, though. You have to corner hard enough to take all the slack out of the suspension in one direction. Then you get (on smooth surface) a sensation of superb stability, and you can feel the tires really bite. The GT-350, in fact, develops so much cornering force that the idiot light came on and the gauge wavered (it has both) on several occasions due to oil surge in the sump. You'd need some baffling if you drove this way regularly.

The car's still a basic understeerer, but nowhere near so much as the regular Mustang. Hard-over-to-hard-over maneuvers, as in a series of esses, produce the most noticeable handling peculiarity. Due to the high center of gravity, the suspension has to deal with some very high dynamic rolling moments under these conditions. When the body rolls from one extreme to the other, it tends to force the tail out momentarily. But with practice, we were able to smooth and tidy up our negotiation of esses considerably.

If you're looking for some justification or excuse for getting a GT-350, it gives very good gas mileage for such a high-performing car. Our best was just under 17 mpg when driving at steady legal speeds on freeways. The low, 11.2, came during performance testing. Average for our 900-mile test was 14 mpg.

The GT-350 is one car that'll never put you to sleep at the wheel. We recommend it as a sure cure for all strains of boredom. It positively exudes character – something rare in this day of follow-the-crowd compromises and design by large committees. The only really unpleasant moment we had with the car was when we had to give it back.

At present, you can have any color GT-350 you want, as long as it's white with blue trim. A no-cost option is 4.11 gears instead of the 3.89 set. The list of extra-cost accessories consists of a set of four custom wheels in place of the standard steel ones. The warranty is the same as Ford's for high-performance stuff: 4000 miles or 90 days. /MT

1965 MUSTANG GT-350

2-door, 2-passenger fastback coupe

OPTIONS ON TEST CAR: Custom wheels
BASE PRICE: $4547
PRICE AS TESTED: $4820 (plus tax and license)
ODOMETER READING AT START OF TEST: 2222 miles
RECOMMENDED ENGINE RED LINE: 6500 rpm

PERFORMANCE

ACCELERATION (2 aboard)
0-30 mph .... 2.7 secs.
0-45 mph 4.5
0-60 mph 7.0

PASSING TIMES AND DISTANCES
40-60 mph .... 2.9 secs., 215 ft.
50-70 mph .... 3.7 secs., 325 ft.

Standing start ¼-mile 15.7 secs. and 91 mph
Speeds in gears @ shift points
1st ...... 56 mph @ 6500 rpm
2nd ..... 82 mph @ 6500 rpm
3rd ..... 108 mph @ 6400 rpm
4th ..... 125 mph @ 6100 rpm (observed)

Speedometer Error on Test Car

Car's speedometer reading	32	47	53	64	74	85
Weston electric speedometer	30	45	50	60	70	80

Observed mph per 1000 rpm in top gear .... 20.5 mph
Stopping Distances — from 30 mph, 24.5 ft.; from 60 mph, 140 ft.

SPECIFICATIONS FROM MANUFACTURER

Engine
Ohv V-8
Bore: 4.00 ins.
Stroke: 2.87 ins.
Displacement: 289 cu. ins.
Compression ratio: 10.5:1
Horsepower: 306 hp @ 6000 rpm
Horsepower per cubic inch: 1.06
Torque: 329 lbs.-ft. @ 4200 rpm
Carburetion: 1 4-bbl.
Ignition: 12-volt coil

Gearbox
4-speed manual, all synchro; floorshift

Driveshaft
1-piece, open tube

Differential
Hypoid, semi-floating
Standard ratio: 3.89:1

Suspension
Front: Independent SLA, with high-mounted coil springs, double-acting tubular shocks, and heavy-duty anti-roll bar
Rear: Solid axle, with semi-elliptic leaf springs, double-acting tubular shocks, control arms, and rebound straps

Steering
Recirculating ball and nut
Turning diameter: 39 ft.
Turns lock to lock: 3.7

Wheels and Tires
14 x 6.0 J, 5-lug, aluminum center, steel rim wheels
7.75 x 14 4-ply Goodyear Blue Spot tires

Brakes
Hydraulic, front, disc; rear, duo-servo
Front: 9.5-in.-dia. caliper discs
Rear: 10-in. dia. x 1.75 ins. wide
Effective lining area: N.A.
Swept drum area: N.A.

Body and Frame
Unitized; bolt-on front fenders
Wheelbase: 108.0 ins.
Track: front, 56.0 ins.; rear, 56.0 ins.
Overall length: 181.6 ins.
Overall width: 68.2 ins.
Overall height: 52.2 ins.
Curb weight: 2800 lbs.

Cobra steering wheel, oil pressure gauge, and tachometer are added. Horn is on dash.

Heart of suspension modification is rear stabilizer bar, Koni shock, and cable strap.

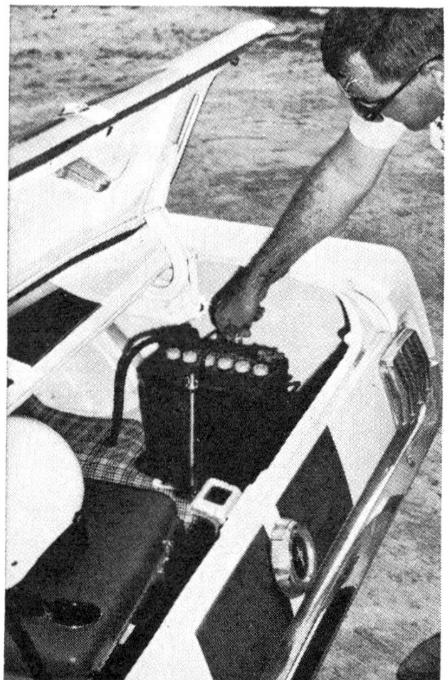

Relocated battery has tray and drain, but sloppy service attendants could fault it.

10,000 MILE ROAD TEST:

SHELBY MUSTANG 350 GT

AD TEST 11/65: *Through 20 States with a fish-eating horse.*

THIS REPORT MARKS A FIRST in magazine road tests. We've long felt that while our usual test procedures were more than adequate for cars that have been in production for some time, they really are not of sufficient duration to catch all the attributes and faults of a completely new car or one that has been changed substantially from previous models.

It isn't easy to get a manufacturer to turn loose a car for the sort of extensive testing we had in mind, but two have done it so far. One was Carroll Shelby whose Mustang 350 GT is the subject of this report; the other was Alfa Romeo with a 2600, the report on which will follow in a subsequent issue. The procedures varied somewhat but the results were the same. In one case we lived with the car for some months; in this case we did an extensive track test, made a shake-down trip from Los Angeles to San Francisco and back and then did a mad, three-week dash that covered 9,000 miles, 20 states and virtually every sort of weather from 110-degree heat through a dust storm to a mountain blizzard. It is easy to grow to hate a car when one is cooped up in the thing for 12 hours a day over a three-week period under these conditions. One can also fall in love with a machine in the same circumstances. We fell in love with the Shelby Mustang 350 GT.

The Shelby Mustang is something of a brute but a totally attractive one. At first blush it looks just about like the standard fastback that has become virtually ubiquitous all over the country, but at first look only. A second glance (and it draws double-takes without fail) shows there is something more, a purposeful, hulking look about the car that insists on further inspection. We have a strong suspicion, based on experience, that one could park one of these things in the middle of the Mojave desert and draw a crowd in something short of five minutes.

What is this car? It is, basically a three-way cross-breed comprised of Mustang two-plus-two, Cobra and Galaxie, with an extra couple of touches that are pure Shelby-out-of-Ken Miles (the organization's talented competitions manager).

Semi-complete Mustangs, less hood, are delivered from San Jose to Shelby American's new El Segundo plant where the cross-breeding takes place. Engines fitted are the 306 bhp "street snake" variety, mated to the Cobra close ratio gearbox. This power unit started life as a 271 bhp Fairlane "High Performance" plant, but has been equipped with a high-rise manifold and larger four-barrel Holley carburetor with equal-sized venturis (1.7 inches), headers and the Cobra cam, all of which combine to produce the 306 horses that make the difference between this and the ordinary breed of Mustang in terms of performance. The close ratios of the all-synchro gearbox make this explosive combination very easy to deal with, even in 15 mph school zones, as long as one remembers that shifting is what a gearbox is for. From the gearbox on back, everything is heavy-duty Galaxie and the differential is equipped with a very positive limited slip unit that lets one know in no uncertain terms when it locks and unlocks — one staff member thought he had broken an axle the first time he felt it unlock! Rear drum brakes are chosen from the scad of options originally built for the rally Falcon Sprints, and are equipped with competition shoes. All this metal is held in place by a very hefty pair of radius arms that ride above the springs and locate *inside* the body. This caused a problem on the very early versions in that fumes from the side exhausts were sucked into the slots for these arms and thence into the interior of the car. This has since been cured on subsequent models, and even on the prototypes and pre-production models, by a very neat rubber boot and fiberglass housing that not only seals everything tightly but stops most of the noise from the positively located rear axles which tended to clunk when the limited slip locked or unlocked. All this beef means unsprung weight which is held in check by

Photos: CHARLENE MEGOWAN

Under heavy acceleration the Shelby Mustang shows slight weight shift to the rear, increasing the bite. Jump is sudden but the tires can be broken loose at half throttle.
Right — Braking is sure and smooth from any speed. As can be seen there is little or no apparent nose drive or rear end lift. In acceleration, braking and flat out the car is totally stable and sure on the road.

MUSTANG 350 GT

heavy-duty Koni shock absorbers that can be set from moderately soft to dead hard. The only time this unsprung weight gets noticeable is on washboard streets when acceleration is wanted. The resulting hopping and skipping is controllable but surprising. The only place we found it really aggravating was New York City, but then that's a place where the only suitable vehicle is a Checker cab anyway — preferably driven by somebody else.

The front end is where the Miles treatment is noticeable. It has been erroneously reported to be decambered but the reworking is much more sophisticated than that. The upper A-arms are relocated downward at their inner pivots, changing the roll center completely. Further changing the roll center is a VERY heavy-duty anti-sway bar. Again all this beef is held in check by Konis. Steering arms and idler arms have been lengthened to increase the steering speed significantly. Brakes on the front are Kelsey-Hayes discs, equipped with very hard pads. Pedal pressure requirements are high but stopping is smooth and firm, with no tendency to lock. The remainder of the handling package comes in the form of Kelsey-Hayes mag-centered, steel rimmed wheels, six inches wide with 7.75 x 15 Goodyear Blue-Streak street tires. The result of these changes is handling that is lighter than the original, far faster, and a change from strong to very light understeer that goes neutral at speed.

At first it seems to want more than a bit of muscle but as one learns to drive the 350 GT this requirement gets to be less and less. For instance we found that while it could be hauled into an early apex in a corner safely, the choosing of a late apex in the same corner made things much easier, faster and smoother. On our test car the ride was, by request, on the hard side but it can be adjusted, as we found when we tried another one with softer settings which would probably be more suitable for the Northeast part of the country. When pressed the Shelby Mustang uses up quite a bit of road, but it is completely predictable, and once one gets the feel of the car this gets to be fun and is easily brought under control. At one point in the track testing we had the car at Riverside and in completely street trim were able to bomb through certain parts of the course including the esses at speeds higher than many all-out racing cars can be run through. At another point during the trip back from New York, in one of the western states, we were able to maintain an average of between 90 and 100 mph over a mountain road for nearly an hour with no strain on driver, passenger or car. It was quite obvious that this car is easily the equal of anything that Europe has to offer in the GT category, even at three times the price.

Inside the car is comfort and space the equal of any two seater Grand Tourer. The seats are of semi-bucket design standard in the tamer Mustangs but are equipped with wide competition-type seat-belts that hold one firmly in place at all speeds. The wheel, a Cobra unit, is flat as opposed to the dished standard item, and therefore sits at a proper distance inviting the "arms out" driving position. An additional instrument group consisting of a tachometer and oil pressure gauge has been placed in the center of the dash cowl, augmenting the rather sparse standard instrumentation. To the rear, the seats have been removed and replaced with a flat fiberglass shelf which holds the spare tire and all sorts of luggage, groceries, equipment and other impedimenta except passengers. As a matter of fact the

(Continued on page **33** *)*

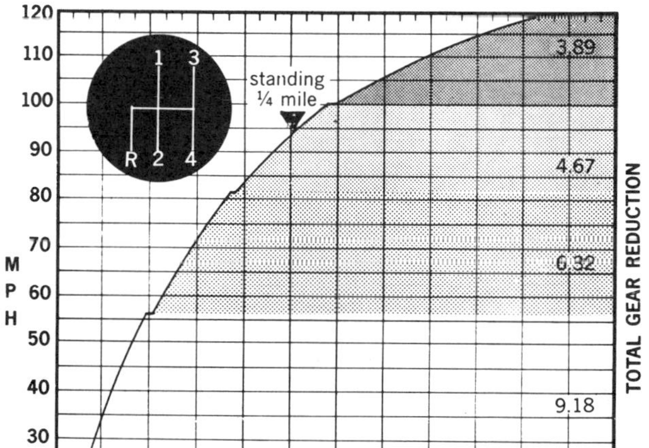

2

3

4

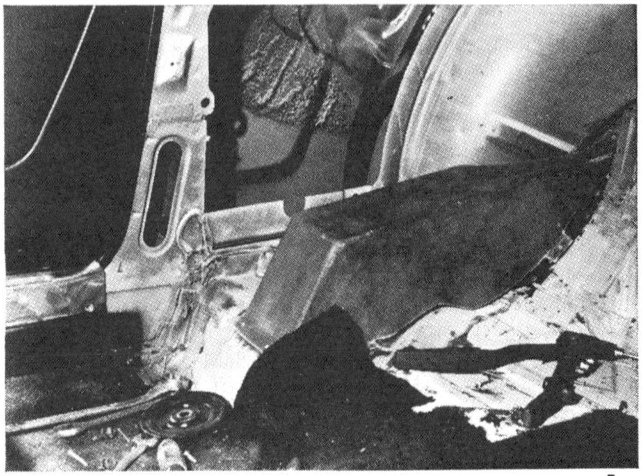

5

Upper left — Under hard cornering, even
wnhill, the GT 350 exhibits little tendency
lean but it does use up road, predictably.
Engine in the 350 bristles with Cobra bits
l pieces is equipped with high rise mani-
d and large Holley 4-barrel carburetion.
Large sump virtually doubles oil capacity
l doubles as an oil cooler, which is needed.
End of radius rod comes through flooring
l attaches inside car. Note large slot.
Slot is sealed by a rubber boot and by
erglass box to keep out water and fumes.
Below — Scotch one canard. The 350 GT
definitely in production, by the yardfull.
Right — Here's how it looks from the seat.
suggest that you try it for yourself now!

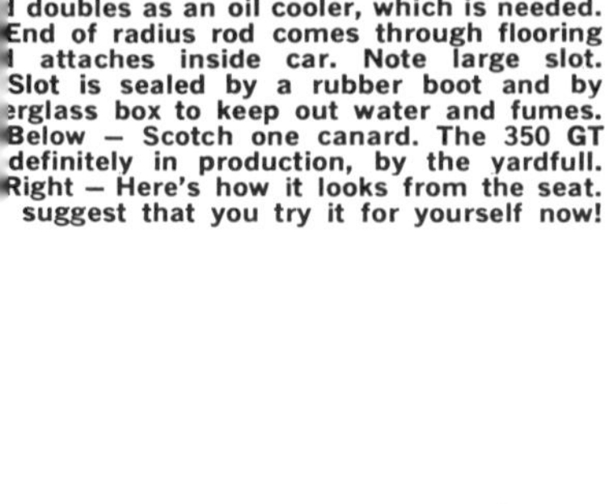

7

ROAD TEST:

SHELBY-AMERICAN GT-350

Back during the Middle Ages, that period between the time Johnny (B.C.) Hart's "gullible simp" invented the wheel and Henry Ford's arrival in the world, people used to turn to sorcerers and magicians for potions to cure their ills and allay their troubles. Most people nowadays don't believe in sorcerers, since quantum physics and new math make them technologically unemployable. But it should be reported that there is one of the old-style wizards still plying his trade, and the magic balm which he dispenses is avidly sought by a few furtive sufferers of tired blood and bland togetherness.

This modern day Merlin practices his black art not in a dank cave or hidden hovel secreted in a dark forest, but beside an ultra-modern time machine emporium. There, thousands of people daily trudge in groups of a hundred or so into the open maws of huge aluminum tubes, emerging four hours later or less in another country, another world, or at the very least another city, unaware that they had almost brushed shoulders with a source of even greater wonderment.

He is Carroll Shelby. He rebuilds cars. Correctly. He takes about 100 Mustangs each month, worth about $2300 each, and gives them his own version of a tune-up that costs another $2200.

This raw-boned Texan does a lot of things to Mustang 2+2 fastbacks that slide under the ministering hands of himself and his apprentices. The cars arrive at his magic laboratory without hoods, rear seats, or grilles, and they're all white with black interiors. When they leave, they've gained a fiberglass hood that's fastened by genuine racing pull-pins, a plastic rear floor for a vinyl-shrouded spare to rest atop, and huge blue stripes along top dead center and bottom rocker panel edge. There's a "GT-350" painted in broad strokes, so there'll be no doubt as to what is being viewed.

The most important thing which is done, because no one

HORSE OF another choler, the GT-350 is a red-blooded, tail-switching range-runner. Shelby-American does the transfusion.

PAUL E. HANSEN PHOTOS

ALL ASPECTS of performance are improved, from braking and acceleration to handling. The Cobra touch is much-evident.

SHELBY GT-350

else does it, is relocating the inner pivot points for upper suspension arms. This little touch alone makes a vastly-needed improvement in the car, raising the roll center at the front so that the car corners flatter and more precisely. Front wheels maintains a more perpendicular attitude toward the road, regardless of body roll and side loadings, because of it.

The Shelby treatment also involves a much stouter anti-roll bar between lower front suspension arms to raise the roll moment and insure the flattest cornering possible. At the rear, a pair of traction bars are fitted to a revised axle pad, with arms running forward to a mounting point directly above the spring hanger centerline – for which portions of body metal must be cut away. Extra sturdy double-acting shock absorbers complete the suspension work.

Under the hood, Shelby works more of his Cobra-derived magic. This can be liberally characterized as blueprinting an already factory-blueprinted engine. Primarily, the passages for inhaling and exhaling are given much clean-up and open-up treatment. Since these modifications, as well as more detailed explanations of suspension changes, are covered in detail elsewhere in this book, it should suffice to say that this Mustang's "350" designation is barely exaggerated from a dynamometer's standpoint.

So set up, this Mustang really rambles. Figures on the data panel illustrate vividly the straight-line performance, but are much too inadequate to describe this beast in action. Acceleration is enough to leave one breathless, although it is necessary to remember that the tach needle should be kept hovering above the 4000 mark for the most spectacular results. This is not difficult to manage, however, with the nicely spaced ratios in its Warner Gear T-10 transmission. The latter, it might be noted, was chosen over the somewhat tougher Ford 4-speed because of its weight-saving aluminum case. With the final drive ratio in the test Mustang the speedometer needle would easily come to rest on the 120 peg. Since the tach was registering 6000 rpm, this trued out to a back-stretch straightaway speed of 105 mph with the fat 7.75-15 Goodyear Blue Dots which were fitted.

Although this was the street version of the GT-350 (there is an even more stark and lightened competition version), it was an exceedingly loud car. A set of thin-wall tubing headers molding into straight-through tailpipe enlargements which masquerade as mufflers was the primary reason for that. So long as one doesn't deliberately bait policemen by revving up the engine in their vicinity, the system is probably legal. The dual tailpipes are bent outward to vent their wrath just in front of the rear tires and this, too, contributes to the overall noise level: The full blast of ripping exhaust assaults the driver from just below ear level.

There was one bother involved with the test car which almost made it mandatory to have a 7/16-in. box-end wrench as standard equipment. That was the tendency for the exhaust headers to work loose with even mild driving. This was attributed to incompatible expansion properties of metals in heads, headers and header bolts. As this appears, the situation will have been corrected, but on the test car at least it was necessary to refasten the headers in place about once a day. Their loosening, it goes without saying, added significantly to the decibel level inside the car.

Noise or no noise, however, this was a car that was designed for the man who wants the maximum from his car. And that pre-supposes a dual-purpose car: Commute to the office with it during the week, then paint numbers on it and race it weekends. As it stands, it runs B Production in SCCA races and will provide a good driver with potential winning machinery. Anyone really serious about winning, however, would prefer the lighter and quicker competition version.

WOOD-RIMMED steering wheel, tachometer and oil pressure gauge cluster are part of S-A treatment; so are huge seat belts.

BLANK grille and horse emblem are characteristic to GT-350. Racing stripes, hood scoop and hood safety locks are standard.

It is on the road course that this car belongs. Shelby modifications have guaranteed that the worst of the production car ghastlies have been eliminated. It still maintains basic understeer characteristics, but the descriptive "near neutral at speed" was never truer. It is a car with great gobs of forgiveness for driver boobery, which is just as well since it insists on one of a Ken Miles stature to extract its full potential. No little measure of this forgiving capacity is due to the huge tread patches with which the Goodyears paw the track. Side loadings of almost devastating force are required to make those tires quit working in the corner. Suffice it to say that none of our test drivers were brave enough to call this bronco's bluff. When we overcooked, as we did once on Riverside Raceway's Turn 7 – a sharp downhill left-hander coming off an equally sharp rise – the tires merely scrubbed off speed, sideways, while we backed off the throttle enough to regain steering control.

This was one car in which there need be no qualms about brakes. Unfortunately, it also was one in which our test crews failed to measure our now-customary deceleration rates. Nevertheless, there was never the slightest concern with stopping power in street use. Pedal pressures were slightly higher than most people now expect, though hardly prohibitive, and fade was largely an academic question. In racing situations, however, that may not hold true although appearances to date have demonstrated no problem. One concern which remained during hard usage, however, was a legacy of the Mustang's most serious vice – rear axle hop. This occurred only in severe race-course braking situations, when weight transfer effect was sufficient to let a lighter loaded rear end lift enough for the huge drums there to lock.

In tight maneuverings on very short courses or exceedingly twisting roads, the GT-350 could conceivably still be a handful. But Shelby's substitution of the quickest gears to yet appear for the Mustang steering has been a great improvement. For those who complain about an absence of road feel, this is the answer. There is definite feedback as part of the price of very quick (for a domestic car) steering and the wood-rimmed aluminum wheel borrowed from the Cobra needs a steady grip to reduce its jerky oscillations. Steering is non-power assisted, but its heaviness is not objectionable.

It might be noted that engineers from both Shelby-American and Ford plunged in to isolate and correct one problem which surfaced as first production GT-350s began racking up mileage. Limited slip differentials, which are actually Galaxie units, seemed to develop an inability to cope with the demands placed upon them. This resulted in some erratic power transmission between rear wheels, which in turn, caused various degrees of rear-end steering. The test car, though afflicted with this trauma, didn't persevere beyond the annoying stage, but it was disconcerting in its presence.

Little in driver comfort has been sacrificed by the GT-350 over that of the off-the-shelf 271-HP Mustang. The same carpeting and bucket seats – among the most comfortable and secure in domestic production – are used. The flatter (and slightly smaller in diameter) steering wheel improves driver position. And the Air Force seat belts, even if less stylish, are vastly greater security than Linus blankets.

Still, despite almost immeasurable improvement in balance and handling and cornering and braking, we came away from the GT-350 with somewhat mixed emotions. It out-performed the last fuel injected Corvette Sting Ray which CL tested, for quite a few dollars less. That, of course, is certainly one recommendation. But the cost is not measurable in dollars to many, and the difference is between gentility and boorishness, between manners and manhandling, between the sophisticated and the strident. On balance, however, it really comes down to the name of the game, and that is To Win. That's our game, too, and Shelby and Ford are playing it with the GT-350. This is one time we'll have to close the textbooks, ignore all pretensions, and shout: I'll take it, I'll Take It! Ol' Wizard Shel has us in his spell.

MUSTANG ROAD TEST

Shelby-American GT-350

SPECIFICATIONS

List price	$4311
Price, as tested	4584
Curb weight, lb.	2790
Test weight	3140
distribution, %	55/45
Tire size	7.75-15
Tire capacity, lb. @ 24 psi	4400
Brake swept area	381
Engine type	V-8, ohv
Bore & stroke	4.00 x 2.87
Displacement, cu. in.	289
Compression ratio	11.5
Carburetion	1 x 4
Bhp @ rpm	285 @ 5000
equivalent mph	102
Torque, lb.-ft.	325 @ 3000
equivalent mph	61

EXTRA-COST OPTIONS

Cast magnesium wheels, Shelby-American tune-up.

GEAR RATIOS

4th (1.00) overall	3.89
3rd (1.20)	4.67
2nd (1.62)	6.30
1st (2.36)	9.18

DIMENSIONS

Wheelbase, in.	108.0
Tread, f & r	56.5/57.0
Overall length, in.	181.6
width	68.2
height	51.2
equivalent vol., cu. ft.	366
Frontal area, sq. ft.	19.3
Ground clearance, in.	5.5
Steering ratio, o/a	19.1
turns, lock to lock	3.75
turning circle, ft.	38.0
Hip room, front	2 x 22
Hip room, rear	n.a.
Pedal to seat back, max.	43.0
Floor to ground	10.0
Luggage vol., cu. ft.	5.5
Fuel tank capacity, gal.	16.0

PERFORMANCE

Top speed (6100), mph	124
Shifts, @ mph (manual)	
3rd (6500)	110
2nd (6500)	82
1st (6500)	56
Total drag at 60 mph, lb.	133

FUEL CONSUMPTION

Normal range, mpg	12-16

SPEEDOMETER ERROR

30 mph, actual	26.4
60 mph	50.4
90 mph	75.1

CALCULATED DATA

Lb./bhp (test wt.)	11.0
Cu. ft./ton mile	157
Mph/1000 rpm	20.3
Engine revs/mile	2950
Piston travel, ft./mile	1410
Car Life wear index	41.6

ACCELERATION

0-30 mph, sec.	2.4
0-40	3.6
0-50	5.2
0-60	6.8
0-70	8.7
0-80	11.2
0-100	19.0
Standing ¼ mile, sec.	14.7
speed at end, mph	90

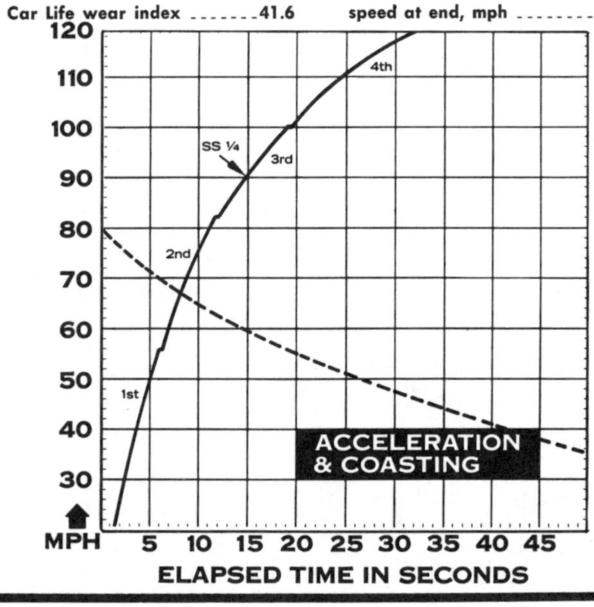

Mustang GT-350

BY THE CARS STAFF

CARROLL SHELBY, with some help from the Ford Motor Company, set the sporty car set back on their heels when he unleashed the Cobra from his Venice, California "Snake Pit". He repeated this performance on the drag strips, when he let his "Dragonsnake" loose to devour A/Sports Sting Rays. These performances also affected the buying habits of the guy on the street. After reading about Shelby's exotic 260-289 Ford machines, many FoMoCo buyers ordered Falcons, Fairlanes and Mustangs with big V-8's and immediately replaced the cheapie chromed or stock tin valve covers with finned aluminum ones with "COBRA-POWERED BY FORD" in big bold letters. Before long "the little old snake maker" introduced a full line or race bred show and go accessories for all Ford small V-8's.

The latest piece of machinery out of the snake shop is a Shelby-ized Mustang that carries a GT-350 name tag and a $4,900 list price tag. At first glance the GT-350 appears to be a dolled-up, hairy version of a stock fastback Mustang coupe. However, a few laps around the back streets and a "hell bent for leather" blast down the Thruway should change all those nasty first off thoughts. The GT-350 in standard street

MUSTANG GT-350

Shelby trim will out-handle, out-accelerate and turn more heads than the hottest factory Mustang machine.

Our old friend Bill Kolb at Larsen Ford in White Plains, New York, gave us a jingle when he received the first of the Shelby-engineered GT-350's. After hearing Bill say, "as soon as we clean it up it's yours for a test", we packed our gear and made tracks for the suburbs. Bill is in charge of racing and the sales of performance cars for the Larsen agency and has been extremely helpful in the past when it came to securing hot models for road tests.

In order to make our deadline, we were forced to use Bill's private test course instead of making the usual drag strip trek. We were not able to record accurate quarter-mile or 0-60 mph times, but we had a pretty good seat-of-the-pants idea of what this baby could do on the strip. It felt hotter than the hottest 271 hp Mustang Ford builds and should check through the traps in grand style. After a few stopwatch runoffs, we felt confident that the GT would run consistent high 14-second, low 90 mph quarter-mile times. These runs were made with the quick disconnect exhaust pipes open and the 4.56 to 1 cogs installed. I'm sure that with a little effort, we could have coaxed seven second flat times on the standard 0-60 mph bashes. The Shelby-modified 289 engine needs tall gears for maximum performance, and the Larsen-installed 4.56's are just the ticket for the quarter-mile go.

The engine Shelby supplies with the GT-350 is a basically stock 271 hp, 289 cubic inch HP model with a few refinements. Setting this one off from the factory stockers is a

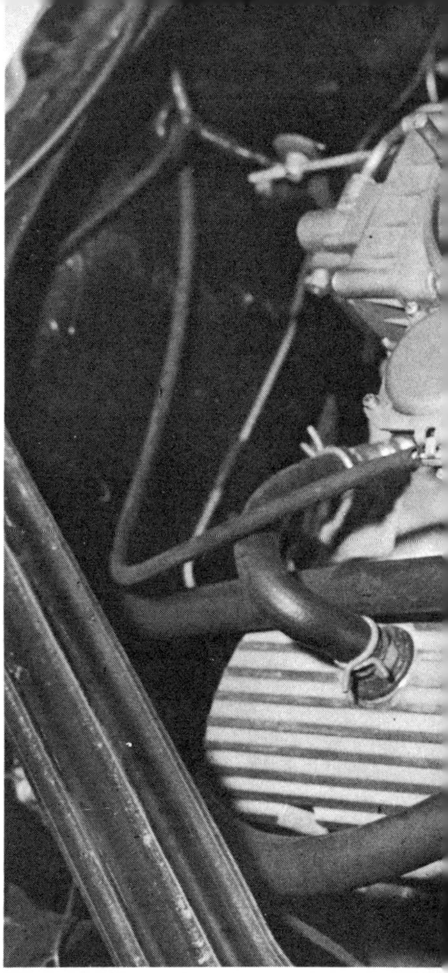

Bill Kolb, left, grabs hold of the special stabilizer that Shelby installs between the front fender aprons. Kelsey Hayes front discs are standard on the GT-350.

GT-350 mill boasts special equipment installed by Shelby's technicians. Huge Holley quad on a new aluminum high riser manifold, Cobra valve covers and tuned tube headers set this mill off from factory stock 271 hp models.

Holley C4AF-9510-DA four-barrel mounted on a tuned, high riser aluminum manifold and a set of tuned tubular steel headers leading to a modified exhaust system which employs quick disconnect pipes and trim fiberglass mufflers. Everything else, including the 450-inch lift, solid lifter cam and dual valve springs, is factory stock. Thanks to a finned aluminum oil pan, lubricant capacity is up to 8½ quarts.

All GT-350's come through with aluminum-cased 2.36 to 1 First gear Borg-Warner T-10 four-speed transmissions and HP 10½-inch balanced clutch assemblies. Bill Kolb's GT came from the Shelby plant with 4.11 to 1 gears, but he felt that 4.56's would get the job done just a little quicker. He changed the gears before we took it out, and shelved the 4.11's for future use.

Shelby installs Borg-Warner T-10 units in place of the factory stock T&C shifter. The B-W unit features a 2.36 to 1 First gear and an all aluminum case. Scatter-shields are stock on all Shelby-ized GT's. The snake charmer himself, Carrol Shelby, right, hams it up a bit at the Riverside GT introduction.

MUSTANG GT-350

Finned aluminum pan holds extra oil, aids cooling. Beefy anti-roll bar is just what the doctor ordered for hard cornering. Bill, below, checks out the optional mag-type sports wheels.

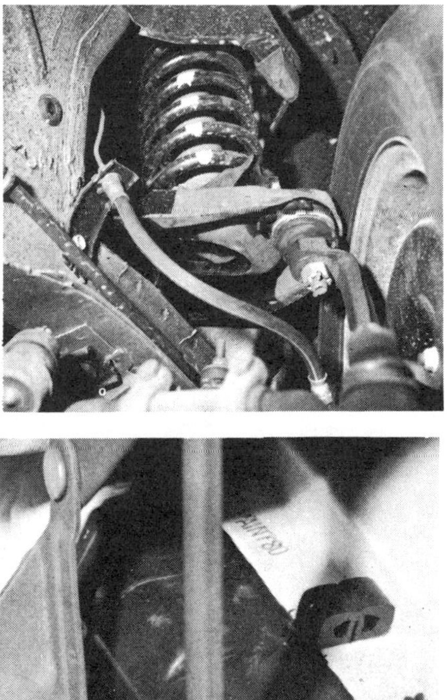

Shelby went one step beyond Ford and replaced the stock Galaxie rears on all GT's with positive locking, gear-type Detroit Automotive units. Anyone who has ever tried to come off the line with a 271 hp stock Mustang at 3,500 rpm can really appreciate Shelby's decision. He offers 3.89 or 4.11 gears as standard equipment on his new striped snakes.

Our test car felt quite strong at the top end of the tach, but was obviously suffering from fuel starvation at any point over 6,000 rpm. Shelby could easily remedy this situation by simply bolting an electric fuel pump next to the gas tank. This modification is a must if serious dragging is in the future.

We were most impressed with the way the GT rode and handled, even on the roughest terrain. The basic optional handling kit available on any Mustang does wonders for the stock product, but Shelby's suspension modifications turn the GT into a thoroughbred, road-hugging stal-

Test car's Detroit Automotive rear, above left, carried 4.56 to 1 gears. Beefy rear shocks are specially calibrated by Koni for the GT. Koni units and special front coil springs, above, are part of the GT's handling and ride package. Special master cylinder and pressure limiting apparatus, left, handles the requirements of the front disc and the oversize conventional rear brakes. This combo works quite well.

The tuned exhaust system can be easily disconnected for short blasts at the drags or road course. Straight through mufflers are used on the street. Exhaust exits in front of the rear wheels.

lion. Regardless of road conditions or speed, the GT stayed in constant contact with terra firma. Responsible for the superior ride, handling and control is a Shelby-engineered suspension which incorporates Koni shocks all around, extra heavy duty coil and leaf springs, sway bars, a pair of traction bars, quicker steering, Kelsey Hayes front disc brakes, and six-inch wide 15-inch wheels. The traction bars aid stability and prevent axle windup under maximum acceleration start conditions, while the Koni shocks and beefier springs take care of ride control. The quicker steering was a welcomed feature and should be installed on all high performance Mustangs.

The ride qualities of the GT should in no way appeal to the pseudo sports type who's looking for a noisy sports machine with boulevard-type ride features. It's far from smooth and it can get a bit hairy on rough roads, such as the ones known as local N.Y.C. streets. Ride characteristics are a heck of a lot harsher than those of a factory Mustang equipped with the optional handling and suspension package, and is only really appreciated by the full-bore enthusiast.

Superior traction is another beauty part of the Shelby machine. Adding to the obvious advantages of the Detroit Automotive differential, are 7.75x15 Goodyear nylon high performance tires good for sustained speeds of 130 mph, wider wheels and a trunk-mounted battery. All these little goodies add up to traction-plus on road or drag courses. Also contributing to the traction ability of the GT is its lightweight and functionally scooped fiberglass hood which replaces the stock steel unit. Positive chromed NASCAR hood fasteners are used to prevent the bonnet from checking out at high track speeds.

Quite a few small styling touches separate the Shelby GT-350 from the factory version. Our test machine featured optional mag-type wheels, dealer-applied bold blue racing stripes, a two-passenger bucket seat interior, extra instrumentation, sexy competition-type seat belts, an open grille and a wooden steering wheel. A customized tachometer and oil pressure gauge, mounted in a contoured bezel on the dash, kept us informed as to what was going on under the scooped hood. Replacing the stock rear seat assembly, which isn't exactly "just what the doctor ordered" for adults anyway is a naugahyde-covered spare mounted on a carpeted fiberglass panel.

As far as we can see, Shelby should find it hard to sell the GT-350's at recommended list price. We don't doubt that the car is worth the stiff price tag for one minute, as the Shelby changes couldn't be duplicated for the (*Continued on page* 33)

Shelby-designed traction bars, Koni shocks and clipped leaf springs help in the traction and handling departments. Traction is also aided by a battery mounted in the trunk compartment over the right rear wheel. Gee dad, just like the Super/Stocks do it!

10,000 MILE ROAD TEST:

SHELBY AMERICAN MUSTANG

G.T. 350

Through 25 states with a fish eating horse

BY JOHN CHRISTY

CARROLL SHELBY'S MUSTANG GT 350 has been praised as fast and good handling in a brutish sort of way and derided as a monstrous joke and a car "more suited to the dropout than the serious scholar."

After 2,500 miles of testing and shakedown, 10,000 miles of cross-country driving through 25 states in virtually all kinds of weather, we can honestly say that this car is far from a joke. And it is hardly suited for dropouts.

When we broke the story of Shelby's super-horse (March, 1965) we knew that virtually every motoring publication in the country would be in print with a road test of the GT 350 within a month or two — it's that kind of car. We could pretty well guess that all the emphasis would be on performance and most of this on the straight line variety.

So what were we to do for an encore? A conversation with Shelby provided the answer. Everybody talked about performance but virtually nobody had bothered to find out about reliability, living comfort and just plain useability. Fast it is and attractive it is, but could one live with it and rely on it under any conditions?

The obvious answer was to put a year's worth of wear and travel on it in as short a space of time as possible under as many weather and terrain conditions as possible. Sebring was coming up, so was the New York International Motor Show the following week. These are two events we always attend, though usually by air. The plan that evolved was to drive from Los Angeles to Sebring via the far southern route, thence to New York, then from New York back to Los Angeles via the northern route. This journey would be a shade over 10,000 miles on every type of paved road with altitude readings from below sea-level (at Indio, Calif.) to more than 8,000 feet in the Sierras. Driving conditions varied from cities and small towns to flat-out desert stretches and constant-speed tollways. Weather turned out to be almost more than we bargained for, varying from 110-degree desert heat and steamy Florida warmth through fog, tropical downpours and electrical storms to mountain blizzards. Road surfaces varied from billiard-table smooth Interstate highways to washboard macadam and lumpy concrete.

How did the car behave under this treatment? Pull up a chair. First off, the car proved absolutely dead reliable throughout the entire journey and as we wrote this chapter we were still using the car for mundane commuting and tooling around town. Only two mechanical troubles have arisen, and one was caused by sheer, painful physical abuse when a cretinous garage attendant in New York took the car off in third gear, slipping the clutch for more than a block to get it moving. How many times he had done this we had no idea — needless to say we made other parking arrangements. The resultant frying took about half the bite out of the clutch, though it stayed with us for the rest of the trip, and is still with us, which says a lot for the high-performance Mustang clutch material.

The other mechanical failure, if it can be called that, was a slight closing of the ignition points. Before we had started out the car would do 131 and a fraction miles per hour and would turn 95 mph in 15 seconds over the quarter. When we returned it would only do a little over 115 mph and power dropped off somewhere between 5500 and 6000 rpm, where it would pull a strong 6500 revs prior to the trip. A little screwdriver and feeler gauge work brought back the 6500 turns in almost less time that it takes to tell about it. The lesson learned is to keep the distributor cam well lubed and check point gap every 5000 miles or so. The points, both primary and secondary should be gapped .018 of an inch, and these had closed to .010 and gotten slightly out of alignment, which cut voltage and retarded the spark. Even so, the car started every time at the first turn of the switch and ran strongly, though without the sharp "bite" it had had at the beginning.

To keep things in order we'll cover the trip area by area. The first area was southwest desert from Southern California to San Antonio, Texas on U.S. highways 80 and 90 with patches of Interstate 10. This route skirts several mountain ranges but does not go over them. In California and Texas it is table-smooth with excellent surfaces and the Mustang could be driven at any speed the law and/or intestinal fortitude allowed. In Arizona and New Mexico it varied from smooth to a kind of scarified pavement peculiar to the Southwest which looks as though someone had dragged a giant comb over it. When combined with washboard it is plain bloody awful.

Sand roads of the deep south presented no problem to the GT 350. The ratchet-type limited slip rear-end would pull the car out of loose sand easily. Here both back wheels are biting fiercely on soft surface. This unit makes noise but it locks very positively.

SHELBY AMERICAN MUSTANG GT 350

This particular GT 350 was set up with shocks on about two-thirds hard. (Konis, remember?) The comb-cum-washboard surfaces left us with the choice of dropping to about 35 mph which was ridiculous or getting up around 80 mph which was illegal, a Scylla-and-Charybdis situation the solution to which we'll plead the Fifth amendment. For anyone living in this area or one with similar roads, we'd recommend setting the car up at about ⅓ hard unless one is willing to accept the odd speeding citation.

In Texas, on that state's smooth roads and windy prairies, we had none of the slow-down-or-speed-up problems but another point became clear. Set stiff and with the larger than stock tires (Goodyear Blue Dots), even with the wider wheels, the car became sensitive to cross-winds, especially those of the quartering variety. An immediate cure was found in using quite high tire pressures, the best combination we found being 40 psi in the front and 38 psi in the back. More important than pressure, however, was equality of pressure on both front wheels and both back wheels. With pressures dead equal from side to side, wind wander became almost nil while a three pound difference would make things quite dicey in a strong wind. High pressures didn't seem to affect the ride at all, which was a rather pleasing discovery. These pressures do tend to accelerate front tire wear, especially in the center of the tread due to a slight bow effect. We would estimate that they would be good for about 15,000 miles and very possibly up to 20,000 if they were cross-switched at about 10,000 miles. The clock on the test car shows 12,500-odd miles at this writing and there is still 3/32 of an inch of tread in the center of the front tires.

Gas mileage was truly amazing compared to most other high performance Detroitware. We have become accustomed to either high-mileage European equipment with figures in the 20 to 30 mpg area or to American equipment which will pass everything but the gas stations. Not so the GT 350, which through this area, got 15.6 miles per gallon while being conducted at very high speeds and booted at full throttle in quick passes of the many trucks that serve the area. Something else proved interesting, and that is that deceleration on the engine in this car is as sharp as the acceleration, and one doesn't have to use the brakes nearly as much as on lesser machinery. Later in the trip, we got rather dramatic proof of this when, headed from Sebring to New York somewhere in the Carolinas, a friend caught up and tailed us for a while. Came a tap on friend's horn and the disturbing information that our brake lights weren't working. Except that they *were* working — we just hadn't had to use the brakes where friend's car did.

From Louisiana to Central Florida we suffered from that Springtime Gulf Coast malady, fog, and so remained slow, though finding that the Blue Dot tires

continued on page **32**

PHOTOS: CHARLENE MEGOWAN

The surrounding photos give an indication to the widely varying conditions and areas through which the GT 350 was run — Southwest desert, San Antonio Texas (where we found the Alamo really does have a back door), the Midwest where it inevitably drew admirers at any pause, the far West, Little America, Wyoming, where it attracted the entire service station crew, mountain blizzards and finally back home with the honest dirt of 25 states caking the sides.

kept a tight grip on wet (though not flooded) pavement. The one point that the weather cleared up on this leg showed us what can happen when one is driving what might be termed an irresistible hazard. We had been extra careful going through towns, pootling along in second cog and remaining some five miles per hour under the speed limit, a wise precaution, especially in some parts of the South. The weather cleared up suddenly in the Florida panhandle and we had just cleared a fairish sized town and got up to cruising speed when up from behind came a pair of super-bright headlights that proceeded to tailgate the GT 350 about 10 feet off our rear bumper. We put up with this nuisance until we got up to a truck at which point we dropped into third gear and romped on it, passing at something like 90 and then dropped back to cruising again. Suddenly there was the tailgater again. We repeated the truck maneuver once more but without realizing it held to a higher than legal cruise (night limit in Florida: 55 mph!) at about 65. Up came the squirrel again but this time accompanied by another pair of lights, unmistakably Dodge, and topped by another light that flashed bright blue—the fuzz had arrived. A few minutes and $25 later, the trooper explained what had happened. We had gone slowly past his station, and he had paid no more than passing attention, letting us go on our way. However, just as we had gone, the squirrel in a standard Mustang had shot by, obviously on his way to take a look at the GT 350, sucking the trooper in his wake. It painfully proved a point. Corvette drivers, already Cobra-shy, generally stay clear, but Mustang drivers of whatever variety all seemed to want to have a go at the Shelby version. Properly driven and with some regard to the law, the GT 350 doesn't attract much attention from the police. What attracts the coppers are the squirrels who want to play Fangio or needle the GT driver into hoofing it.

Once arrived at Sebring we used the GT 350 for a general utility car, transporting SCG staffers and photographic gear around the course, lugging through infield dirt and through the crowded paddock area. On race day, just getting to the track was two hours of misery in stop-start and just plain stalled traffic. Not once did the car even look like it wanted to overheat, even though ambient temperatures were in the 90's. It just chugged along and stayed at 180 degrees regardless of idling, lugging and churning along in low gear nor did it want to load up.

Two things of note happened with the car at Sebring. First the car is irresistible to the curious and some of the curious get light fingered. The first night the car was neatly opened without damage and the gearshift knob stolen. Two nights later someone who knew that the car was carrying roughly $2,000 worth of photographic gear made an attempt to get it. They broke a vent window and tried for the trunk through the cockpit and failed. They tried the trunk with a jimmy bar but failed to break or even loosen the sturdy lock, succeeding only in leaving some gashes in the trunklid, a mess of broken glass and several nice, useable fingerprints which were found by the efficient Sebring town police. Think what you like about small town police, but this is a polite, sharp heads-up force.

The second thing that happened was RAIN. It came pouring down on race day, and the day after as though someone upstairs had turned over the bucket. In spite of the damaged windwing (hastily repaired with box board) not one drop got into the cockpit or the trunk area. We were to see a lot more water during the course of the trip, but this was the heaviest and neither then nor later was there evidence of leaks. The windshield wipers worked well in Florida, but we found out later that at any speed over 60 mph they tended to skip. Since this is a car that can be driven at speed in the wet, we humbly suggest stronger springs, aerofoil arms or both be used on the GT 350.

North from Sebring, U.S. 301 is a main tourist and trade route that touches every Atlantic seaboard state from Florida to Maryland where it runs into U.S. 40 and then the New Jersey turnpike and is typical of most eastern two-laners, varying from fairly smooth to moderately rough. We were worried at first that the older concrete slab section would get up an oscillation in the firmly suspended Mustang, an action we had noted in other first sports-type cars. However, the Mustang, either due to wheelbase dimensions or other factors, did not show this trait. The tar expansion strips could be felt but nowhere near objectionably, and there was no bounce or yaw even on turns. At several places, we took off on side roads, some of them sand or dirt. Our test car was equipped with a ratchet type limited slip differential that is sometimes noisy but has one unquestionable advantage in that it is totally positive. On sand, dirt or even pavement, it took hold and there was no stopping the car. It might go "clunk", but it sure works. The trip north was totally uneventful except for that bit mentioned earlier about the brake lights – no undue attention from police, no thieves and the usual horde of onlookers at stops. One thing about this car is the immediacy of the service one gets at gas stations; people just can't stay away from it.

Arrival in New York was a bit different. The car poked its blue and white nose out of the Lincoln tunnel just in time to nuzzle New York's ghastly noontime traffic. New York at noon – or most any other time – is no place for any vehicle other than a Checker cab or a Mini. Just getting across town and up the few blocks to the hotel in the GT 350 was torture, and we made an immediate vow not to take it out unless we absolutely had to. It was a good thing we did. In New York City you don't park your car, not if you are a visitor. Attendants do, and to a man they are a surly, insensitive lot, most of whom seem to have graduated from a home for the mentally retarded. New York is not the place for a Mustang GT 350; it's about the only place we found where the car isn't suited or readily adaptable to the local scene. If you are a New Yorker and are desirous of a 350, move.

Five days later we headed west into a drizzling rain. From there to a point 20 miles west of Chicago, we remained on Interstate toll roads, a distance of about a third of the way across country as the crow flies. As expected, gas mileage came up to 16.4 miles per gallon due to the steady throttle settings and constant speed without need for accelerating. Though the front tires were beginning to show some wear (8,300 miles on the odometer) handling in wet or dry was excellent with no apparent change in bite.

Through Ohio, Indiana and Illinois we were blessed with clear, sunny weather and no incidents whatsoever except that we were to read two days later that we had missed the destructive tornadoes by one day. It was the only piece of weather we did miss. From then on we got it from every angle.

It started with a drizzle in western Illinois. In Iowa it was, as anybody who reads newspapers knows, FLOODED. The signs kept coming: *Bridge Out, Detour, Watch for Trucks, Emergency Road Repairs, etc., etc.* Virtually every river, tributary and creek in the state was overflowing and much of the overflow was spilling on the road. The Mustang took on the universal brown of the Iowa mud but it kept going strong and stayed well glued to the road even where the mud had overflowed.

Nebraska was just the opposite, cold and very dry – at first. Then a warm wind came up from the south, stronger and stronger, straight from the left side of the car. Soon it began to pick up the farm land and river bottom from south of road and send it horizontally to the hills to the north. It was, in brief, a Nebraska dust storm. Like a thick, brown fog it was – we couldn't take a picture of it for fear of jamming up the cameras with the stuff. Then the wind shifted to come from the north, and the sandblasting started all over again from the other direction. In each case, the wind was strong enough to force the driver to put opposite lock on the wheel. This probably contributed to the earlier noted tire scrub. This stuff lasted for about four hours, clearing up just as we approached Wyoming.

Wyoming greeted us with an electrical storm of awesome proportions, thick, black, shot with gigantic bolts of lightning and carrying huge drops of rain that the poor windshield wipers couldn't really cope with. This shortly turned to snow and which, fortunately, didn't stick to the road. The next day dawned bright, sunny and cold. The snow had stuck in the high (6,000 feet) plateau country, but not on the road itself. This was the last fine weather we were to have. We barreled over Wyoming's excellent roads and enjoyed the scenery. Set up on the lean side to begin with, the car went better above 4,000 feet than it had at sea level; high altitude starvation didn't seem to appear other than for a slight sluggishness at top revs.

The only thing we hadn't encountered (other than the tornadoes) was a blizzard. Utah obliged. We were attempting to get through that state and into Nevada before dark, but Utah weather did its best to keep us there. Determination and the Mustang got us through, but it took just about twice as long as a normal trip over the same route.

The weather in Nevada handed us something between rain and snow that could only be called snush – any other words for the stuff cannot be printed in a family magazine. This stuff finally cleared and turned into intermittent rain and then cleared completely. Nevada has no speed limit as such – you can be nailed for reckless driving if you overstep the bounds of decency, but on a clear road nobody is going to clock you as long as you don't act like a squirrel. The GT 350 can go tremendously fast through highway bends without getting out of shape or looking the least bit squirrely, so we kicked up the pace, *averaging* in the neighborhood of 100 mph in several areas where the roads were clear of traffic, making sure in the areas that weren't so clear to avoid taking advantage of Nevada's permissive attitudes. We like Nevada the way it is, and don't want to see it changed.

As we wound up the mountain to Virginia City (where we wanted to discuss a Fall hillclimb with the powers that be) it began to rain. Up higher the rain turned to slush and then to snow. We had taken the Gold Hill road, site of last year's Martini & Rossi Hillclimb, and the last few yards got very dicey indeed on the new snow. We came over the Divide sideways, got the car straightened out but not stopped until the middle of town. Great in the wet, the Blue Dots leave something to be desired in snow. For this use we suggest prospective owners invest in a set of Pirelli Cinturatos and chains, a combination we've tested and found hard to beat.

We've wondered what it would be like to be marooned in Virginia City with a legitimate excuse to stay over. We found out. The snow kept up and, what do you know, we were snowed in for a whole day!

The next day we set out over the most mountainous part of the journey, U.S. 395 over the spine of the Sierras and including three 8,000-foot passes then down through the Owens Valley to the Mojave Desert and into Los Angeles.

The weather, happily, was clear as was the road. This piece of road is wide though quite curvy, and is a good test of suspension and general roadability. We borrowed a technique from the Shelby Driving School (more about which next month), using a very late apex on all turns. Using the full width of our own lane, we apexed each turn almost at the exit. This cuts down the usual amount of drift that an understeering car normally has on the exit of any turn, and also requires considerably less steering effort. It really works. The car could do the twisty road without effort at an illegal rate speed in an area where most cars find it hard to even maintain a legal speed. Suffice it to say that the run from Virginia City to Los Angeles was made in under eight hours, not so much a result of high top velocity or bursts, but a steady maintenance of top legal speeds.

As we mentioned last month, one can grow to hate a car under these conditions. How did we feel about it? The fact that we ordered one of our own immediately on return should be enough said.

MUSTANG

continued from page **18**

luggage space, including the trunk, will hold enough luggage and gear for two people to cover a month's worth of grand touring in the grand manner.

As we said, the 350 GT will draw a crowd anywhere, any time. Part of this crowd appeal is due to the exhaust system which emerges just in front of each rear wheel and announces the presence of the wild oats in rather insistent terms, especially when warming up in the morning. One passenger made the comment that it felt at first as though the pipes came in at each window and plugged in, stethoscope style, at each ear. This is a bit exaggerated but the noise level *is* on the high side until the car gets up to about 70, when the Cobra-like roar turns into a steady buzz for some reason or other, possibly due to distributor advance reaching peak. It was noticed by *Motor Trend's* test crew that their car became noisy when the ignition points closed up, retarding the spark.

All in all, the Shelby Mustang proved to be one of the most reliable sports cars we have ever tested. It went through every conceivable kind of weather, over every conceivable kind of road, with absolutely no trouble of any kind and it is, purely and simply, a sports car in every sense of the word. No car is perfect, at least not in pre-production form (ours was number seven) and the 350 GT was no exception. Most of the faults were of a niggling nature having to do with convenience rather than actual operation. The most glaring was the lack of ability of the windshield wipers to cope with the speeds with which the car can be driven in the wet. At anything over 60 mph the wipers skipped abominably, though the car was stable and secure at far higher velocities on wet pavement. However, how the car behaved under various conditions is the subject of the following episode, as are the adventures of those who would take what could be aptly termed an irresistible hazard into all parts of the country. Next month we'll tell you about that – fuzz, rain, wind, snow, thieves, idiots and all. Don't miss the next episode of the fish-eating horse. *–John Christy*

"I lost my hacksaw!"

MUSTANG GT,

CONTINUED FROM PAGE 27

price differential between the HP Mustang and the GT-350. However, all these modifications are not really necessary for street use. Most performance buffs would be satisfied with the suspension goodies and the Detroit Automotive limited slip rear end. I'm sure it won't be long before Shelby starts to offer GT-350 do-it-yourself conversion kits for those who already own Mustangs or who can't afford a ready-to-run GT-350.

If you feel that you need a two-seater interior, Shelby instrumentation, oversize wheels, a sexy exhaust system, quick steering, a trunk-mounted battery, the complete suspension and a fiberglass hood, then by all means honk on over to your local Cobra dealer and sign on the dotted line. We felt it was one hell of a car and found it quite difficult having to turn it back to friend Bill at Larsen Ford after testing. He did promise that we could take it to a local strip for a day of racing fun and games as soon as he took full display advantage of it.

We brought the test car back to the showroom at 11 PM, long after the place officially closed, and was greeted by dozens of sports car and rod enthusiasts waiting to soak up some of that Shelby Cobra jazz. Maybe we're jumping to conclusions as far as that list price jazz goes. It may be just what they're waiting for!

The Ultimate Total Performance car

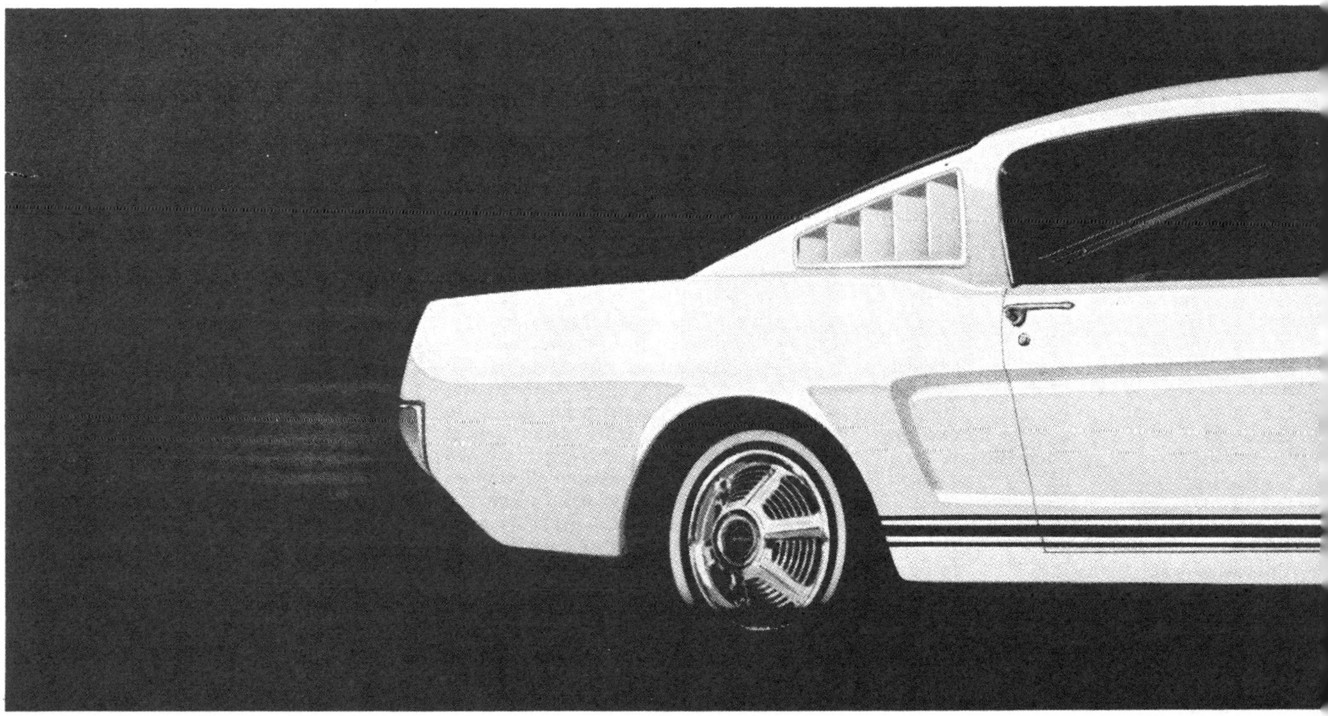

The car it inspired

Here's Mustang's brilliant new package for the driver who knows what Total Performance really ought to mean. Who else delivers exhilaration like this: a 225-horsepower V-8, fueled through a four-barrel carburetor. A 3-speed all-synchronized floor shift. A straight-through dual exhaust system with chrome "trumpet" extensions. Front disc brakes. Built-in fog lamps. The cornering power of Mustang's Special Handling Package. A GT insignia on the front fenders and the new symbol of Total Performance—the

—Ford GT

—new MUSTANG GT

GT stripe—above the rocker panels. And inside, a custom GT five-dial instrument cluster including ammeter and oil pressure gauge.

That's the basic GT package. It will light up most drivers' veins like neon tubing. But for even lustier appetites, there's a super-option—Mustang's 271-horsepower solid-lifter V-8 and the 4-speed manual shift. Both packages are available on Hardtop, Convertible or 2+2.

Happy news—and it's all spelled out at your Ford Dealer's. Take a taste. Soon!

MUSTANG

Unique Ford GT stripe—badge of America's greatest Total Performance cars!

MUSTANG WITH

text and photos by Eric Rickman

Carroll Shelby puts his Cobra's sting in America's most popular sports car — with jarring results!

There's a story that's been around awhile to the effect that if you cross a parrot with a tiger there is apt to be some doubt about just exactly what you'll get ... but when it talks, you'd better listen! This is somewhat the position that Carroll Shelby finds himself in after crossing a Cobra with a Mustang; the Cobra venom has given Ford's sporty little workhorse considerably more authority, enough to call it a race car.

To have a car "homologated" (accepted and classified) by FIA, the international governing body of automobile racing, certain rules must be observed. To establish a car as a production model, a hundred or more must be built and sold. Once the car has been homologated, changes are permitted in the engine and drive train, or the chassis, but not in both. This meant that Shelby had the choice of building a hundred all-out race cars, and trying to sell them, or building a few race cars and making the remainder street machines to be sold to the lucky public. With chassis or engine changes being the only alterations permitted, he chose to produce all the cars with a racing chassis and concentrate only on engine changes between the street and racing versions.

The street machine provided HRM is rated at 306 hp, packing a 289 cubic inch Cobra engine. The competition version pulls 350 horses from the same engine with the installation of an optional cam, increased compression, port modification, etc., plus some mighty fine tuning that adds 44 hp. The bare makin's for both the street and competition versions are sent to Shelby's Los Angeles production line less hood, engine and rear seat. Stock high-performance engines arrive in crates and go directly to the engine department where they are converted to Cobras: High-riser

FANGS...GT-350

manifold and center-pivot Holley carburetor are fitted; the latter eliminates stall on turns and is rated at 718 cubic feet per minute air flow. Ford dual-point high-performance ignition with centrifugal advance is next, along with the Cobra 6-quart, finned, cast aluminum oil pan, and rocker arm covers of similar material and motif. A set of 1½-inch i.d. headers that mate to a 2-inch exhaust system completes the engine package. This Cobra treatment raises the stock engine's horsepower from 271 to 306.

Chassis and body modifications are more involved. Beginning at the front, the upper A-arms are lowered an inch to drop the roll center and effect slight negative camber for high-speed turns. A stabilizer bar of 1-inch diameter is installed to reduce body lean under the same conditions. Longer pitman and idler arms reduce steering wheel travel from four to 3½ turns lock to lock (19:1 ratio). Ford's Kelsey-Hayes spot brakes with DS-11 lining are used on the front wheels.

The rear end really gets a working-over. Shelby's Mustangs arrive with the Galaxie high-performance rear end and brakes already in place, but they are switched to 2½-inch wide metallic brake shoes and a No-Spin center section. This special center section has no differential action, working on an actuating ramp and dog clutch principle. When the load from an axle over-rides the power input from the pinion gear, such as encountered during slow cornering when the outside rear wheel must travel farther, a ramp mechanism slips and disengages the dog clutch teeth on the over-riding axle, allowing it to turn free. When cornering under power, however, and the pinion gear is feeding more power than the outboard axle, the ramp setup can't slip and the clutch remains engaged, effecting a locked rear end: Ideal for racing, but not street use.

To absorb torque and prevent spring wrap-up, traction bars are in-

BELOW—Even in a full broadslide, Mustang GT has all four wheels planted squarely on the ground. The 1-inch front stabilizer bar holds body roll to a minimum, and the oversize Goodyear 7.75x15 Blue Dot tires show no roll-under.

BELOW RIGHT — The added cowl brace, heavier shock mounts and upper cross bracing are the first things to note when looking into the goody box. The rest of the compartment is filled with pure Cobra. Removing chromed air cleaner will reveal a center-pivot Holley four-throat carb on a high-riser manifold. Hood scoop ducts air directly to top of the carb. Ignition is Ford dual-point HP unit.

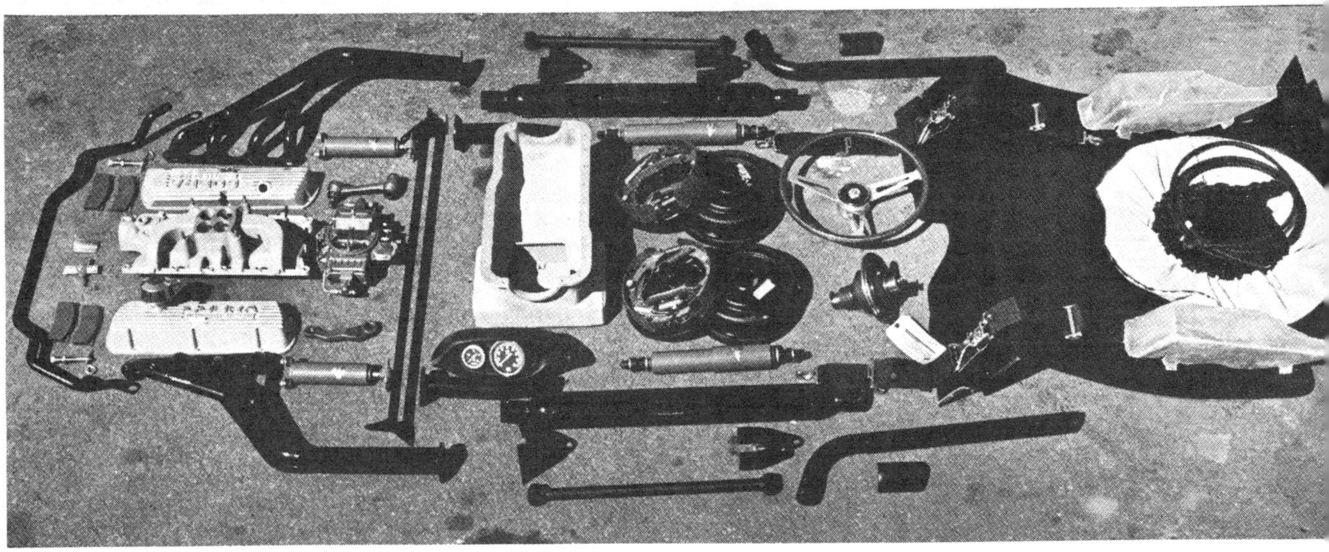

ABOVE – All this and the labor to make the conversion account for the $4547 price tag of the street racer. The latest version will retain the rear seat if you're a family type. Shelby-Cragar wheels optional.

FAR LEFT – The finned aluminum Cobra pan serves as oil cooler as well as providing extra capacity. Engine location remains unchanged.

LEFT – Cobra six-quart oil pan has interior baffling at sides as well as at mid-point. The check-valve-type flappers ensure oil supply to the pump even in severe cornering.

MUSTANG WITH FANGS...GT-350

stalled above the rear axle housing. To limit body lift, a steel cable attached to the body is looped under each axle housing. Slowing this movement also are heavier Koni shock absorbers on all four wheels. As an option, Shelby-Cragar 15-inch aluminum wheels are available. In either dress, all GT's are equipped with Goodyear Blue Dot High Performance 7.75 x 15 tires.

Body changes, along with a grille face lift, include a cowl brace and reinforcement of the shock tower area. The battery is moved to the trunk, and a fiberglass hood incorporating an air scoop is fitted over the engine compartment. Interior alterations consist primarily of the removal of the rear seat and the installation of a fiberglass deck carrying the spare tire (the rear seat is removed to conform with the Sports Car Club of America's definition of a sports car). To keep the driver informed, a supplemental instrument cluster with tach and oil gage is bolted to the upper edge of the dash.

The finished product weighs 2800 pounds at the curb, with 53/47 per cent distribution respectively, front and rear. Price out the door for the street car is $4547 plus tax and license, while the fire-breathin' competition job is definitely in the ball park as far as racing jobs go at $5950 plus tax. And if you do go for the race model, here's their exciting list of extras, in addition to the aforementioned group: blueprinted engine with racing cam, engine oil cooler, cold air plenum chamber between carbs and hood scoop, extra-capacity radiator, even larger straight-through exhausts, front and rear brake cooling ducts, 34-gallon fuel tank with 3½-inch quick-fill cap, electric fuel pump, magnesium wheels with 7-inch rims, roll bar with shoulder harness, fully instrumented dash, wrap-around bucket seats, fiberglass bumpers and apron, plexiglass windows, fireproof interior with extinguisher, and several other refinements. Even with this added, the competition version has a curb weight of only 2550 pounds.

With lots of rubber on the ground, and a near 50/50 weight distribution, the GT-350 corners as though it were on the proverbial rails; rather than spin it prefers to move into a four-wheel drift. The feeling is one of being in full command at all times. Shelby's street version is docile enough for everyday driving, with the possible exception that it steers a bit heavy and rides rather stiffly. It's long on performance, idles down smoothly, and still punches out from zero to 60 in 5.7 seconds. The brakes, in combination with the big tires, bring the GT down in one big hurry. It does take a heavy foot, however, since there is no power assist to detract from the true feel of what's going on at the wheels.

One final incentive to some is the fact that NHRA has placed the GT-350 in B/Sports category, so you're all ready to drag. We guarantee you won't fall asleep while driving this one, and it'll give anything you're liable to encounter fits, particularly those in the "other" camp. ■ ■

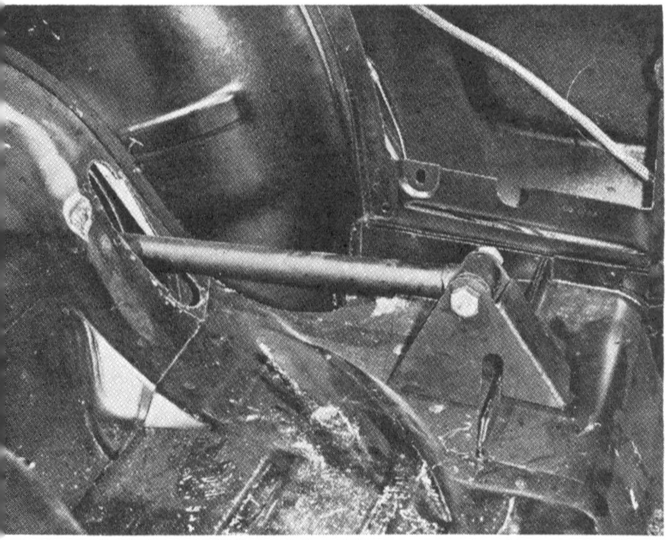

ABOVE – Forward mounting of traction bar is on top the frame as in a Corvette. The pivot point of the mount is directly over the forward spring eye, thereby giving correct parallelogram geometry and avoiding any binding action as axle moves.

ABOVE RIGHT – The rear traction bar mounting is a yoke welded to the axle housing. Again the pivot point is directly in line with center of the axle. Heavy-duty Koni shocks and a steel cable tiedown limit the body's vertical travel.

RIGHT – Front binders are made by Kelsey-Hayes for Ford. This is the optional unit available on stock Mustangs. The GT features DS-11 brake pucks which are harder and contain some sintered metal. These will withstand severe use.

BELOW RIGHT – The stock brake appears at left. Two and a half-inch wide replacement uses metallic linings. Heavy foot is needed as there is no power assist. Braking of light car is fantastic.

BELOW – NoSpin rear end has no differential action. It is basically a dog clutch assembly. Over-riding axle torque causes ramp system to disengage axle. If drive torque is greater, unit locks up.

This is a Hertz?

HERTZ TO RENT HIGH-PERFORMANCE CHAMPIONSHIP SHELBY GT-350-H WITH COBRA ENGINE

The newest, liveliest sports car in the U.S., the sleek Shelby GT-350, have hit the highways of America in an exclusive version available for rent only from Hertz.

Powered by the celebrated engine that has propelled Shelby's Cobra racing cars first across many finish lines here and in Europe, the spirited new auto is a roomy, four-seat vehicle with racing-car safety, disk and drum brakes and brisk handling characteristics. It holds the Sports Car Club of America's National Championship for B Production class and many industry observers consider it the hottest sports car produced in the U.S.

In the exciting GT-350-H version available exclusively through Hertz Rent A Car in more than 60 cities coast to coast, the car is readily identifiable by its jet black body and vivid twin gold racing stripes from grill to rear bumper. Other colors are also available from Hertz.

The addition to its rental fleet of the new "H" model — a "first" in sports car history — represents a sizeable expansion of Hertz' sports car operation which began two years ago in only seven U.S. cities, renting high-performance U.S. and English autos. Typically, Hertz reports, these cars are rented by business travelers who want a change-of-pace motoring experience, by sports-car owners away from home and by vacationers who consider driving these cars an enjoyable holiday sport.

The new GT-350-H is available with high-performance three-speed, competition-type automatic transmission on the same basis as any other car in the regular Hertz fleet of late-model vehicles. A motorist who wishes to rent the 4-speed manual-shift GT-350-H is given a familiarization drive through the Hertz Sports Car Club. He then receives a card which enables him to rent the vehicle from Hertz at any other club location. The same procedure is followed for the other manual-shift cars in the Sports Car Club fleet.

The new car is produced in Los Angeles by Shelby American, Inc., owned by famed race driver Carroll Shelby. The company develops Ford's 2-plus-2 Mustang fastback into a true track-bred, high-performance vehicle using racing components in suspension, styling and power plant.

In Hertz' GT-350-H version, side ventilators are replaced by quarter windows, giving a "wraparound" effect. Side air scoops are added for the rear brakes. The honeycomb grill is replaced by a horizontal bar assembly. The track-style suspension utilizes front sway-bar and anti-torque and stabilizer bars. Added instrumentation includes oil-pressure gauge and tachometer. A 306-horsepower, 289-cubic-inch Cobra engine with aluminum high-riser manifold and tuned header-exhaust system are among the other modifications in the car.

When Shelby unveiled the GT-350 last year, the new car drew enthusiastic notices from motor publications, which termed it: "easily the equal of anything that Europe has to offer in the GT category . . . the most exciting car to hit the domestic market . . . positively exudes character, something rare in this day of follow-the-crowd compromises and design." The GT-350 and the world champion Cobra are distributed internationally through a network of franchised Shelby dealers. ■

G.T. 350

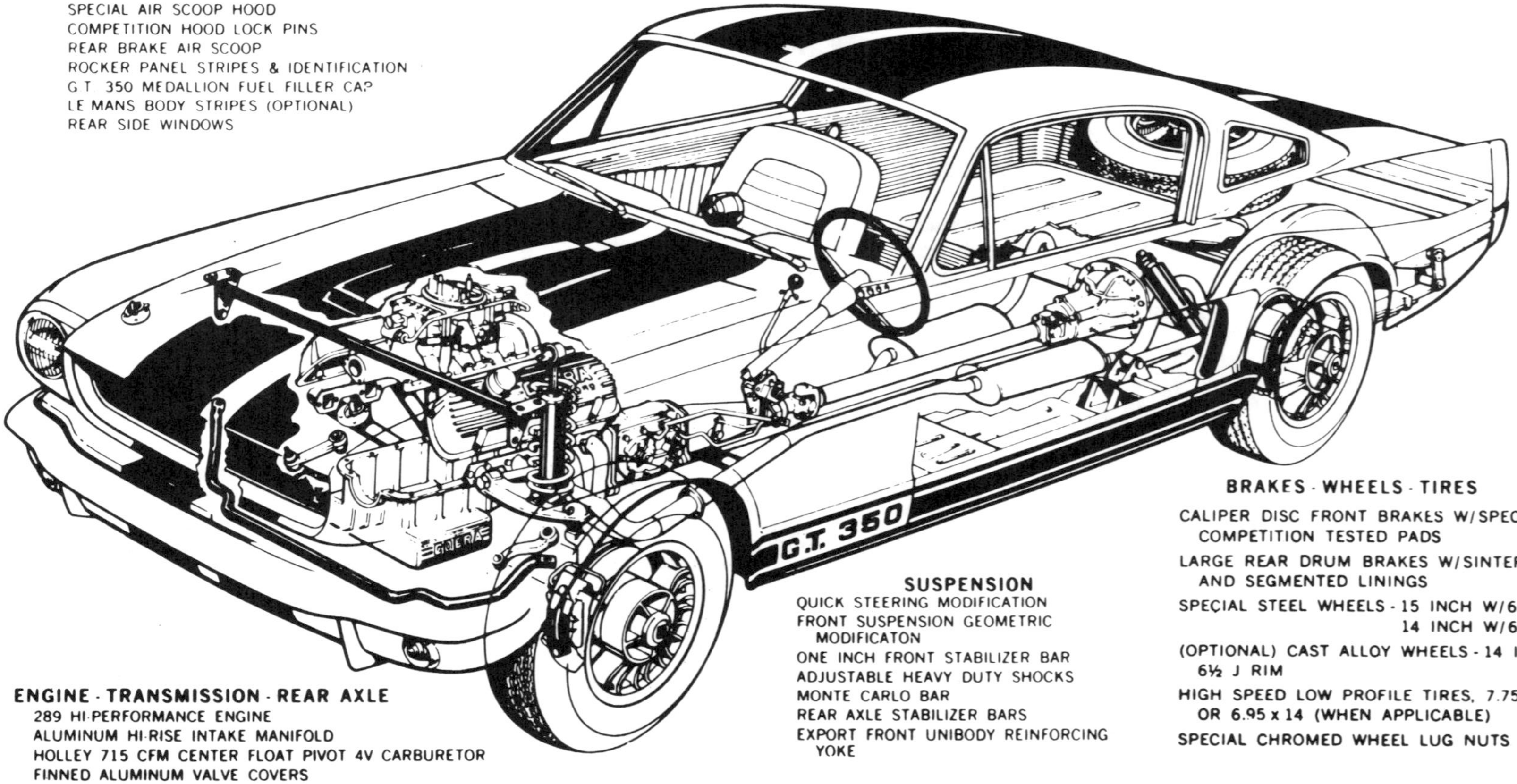

INTERIOR
REAR LUGGAGE PLATFORM
FOLDING REAR SEATS (OPTIONAL)
TWIN BUCKET SEATS
COMPETITION 3 INCH WIDE SEAT BELTS
FIVE DIAL INSTRUMENT PANEL
COWL MOUNTED TACHOMETER
SIMULATED WOOD RIM STEERING WHEEL

EXTERIOR
SPECIAL AIR SCOOP HOOD
COMPETITION HOOD LOCK PINS
REAR BRAKE AIR SCOOP
ROCKER PANEL STRIPES & IDENTIFICATION
G.T. 350 MEDALLION FUEL FILLER CAP
LE MANS BODY STRIPES (OPTIONAL)
REAR SIDE WINDOWS

ENGINE - TRANSMISSION - REAR AXLE
289 HI-PERFORMANCE ENGINE
ALUMINUM HI-RISE INTAKE MANIFOLD
HOLLEY 715 CFM CENTER FLOAT PIVOT 4V CARBURETOR
FINNED ALUMINUM VALVE COVERS
TUBULAR EXHAUST HEADERS DUAL MUFFLERS
EXTRA CAPACITY BAFFLED ALUMINUM OIL PAN
4-SPEED CLOSE RATIO ALL SYNCHRO TRANSMISSION
HEAVY DUTY CLUTCH
(OPTIONAL) 3-SPEED AUTOMATIC TRANSMISSION
HEAVY DUTY DIFFERENTIAL & REAR AXLE
(OPTIONAL) NO-SPIN SELF LOCKING DIFFERENTIAL

SUSPENSION
QUICK STEERING MODIFICATION
FRONT SUSPENSION GEOMETRIC MODIFICATON
ONE INCH FRONT STABILIZER BAR
ADJUSTABLE HEAVY DUTY SHOCKS
MONTE CARLO BAR
REAR AXLE STABILIZER BARS
EXPORT FRONT UNIBODY REINFORCING YOKE

BRAKES - WHEELS - TIRES
CALIPER DISC FRONT BRAKES W/SPECIAL COMPETITION TESTED PADS
LARGE REAR DRUM BRAKES W/SINTERED AND SEGMENTED LININGS
SPECIAL STEEL WHEELS - 15 INCH W/6 K RIM
14 INCH W/6JK RIM
(OPTIONAL) CAST ALLOY WHEELS - 14 INCH 6½ J RIM
HIGH SPEED LOW PROFILE TIRES, 7.75 x 15 OR 6.95 x 14 (WHEN APPLICABLE)
SPECIAL CHROMED WHEEL LUG NUTS

IMPORT

Shelby Mustang GT 350 EGO-BOOSTER

The Mustang with the big sting — Shelby's GT 350 "street" version, turns up Down Under. There should be more high performance American machinery here in 1966.

A GLIMPSE of things to come? Shelby's GT 350 Mustang for street-and-strip here reposes in uncharacteristically sedate surroundings. We see all too few examples of high performance "yank iron" in Australia because of the enormous cost of importing such vehicles. Government restrictions are overwhelming and even when the cars are finally landed here there is the large additional cost of conversion to right hand drive.

Inevitably when the odd car does slip in, it attracts a justly large share of public attention. So it is with the Shelby Mustangs. The Shelby GT 350 is not hard to pick. All GT 350 Shelby cars have white main color exterior paintwork with two ultra-wide blue GT stripes running the full length of the body. Bonnet scoop, smart airducting, enormous polished and chromed magnesium wheels, sill-stripes, competition bonnet safety latches, complement the suave all-black interior trim, with such accessory niceties as wood-rimmed steering wheel and tachometer-oil pressure binnacle.

This is the '65 model imported just at the close of last year by Laurie O'Neill of Peterbilt Truck Sales. It came to us by person of Frank Matich who happened to become infatuated with it, soon

The interior is fitted out with two buckets only, and the back seat area has been sacrificed to the luggage department to house spare wheel and provide a large platform.

Passenger and driver are equipped with an enormous pair of lap strap safety belts with stirrup-type spring-ball and lever fasteners. Each nylon strap is about seven inches wide, and is tipped by a large square of tanned leather on which the steel fasteners are secured. Gear-lever is the normal Mustang pattern (four speed conventional) with lift-lever reverse lock. The wood rimmed wheel is brass-studded and extremely light and sensitive.

The boot is lined and the battery stored there. Probably the least impressive part of the whole car is the engine bay in which the big 350 engine seems dwarfed. COBRA lettering caps all engine covers — tappet box, sump etc, and there is a strong torque rod running laterally between the guards — a special addition to the Shelby Mustang. This healthy unit however produces a neat 285 bhp at 5000 rpm with 325 lb/ft of torque at 3000 rpm. Top speed has been tested in the States at a figure in excess of 120 mph and the car will blast down a quarter mile in just 14.7 seconds.

But this car will probably appeal in a far more obscure way to most people . . . its stunning exterior face makes it an ultimate car as an ego-booster. #

after first seeing it on the wharf. The 1966 version, slightly re-styled, can be seen in the accompanying story of Shelby and the '66 prospects.

The car can be readily identified from all angles as the GT 350. From the front the honeycomb-wire mesh grille (replaced by horizontal slats in '66 model) and bonnet scoop make it easily recognisable. From the sides, there are the magnesium wheels, identification stripes; from the rear and above, the big GT stripes, and from below (slow pedestrians only) the twin protruding exhausts just forward of the rear wheels.

Tinted glass all round is standard. A very smooth touch is the special type bonnet catch. Two vertical poles simply protrude through the glass fibre bonnet. Through each pole slips a retaining pin, and hook-over ring. It is the simplest and safest method we've seen and possibly the only one that could be employed with the fibre bonnet. Doors and deck lid are all steel as on standard Mustangs.

FOUR ASSORTED MUSTANGS

One of the most controversial stories in the first issue of ROAD TEST was about the then-new Mustang in which we found several deficiencies along with a great many attractive features. The staff rightly opined that it would be a popular seller in spite of these negative values and we have kept a continuous check on it because of high public interest.

To date the magazine and staff members have owned six Mustangs: 1965 Six (170 cubic inch), 1965 V8, 1965 V8 High Performance, 1966 Six (200 cubic inch); 1966 V8 High Performance; 1966 Shelby GT-350. This group of cars, in daily use, for the most part, has furnished a considerable accumulation of data and some rather firm convictions about their design, manufacture and performance which could not have been acquired with less familiarity.

An intriguing feature of the Mustang is that it can be almost anything you want. Ranging from the basic, lowest priced Six (around $2,600) up to the Shelby GT-350 (around $4,500), there is a near-infinite number of options or alterations which can be made to tailor it to your personal idea of cost and performance.

nd, if Ford Motor Company can't supply what you ant there are many independent manufacturers of peed equipment and accessories who will be glad).

The four cars reviewed ere are examples of this ıultiplicity of choice.

1: 1966 Six Cylinder, utomatic transmission, ower steering, de Luxe ıterior. Basically suited the person who wants Mustang styling but not ecessarily outstanding cceleration or speed.

2: 1966 V8 (200 hp) utomatic transmission, ower steering, Special Handling Package, disc rakes, air conditioning. Equipped in manner which eems to be preferred by he young executive who vants performance but insists on keeping the creature comforts in.

3: 1965 V8-HP (271 hp), four speed transmission, handling package, disc brakes, wide base alloy wheels. Set up by Dean Moon, this car represents the Mustang tailored or the knowledgeable car enthusiast who does not go o far as to compete in porty-car events.

4: Shelby GT-350, V8, (306 hp) 4 speed gearbox, vide base 15″ wheels, Pirelli Ciuturato tires, disc brakes, metallic rear brake inings, fast steering (manual), modified suspension front and rear, Koni shocks, competition fuel tank, plus special rally modifications. Strictly for the rally enthusiast or high speed touring in no-speed-limit countries.

MUSTANG SIX

This is our second six cylinder model and represents a great improvement over the 170 cubic inch engined 1965 model. As regular readers of ROAD TEST know, it was never a wholly satisfactory machine (see RT, July 1965) but after 25,000 miles we can report that it has not required any major repairs. Once the driver becomes used to the thumps, whines, growls and shakes plus the fact that the engine overheats when the air conditioner is used in hot weather, it isn't so bad.

Overall gas mileage has been close to 18 mpg for city and freeway driving. Long trip mileage has varied from 18.7 to 22, depending on whether the air conditioner drive belt was removed or not. The 170 cubic inch six just doesn't have the torque to cope with the air conditioner and we found that it was far better to free it up on long trips.

The 1966 model is down slightly from these mileages but to judge from the performance of the 1965, consumption will improve as the car grows older. This is also true of oil consumption. During the first 6,000 miles, the average was 1,500 miles per quart. Thereafter it used a negligible amount.

From its present indications the Six, driven in its normal manner, will have the advantage of about two mpg over the V8 automatic driven as its performance encourages one to drive.

Sixes with the basic three-speed or four-speed manual transmission are extremely rare and ROAD TEST has not been able to

Six cylinder Automatic. Zero to 60 mph: approximately 15 seconds. Fuel economy: 18-22 mpg. Price: $2870.

V8 Automatic, 200 hp. Zero to 60 mph: approximately 12 seconds. Fuel economy: 16-20 mpg. Price: $3,353.

V8 4-Speed, 271 hp (modified). Zero to 60: approximately 8.5 seconds. Fuel economy 15-17 mpg. Price: $3,650 plus.

350-GT 4-Speed. Zero to 60 mph: Approximately 6.6 seconds. Fuel economy 12-14 mpg. Price: $4,500 plus.

gather enough statistical data to indicate whether there is any economy advantage in shifting for yourself. We doubt it. The Ford three-speed automatic is a pretty efficient box with well-chosen ratios and the final drive ratio (2.83-1) with the automatic is matched fairly to peak torque vs. cruising speed. Moreover, the automatic, which costs about $175 as an option, brings $150 over the manual three speed from the used car appraiser at your friendly Ford dealership.

Power steering in this car is a concession to the woman driver. The Six is 250 pounds lighter than the V8 and some 200 pounds of this is removed from the front wheel burden, so it isn't unpleasant with manual steering (which also deletes another 28 pounds). When resale time comes, power steering is worth an extra $100 and it only costs $85, so why not make $15?

Lighter though it may be than the V8, the Six is not so much lighter that its 120 horsepower will make it a ball of fire in performance. Acceleration rate is comparable to that of the Falcon, Chevy II or Rambler American with similar engines. The Mustang fastback weighs only 20 pounds less than the Falcon 200 cubic inch six 2-door club coupe, a fact which surprises almost everybody. The 7 main bearing engine is highly reliable and notably smooth. However, the man who assented to installing a multi-blade fan instead of increasing the radiator capacity won't get any vote of confidence from ROAD TEST. (V8-engined cars have 15-quart capacity cooling systems and the Sixes only 9.5-quart.) With a larger radiator the Six could use a viscous-hub fan and lower the noise level inside the car by about 40% in addition to picking up a few horsepower. At 70 mph conversation is a bit strained because of the fan-created roar . . . which is not true in the stock V8.

The de luxe interior includes simulated wood paneling, simulated wood-rim steering wheel and embossed vinyl upholstery. In the case of this Six, the optional center console was installed. Taking these benefits one at a time: The paneling is superior to that in some competitive makes, such as Pontiac, at least as far as ROAD TEST is concerned because it shows no inclination to peel off. The steering wheel intrigues everybody at first and satisfies some wholly. However, according to some members of the staff, after a couple of hours of driving it begins to seem awkward and not as comfortable as the regular wheel. The upholstery is attractive in its running horse motif but it seems likely that after a long, hot ride you might have a Mustang embossed on your back. The console has no compartment in it and its chief function, aside from decoration, seems to be to provide another bit of chrome which can dazzle you when the sun is right.

When delivered, this car was in pretty good shape. Only the sticking right door and a hard-to close deck

Four Mustangs at same speed in same corner at Willow Springs Raceway exhibit different handling characteristics. Six has more roll but less understeer than 1966 V8. Moontang's wider base wheels make it more neutral. Shelby 350 is loafing through turn where others are working.

lid marred perfection. In spite of greatly increased production, Mustang quality control seems to be higher this year.

In summary; the Six is smooth; has adequate performance for non-sporty driving; can be expected to be relatively trouble free and more economical than the eight and is just as much of a Mustang as the others as far as your neighbors will be concerned. You can save about $250 by selecting it instead of the basic V8. But, you'll have to endure a higher noise level at highway speeds.

MUSTANG V8, 200

This is the big seller of the nation's most sensational seller, so it deserves a close scrutiny.

The advantages of the 200 hp V8 Mustang over the Six is in noticeably better performance—as regards acceleration and top speed. But, and this is an important factor, it doesn't handle as well—which shoots it down right there. Our experience with the early V8, which we found to require special shock absorbers before it was at all roadable, indicated that the car needed stiffer suspension. Having bought another 1965 with the optional Special Handling Package, we wouldn't consider the standard V8 in 1966 without the same equipment and we wouldn't recommend it to anyone.

The special handling package is a matter of $30 and must come on all cars equipped with disc brakes anyway, so there is no excuse for not including it in your order.

Drum brake capacity on the V8 is increased over the Six by almost 20% through the use of larger shoes but effectiveness is reduced by increased weight transfer to the front wheels.

Discs are absolutely mandatory as far as the staff of this magazine is concerned. At that, the Six still establishes a superior deceleration rate on its initial stops from 70 mph with 27 ft./sec^2 compared with a maximum of 25 ft./sec^2 posted by the V8. This is due, of course, to the better weight distribution of the six which tends to decrease the onset of rear wheel lock up . . . the limiting factor in deceleration with these cars. In repeated panic stops and in flooded streets the discs proved to be much superior, naturally.

The weight distribution factor is strictly responsible for the stock V8's inferior handling in comparison with the Six. However, measures can be taken to improve it beyond those which can be carried out in the Six.

The V8 has five-bolt wheels with 5 inch rims on which can be mounted 7.35 x 14 tires. The Six has four-bolt 4½ inch rims that will only take the standard 6.95 size. You can get 100 level Nylon or 120 level red-strip Nylon optional 6.95 tires from your Ford dealer when you order, but the extra tread width (almost a half inch) on the fatter tires is a much better buy for handling.

Air conditioning unit and floor-shift for automatic eliminate center of front seat as adult passenger spot.

Unyielding, sharp edged padding for top of Mustang dash is poor design feature, offers an unnecessary hazard.

Disc brakes, available again on Mustang after being discontinued early in model year, are highly satisfactory.

For what it is worth, the tires which came with this particular Mustang were B. F. Goodrich and exhibited far poorer traction in the wet than the Firestones on the Six. Heavy rain also proved that a single-speed wiper is not the answer. Although the 2-speed is listed separately as an option, and is being included as "standard equipment" on some vehicles, you can't get it on the Mustang without buying the whole "visiiblity group" ($30) which includes an inside-outside mirror of dubious value. Better to take the mirror and ignore it, we now feel.

The new optional bench seat with the folding center armrest was specified for this car. However, carrying three adults in the front of a Mustang isn't really practical so aside from the armrest its value is comme-ci, comme-ca. A long trip in the back seat is, to say the least, not something to look forward to either. Interestingly enough, in the wonderful world of options, the extra-cost bench seat ($25) cannot be had with the de luxe interior decor group.

Instrumentation is improved in the '66 by virtue of the fact that an ammeter and an oil pressure gage are used, rather than warning lights. Still, the sharp-edged "padded dash" is as sharp-edged as ever and constitutes a hazard in our opinion. The seat rake is adjustable (by screw), which is not the case with many cars and can be made more comfortable for the outsize driver. (No six-way power seats for Mustang buckaroos.)

You can expect around 17 mpg in the long run if you aren't a leadfoot. Use of the air conditioner brings this down a couple of mpg during hot weather but the engine has no problems in coping with the load.

This car was delivered in good condition with the exception that the padded vinyl top had been polished with the rest of the paint which left it a trifle murky. It has developed a creak in the right door and a slight whine in the rear axle but no major problems seem in the offing. To judge by experience with 1965s, there will be low upkeep if nothing has turned up in the first five thousand miles.

MOON-MODIFIED V8-H.P.

Engine options available for the Mustang above those discussed include a 225 hp and a 271 hp V8. These are 289 cubic inch displacement, the same as the base V8 with some hotrod techniques applied: longer cam timing, larger carburetors, more compression, etc. The 225 costs a little over $50 above the 200 but the High Performance 271 calls for a whopping $330 more, or, figured in with the GT equipment group is $275 more (the GT group adds $151 to the base price).

Many knowledgeable owners prefer to make their own modifications and put the nearly $500 difference toward some proprietary items to improve handling and performance. One of the magazine's 1965 Mustang's was sold to Bill Eggleston, a friend of staff

Substitution of Moon quick-fill gas cap for stock Mustang cap ended thefts of this attractive item.

Moontang owner Bill Eggleston checks tread depth. Tire mileage has been excellent.

Additions to 1965 HP V8 by owner include Moon tachometer mounted on steering column. Car has been used as rolling test bed for speed equipment.

member Dean Moon who heads one of the country's largest speed equipment firms. Moon has used Eggleston's car as a rolling test bed for numerous carburetor, ignition and camshaft experiments seeking maximum output with tractability. The "Moontang" as the bright yellow machine has been dubbed, is in daily service as drive-to-work car and now has 30,000 miles on the odometer.

With the presently installed Potvin camshaft, the Moontang 289 delivers 285 horsepower but has no objectionable idling or low speed characteristics. It is smoother and has better low end torque than the GT-350 which gets its peak 329 lb./ft. of torque at 4,200 rpm. The Moon modified engine is comparable to the regular 271 in its streetability. The 4-speed gearbox in the Eggleston car is a Ford design with slightly different gear ratios than the Warner T-10 in the Shelby version. The HP Mustang has 3.50-1 final drive ratio, the 350-GT runs 3.89s. The combination being run in each car is near-ideal.

A marked improvement over the car's original handling was obtained by merely mounting Keystone wheels with six inch rims which adds quite a bit of base to the tires. Tire wear has been good, with an estimated 5.000 miles left on the original tread.

The only problems encountered in more than a year of vigorous driving have been a bent clutch release lever and a noisy bearing in the transmission.

SHELBY GT-350

Modifications made to the fastback Mustang by Carroll Shelby's organization include almost everything but the sheetmetal . . . and even the hood hasn't been untouched; it is fiberglass. The engine has individual-pipe headers, open exhaust, hi-rise manifold, Holley 4-barrel carburetor, extra capacity sump and dress-up rocker covers to distinguish it from the production model. Suspension has been changed in several details such as: moving the inner anchorage of the front upper control arm down to raise the front roll center; substitution of a one-inch stabilizer bar for the stock .84 inch; mounting torque-reaction arms at the rear, using Koni adjustable shocks all around along with 6" rims and 7.75 x 15 Goodyear Blue Dot tires. A heavy-duty limited slip differential is used; metallic rear brake linings are fitted and the power steering gear ratio is used without power assist. An instrument cluster is mounted on the dash. Distinguishing stripes, wood rimmed wheel, etc. are dress-up items. A 35 gallon fuel tank and electric pump are available for the competition-minded.

This is basically the car bought by staff photographer Bob McKay who participates in several international type rallies each year. His analysis of the car is that it is strictly for such a purpose; that it is too heavy in steering effort and too stiff in suspension to be acceptable as an every day street machine except for "the absolute nut." As a GT car and rally material, it is something else. His story follows.

Rear seat space in Shelby 350 is usurped by spare tire moved from luggage compartment to make room for big tank.

Fiberglass hood on 350 is secured by pin-lock of type used on racing cars. Scoop is functional, stripes cost extra.

Straight-through mufflers and short exhaust pipes ending in front of rear wheels are part of Shelby package.

Braking action of Mustangs with three different suspensions and weight distributions shows considerable variation. Six has better deceleration rate on first stop than V8, but V8's disc brakes are better when hot or wet. GT car showed highest deceleration rate and best balance between front and rear braking power.

THE MUSTANG GT-350 AS A RALLY CAR

By
Bob McKay

Although the emphasis in rallies today is more and more on regularity and accurate navigation, there are still a few events in which the car and the driver are put to a real test. You might easily win a typical SCCA rally in an old Hudson with the right navigator and equipment, but the high speed, late-penalty type of rally such as the 24-Hour go in Mexico calls for a real GT-prepared automobile. Having participated in this International event in 1963 (driving a Corvette), I knew that my Buick Riviera wasn't exactly the vehicle to enter in the Class A Gran Turismo category.

Approximately 95% of the route of the IX RALLY INTERNACIONAL DE LAS 24 HORAS is over mountain roads. Altitudes range from sea level to 11,000 feet and, while it is all paved, the highway system can't be compared to anything the domestic driver encounters except in some parts of the Appalachians. There is every expectation that bad weather will be encountered and the fact that the event is run at high speed over open roads where people and animals move at a slow pace, makes additional demands on the car. What is needed, in short, is an honest full-bore GT machine — if you are going for Class A. There are smaller-classification vehicles eminently suited to this terrain and these road conditions, but the real challenge is in the top category where entrants are allowed only 22 hours to complete the rally as against 26 for the small cars.

This situation calls for something on the order of the Corvette, Ferrari or Aston Martin. I'd had considerable experience with the 'Vette, in the 24 Horas and the 2,000 mile Rally Centro-America-Mexico, (which I co-drove with OCee Ritch in 1962), and was tempted to buy another for this event. But, having read a great deal about the Shelby-350 Mustang, I decided to try it. I did, liked it and made arrangements to take delivery on one from Lew Spencer's High Performance Motors in Los Angeles.

There are a number of characteristics of the car which are evident only under the stresses of many miles of competition driving and practice. These conditions are hard to impose artificially and, like a sail-

Driver Bob McKay and navigator Paul Martel led Class A GT category and held First Overall spot for 14 hours of 24 Hr. Rally.

Mustang 350 was first tested on "Mil Cumbres" (Thousand Peaks) section of rally route where road is never straight for as much as 100 yards.

Interested spectators of rally practice trip question owner McKay about technical details of Shelby 350-GT.

Although well suited to open-road, high speed travel, 350 is annoying to drive in crowded streets such as this, in Guadalajara.

Roads are paved throughout Mexico but can change abruptly from wide highway to narrow lanes over ancient stone bridges.

ing vessel, the only way to find out if it can sail around the Horn is to go round the Horn. The biggest apparent item in the GT's favor is its modified suspension which makes it a handy thing to haul around mountain roads. Some of the options, such as the big fuel tank, are also bonuses for the serious rallyist. There are also only a few deficiencies (which can be corrected at relatively small cost). Therefore, it forms a good basic package at a fair initial price.

No production car has sufficient headlamp intensity to permit the sort of driving called for in the all-up rally, so my first move was to install Marchal Optique lamps in the place of the regular sealed beams. These lights fan out enough to illuminate the verge of the road but carry for several hundred yards. They will get by for highway use in the United States if used on low beam only. On high beam they'll blind any oncoming motorist. These lamps would probably be sufficient for domestic road conditions, but in Mexico or Central America, the prevalence of cattle, ambling pedestrians carrying huge loads and slow-moving carts makes even more illumination necessary. I mounted two Marchal Quartz-Iodine long distance lights in the grille opening. These lamps are usable only where there is no oncoming traffic to worry about. The combination was tremendous and enabled us to drive flat to the boards during the darkness. The only addition I'd make would be to add two fog lamps mounted as low to the road as possible. During the foggy stretches we were somewhat dazzled by the reflected candlepower.

My first get-acquainted drive with the car convinced me that it needs even more rubber to take advantage of the suspension. A set of 15″ x 6″ Cragar Super Sport wheels was fitted and I chose Pirelli Cinturato tires because my navigator Paul Martel, (a Frenchman living in Mexico), wrote me to: ". . . be sure we have tires for the raining roads . . ." My past experience with Cinturatos showed that they will stick well in the wet and yet live under the scrubbing they get from volcanic rock roads. The Cragar wheels, in addition to improving the looks of the car a great deal, are beautifully balanced and are held to close tolerances on run-out and roundness. These items are a wise investment for the competition minded. I switched back to the original Blue Dots for occasional street use but they are badly checked and cracked in 4,000 miles.

I didn't have adequate time to check the car out before leaving Los Angeles for Mexico City, so I drove it to Tucson, trailered it from there to Zamorra and drove in through the **Mil Cumbres,** ("Thousand Peaks"), section which was to be part of the Rally route. Here I discovered that hard cornering, rally style, produced some nerve-shattering thuds, grinding noises and whumps from the rear axle. On examining it in Mexico City I found that the trailing arms were fouling the remnants of the rebound-cushion bracket which had been torched off during the suspension modification process. We did some more flame-cutting, frame-pounding and ground off part of a bolt head and managed to cure the clearance problem.

Aside from having carried a good supply of the wrong heat range spark plugs, there were no other obvious foul ups. If I had been a bit more thorough in my under-car examination, I might have found the assembly miscue which put us out of the rally while leading, however.

In re-locating the battery to the rear compartment, a long cable is required. In this particular GT-350, the cable was draped over the frame and not clipped, or given extra wrapping where it went from the frame to the compartment. The cable, swinging and flopping as we pounded over the rough roads, eventually frayed and shorted, causing the fire which ended our ride.

Prior to this occurrence, though, the GT-350 proved that it is a great rally car.

After leaving the start at **Ciudad Satillite**, suburb of Mexico City, the Mustang's superior acceleration, brakes and cornering ability let us gobble up the narrow, 2-lane highway at a great rate. We could pass several cars at a time and safely make maneuvers which less agile automobiles simply couldn't match. On the initial stretch we quickly moved past the Pontiac GTO which had started 2 minutes ahead and repeatedly demonstrated that the GT-350 could outperform it on any type of road. (We were back and forth often because of intermittent regularity zones which placed us at the 2-minute interval again.)

The road from Mexico City to Tamazunchale, (where it seems there isn't a single stretch of straight road 100 yards in length), was built for the 350. I have never been so confident, or had such precise controllability, in a car on this type of road before. The fact that we could drive extremely fast through here enabled Paul to con us through the regularity runs with ease.

As an indication of the rate at which we travelled, we covered a section of highway in 4½ hours which took me almost 10 hours last summer in my Riviera.

Through the first 14 hours of the rally the car went superbly in driving rain and blackest night and I feel that we would have held onto First Overall (all the other Class A GT cars had retired) if it hadn't been for the shorted battery cable. However, that's rallying. We proved the essential worth of the car by taking it around the Horn.

If you aren't a hot-blooded rallyist willing to put up with discomforts, forget it. There are some things about the 350 that you won't appreciate and even some which I intend to change before my next event. (The quick-fill gas cap, for one. It does not seal completely and you have to live on a mixture of air and gas fumes.) But, as an enthusiast, you'll have a hard time to find a big GT machine which will perform as this one does for this kind of money.

Free-roaming cattle, lumbering buses and omnipresent road construction were hazards to rallyists who traveled at high speed on roads filled with normal traffic. Rally was excellent test of Mustang which McKay rates as admirably suited to purpose.

McKay shows Rally official broken battery cable which caused Mustang to retire. Shorted-out cable started fire adjacent to fuel tank, caused considerable excitement.

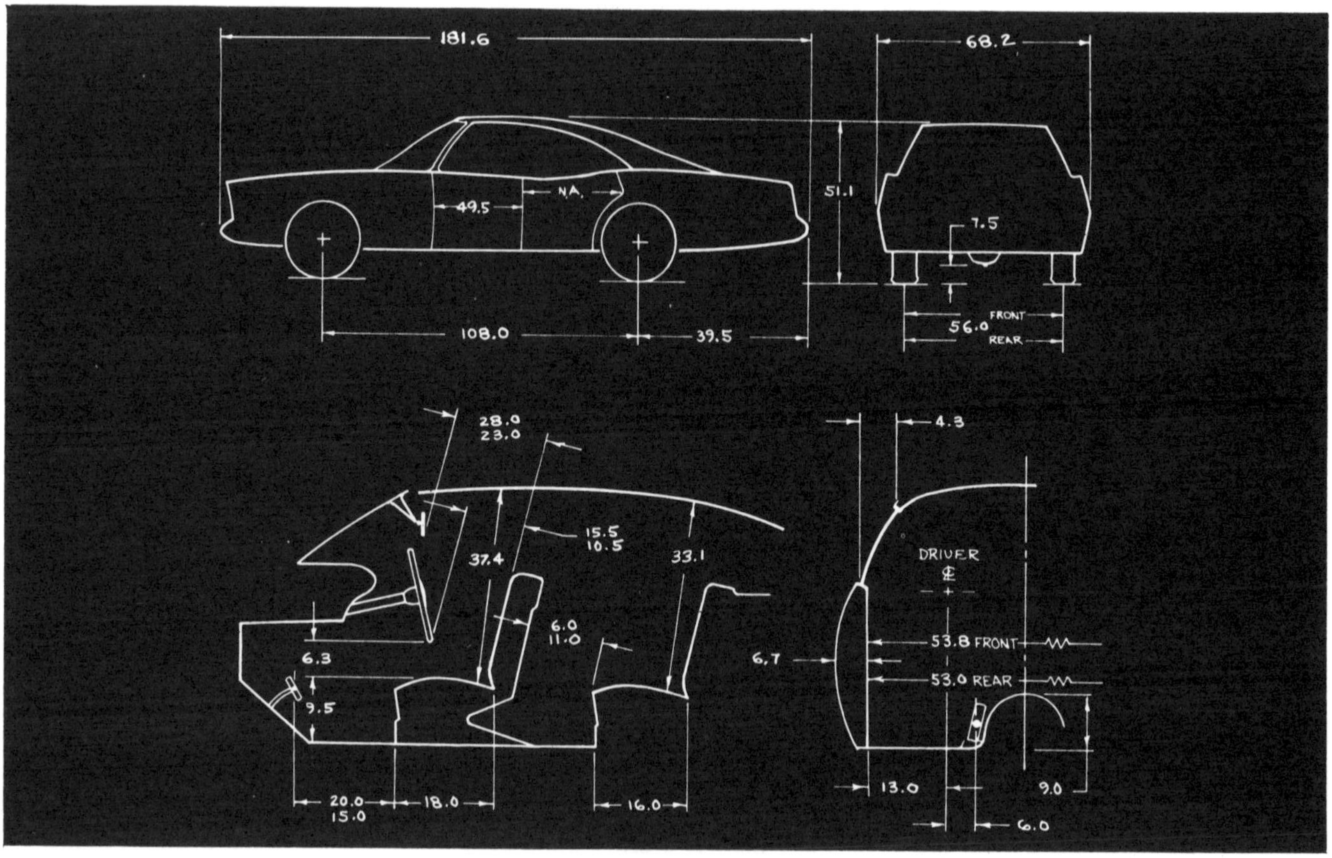

SPECIFICATIONS

Curb weight	2856 (8-H.T.)
Weight dist.	56/44
Brake type	drum
Swept area, sq. in.	212
Brake type	drum
Swept area, sq. in.	251
Brake type	disc/drum
Swept area, sq. in.	330
Tire size	7.75 × 14

Steering, turns	
Manual	4.64
Power	3.73
Turning circle	41.2
Clutch dia.	(6)8.5
Clutch dia.	(8-2V)9.0
Clutch dia.	(8-4V)10.0
Clutch dia.	(8-HP)10.4

Std. ratios,	
4th	
3rd	1.00:1
2nd	1.83:1
1st	3.29:1
Automatic:	
Drive	1.00:1
Drive	1.46:1
Low/drive	2.46:1

Optional transmissions:

	200-6	V-8 2V	V-8 HP	350 GT
4th	1.00:1	1.00:1	1.00:1	1.00:1
3rd	1.41:1	1.36:1	1.29:1	1.20:1
2nd	2.21:1	1.93:1	1.69:1	1.62:1
1st	3.16:1	2.78:1	2.32:1	2.36:1

Differentials:

(6)	3.20:1
(6)	2.83:1
(8-2V)	2.80:1
(8-4V)	3.00:1
(8-HP)	3.50:1

ENGINE

Type	6 cyl ohv
Bore	3.684
Stroke	3.13
Disp. cu. in.	200
Comp. ratio	9.2:1
BHP @ rpm	120 @ 4400
Torque @ rpm	190 @ 2400

ENGINE

Type	90° V-8 ohv
Bore	4.005
Stroke	2.87
Disp. cu. in.	289
Comp. ratio	9.3:1
BHP @ rpm	200 @ 4400
Torque @ rpm	282 @ 2400

ENGINE

Type	90° V-8 ohv
Bore	4.005
Stroke	2.87
Disp. cu. in.	289
Comp. ratio	10.1:1
BHP @ rpm	225 @ 4800
Torque @ rpm	305 @ 3200

ENGINE

Type	90° V-8 ohv HP
Bore	4.005
Stroke	2.87
Disp. cu. in.	289
Comp. ratio	10.5:1
BHP @ rpm	271 @ 6000
Torque @ rpm	312 @ 3400

ENGINE

Type	90° V-8 ohv 350 GT
Bore	4.005
Stroke	2.87
Disp. cu. in.	289
Comp. ratio	11.5:1
BHP @ rpm	306 @ 6000
Torque @ rpm	329 @ 4200

CAR and DRIVER ROAD TEST

SHELBY MUSTANG GT 350H

Yipes! Stripes! Let Hertz Put You In the GT 350's Seat

Cigarette butts in the ashtray of a rented car? Not when we rented a Shelby Mustang GT 350H from Hertz. No sir—we found cigarette butts in the rear brake scoop. What other rent-a-car outfit can make that claim? One of the technical boys at Shelby once hinted that those distinctive scoops on the flanks of the Cobra-ized Mustang are more decorative than functional, that the rear brakes don't need any more fresh air, except when racing. Apparently, nobody realized that Shelby's ingenious crew had created the *world's first external cigarette receptacle.*

Seriously, the alliance between Shelby and Hertz has eliminated the necessity of owning a sports car. Now the enthusiast can have his cake and eat it too. Rates vary according to locale: in the New York area it was a moderate $17 per day (or $70 per week) and 17 cents per mile. Hertz wouldn't rent us one when there was snow on the ground; said they didn't have snow tires for it. (New York has a law that says you get towed off the roads in a "snow emergency" if you don't have approved snow tires.)

Renting a Shelby Mustang is bound to be more ego-gratifying than a no-go small-bore import or a big, blowsy "sport-type" American sedan. The GT 350 is a real *guts* sports car, with hair on its chest—all the way down to its navel. Shelby has contracted to supply Hertz with one thousand GT 350s, designated the GT 350H ("H" for Hertz). Most of these special GT 350s will have the new high-performance automatic transmission, although a limited number will be available with 4-speed manual transmissions for the do-or-die purists. Said *puristi* will have to join the Hertz Sports Car Club, the qualification for membership being a demonstration of your ability to operate a manual gearbox. Hertz then gives you a little card, so the next time you want to rent a stick shift Shelby Mustang, you just flash your smile and your HSSC card.

There isn't any significant difference between the GT 350 you can buy and the Hertz version. The standard GT 350H color scheme is black with two broad gold stripes—a sensational, crowd-stopping combination. Other color schemes are available, including the regular GT 350's white with two broad blue stripes. A trio of narrower stripes along the rocker panel are interrupted by a "G. T. 350H" nameplate behind each front wheel. All the Hertz cars have the "occasional" rear seats and "mag-type" wheels that are options on the GT 350. Incidentally, the "H" might well stand for "Homologated" if Shelby—or, for that matter, Hertz—wanted to race the car as a Group 2 sedan; the 1000 examples Shelby will produce for Hertz fulfill the FIA's minimum production requirement.

Nor is there any significant difference between this year's Shelby Mustang and last year's, although there is a whale of a difference between Ford's stock $3500 Mustang and Shelby's version. Standard

high-performance, 271-horsepower Mustangs are delivered to the Shelby American factory near Los Angeles International Airport, where they are rebuilt to GT 350 specifications. Wide-base wheels and 130-mph 7.75 x 15 Goodyear Blue Streaks are fitted, as are trailing arms at the rear, a one-inch anti-sway bar at the front, and Konis all around. The front suspension geometry is altered, the steering speeded up, and a chassis brace is installed across the engine compartment between the upper shock mounts. The front brakes are 11-inch Kelsey-Hayes discs with heavy-duty pads; the rear brakes are 10 x 3-inch drums (3/4-inch wider than stock) with sintered metallic linings. Finned, cast aluminum rocker covers and sump are bolted on, as is a high-riser intake manifold and a big 4-barrel carb with 1.7-inch venturis and center-pivot floats so it won't cut out in turns. This, along with fabricated steel headers and low-restriction mufflers, boosts the horsepower figure by 35, to 306 @ 6000 (vs. stock 271 @ 6000), and the torque from 312 lbs-ft. @ 3400 to 329 lbs-ft. @ 4200. Final touches include the rear brake scoops, a new hood with a big air scoop and NASCAR-style hood pins, plexiglass rear quarter windows in place of the regular Mustang fastback's vent panels, a cleaned-up grille with the Mustang emblem offset to the driver's side, and the stripes. There are about a million options for racing the GT 350 in the SCCA's B Production category, many of which were proved in competition by the Cobra 289 and Ford's racing GT 40, both of which also use the basic Fairlane V-8. Last year, in their first season of racing, the GT 350s won several National Championships, including the final play-off at Daytona.

Last year, one staffer characterized one of the first GT 350s as a "brand-new clapped-out race car" and likened it to a World War II fighter plane. For '66 the car has been considerably refined, though it's still a tough, for-men-only machine, requiring strong arms to twist the steering wheel, strong legs to push the pedals, and strong kidneys to survive the ride. The exhaust pipes, which used to end just ahead of the rear wheels (stock-car style, and right under your ear) have been lengthened and rerouted to terminate aft of the rear axle. This change has made the noise level more bearable, and almost solved the problem we mentioned last year of exhaust fumes seeping into the cockpit. The ride seems more supple, though still what the British call "gratifyingly stiff," and the noisy, ratcheting-type, limited-slip racing differential is gone (thank heavens; it used to scuff the inside rear wheel around a turn and then unlock with a crack like a breaking suspension member). A regular, street-type limited-slip is optional, but none are fitted to the Hertz cars that we know of.

Visually, a '66 is distinguished from the '65 by the rear quarter windows and the brake scoops, which weren't on last year's model. The hood and its scoop look the same as before, but last year it was molded fiberglass, this year stamped steel. Inside, the change is more pronounced. The '65 GT 350s had standard Mustang instrumentation plus a pod atop the dash housing a tach and oil pressure gauge. The steering wheel had a wood rim and considerably less "dish" than the standard wheels, and the horn was operated by a spring-return toggle switch on the fascia. This year, the pseudo-woodrim with phony rivets, and the 5-dial instrument panel of Ford's Mustang "GT" have been adopted. The "GT" instrumentation includes an oil pressure gauge, so the tach now sits alone on top of the dash. The automatic transmission is a new feature, as are the optional rear

(Continued on page 57
Specifications Overleaf)

SHELBY MUSTANG GT 350H

Manufacturer: Shelby American, Inc.
6501 W. Imperial Highway
Los Angeles, Calif.

Price as Tested: $17/day or $70/week, plus 17¢/mile

ACCELERATION

Zero To	Seconds
30 mph	2.1
40 mph	3.1
50 mph	4.9
60 mph	6.6
70 mph	8.9
80 mph	10.8
90 mph	14.2
100 mph	17.9
Standing ¼ mile	93 mph in 15.2

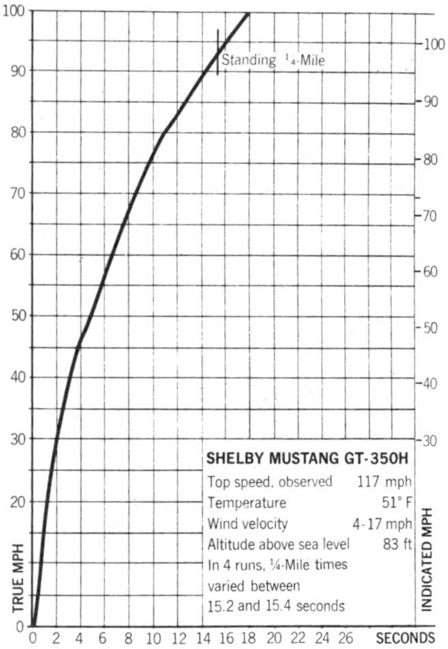

ENGINE

Water-cooled V-8, cast iron block, 5 main bearings

Bore x stroke..4.00 x 2.87 in, 101.6 x 72.9 mm
Displacement..............289 cu. in, 4727 cc
Compression ratio................10.5 to one
Carburetion....Single 4-barrel, 1.7-in venturis
Valve gear. Pushrod-operated ohv, solid lifters
Power (SAE)............306 bhp @ 6000 rpm
Torque..............329 lbs-ft @ 4200 rpm
Specific power output.....1.05 bhp per cu. in, 64.4 bhp per liter
Usable range of engine speeds.800-6000 rpm
Electrical system..12-volt, 55 amp-hr battery,
Fuel recommended...........Super premium
Mileage..............................6-12 mpg
Range on 16-gallon tank.........96-192 miles

DRIVE TRAIN

Transmission...3-speed automatic with torque converter

Gear	Ratio	Overall	mph/1000 rpm	Max mph
Rev	2.80	10.29	−7.0	−42
1st	2.46	9.57	7.9	47
2nd	1.46	5.68	13.4	80
3rd	1.00	3.89	19.5	117

Final drive ratio....................3.89 to one

CHASSIS

Platform steel frame, semi-integral steel body
Wheelbase.............................108.0 in
Track......................F: 57.0, R: 57.0 in
Length...............................181.6 in
Width.................................68.2 in
Height................................51.2 in
Ground Clearance......................5.3 in
Curb Weight.........................2884 lbs
Test Weight.........................3158 lbs
Weight distribution front/rear.........52/48%
Pounds per bhp (test weight).............10.9
Suspension F: Ind., upper wishbone, lower control arm and drag strut, coil springs, anti-sway bar
R: Rigid axle, semi-elliptic leaf springs, trailing arms
Brakes, F: 11.3-in Kelsey-Hayes discs, R: 10 x 3-in drums, 408 sq in swept area
Steering.......................Recirculating ball
Turns, lock to lock..........................4.0
Turning circle..............................4.0 ft
Tires and wheels..........7.75 x 15 Goodyear Blue Streaks on 6JK rims

CHECK LIST

ENGINE

Starting	Very Good
Response	Excellent
Noise	Fair
Vibration	Very Good

DRIVE TRAIN

Clutch action	—
Transmission linkage	Good
Synchromesh action	—
Power-to-ground transmission	Very Good

BRAKES

Response	Very Good
Pedal pressure	Poor
Fade resistance	Very Good
Smoothness	Good
Directional stability	Good

STEERING

Response	Good
Accuracy	Good
Feedback	Good
Road feel	Good

SUSPENSION

Harshness control	Fair
Roll stiffness	Very Good
Tracking	Excellent
Pitch control	Very Good
Shock damping	Very Good

CONTROLS

Location	Good
Relationship	Good
Small controls	Good

INTERIOR

Visibility	Good
Instrumentation	Good
Lighting	Good
Entry/exit	Good
Front seating comfort	Very Good
Front seating room	Very Good
Rear seating comfort	Fair
Rear seating room	Poor
Storage space	Very Good
Wind noise	Fair
Road noise	Fair

WEATHER PROTECTION

Heater	Excellent
Defroster	Excellent
Ventilation	Good
Weather sealing	Good
Windshield wiper action	Very Good

QUALITY CONTROL

Materials, exterior	Very Good
Materials, interior	Fair
Exterior finish	Very Good
Interior finish	Fair
Hardware and trim	Fair

GENERAL

Service accessibility	Very Good
Luggage space	Fair
Bumper protection	Fair
Exterior lighting	Very Good
Resistance to crosswinds	Very Good

SHELBY MUSTANG *(Continued from page 55)*

seats. A carry-over feature is the use of USAF-style, three-inch-wide seat belts with-metal-to-metal buckles and quick release mechanisms, just like the racing cars.

The changes for '66 have made the GT 350 more civilized, and we still think it's a great sports car in the classic tradition, but there are some aspects of Shelby's metamorphosis that we criticize. Prime among them is the car's interior appearance—which, after all, is what the driver spends most of his time looking at. It looks too much like any run-of-the-mill Mustang that half-a-million average Americans are using for utility transportation.

We also found fault with the rear seats and quarter windows. Construction of both was rough-and-ready, and in poor condition after only 5000 miles. The upholstery was beginning to tear, the trim was starting to come loose, and rain leaked around the plexiglass window. The rear seats are none too comfortable, but, surprisingly, better than the last notchback Mustang we drove. But it's sports car, not a bloody bus, and besides, few of our complaints would bother the man who only rented the car.

The staff was most impressed with the looks of the GT 350H. It makes any stock Mustang look sick, particularly with those oversize tires and stylized wheels. We were also favorably impressed with the overall performance and general roadability of the car. The engine is lively and responsive, and does a much better job of getting its power to the ground than any of the six "Super Cars" we tested in the March issue.

The automatic transmission felt a bit like the one used in the Ford Fairlane GT/A of our "Super Car" test, but better on part-throttle acceleration. Shelby American modifies this transmission to shift at 5500 rpm at full throttle, although the driver can hold it in each of the three gears with the shifter. Our shifts were made at 6000 rpm, and the acceleration times compare favorably with those for the 4-speed. Starting without wheelspin, the automatic actually has an edge on the 4-speed up to about 45 mph. Wheelspin can be induced by "pumping up" the converter (using brake and throttle simultaneously), but even then the tires have the situation well in hand—there's no fishtailing or useless clouds of smoke. Shifts are crisp and clean, probably faster than anybody but a drag racer could achieve with a manual. Fuel economy doesn't seem to suffer much, and it loses only two miles an hour of top speed. Flat-out, the car has a high, hammering note. At highway speeds, the changing pitch of the automatic shift jangled some people's nerves. The idle is lumpy, but in a way that promises good things to come.

Good things do come for the driver of a GT 350. Its cornering ability is a lovely mixture of the beast getting the better of you and you keeping hold of the tiger's tail. The taut suspension, well-controlled geometry, and big tires suffice to keep it on the road at insane speeds, and when it starts to slide, you can wrestle the slide in the direction of your choice. Its steering characteristic is pretty neutral, tending toward understeer. Driven too deep and too fast into a corner, the car can be "saved" by backing off (which scrubs off speed), by giving it part throttle (which squats the tail down for a better bite), or by full throttle (which gets the tail out and tightens up the turning radius). If steering with your right foot gets boring, you can always give the steering wheel a wrench to make the front end perform appropriate maneuvers. Needless to say, the GT 350 is more fun to drive than anything since the Mini-Cooper S. Maybe more fun than the Mini; at least when you floor the Mustang, it *moves*.

The level of creature comfort on long trips is amazingly high. There is a screw adjustment for the seat back rake angle which we unbolted to let the seat flop as far backward as it would go. Head-, foot-, and hip-room are good, but the door is a little close to the shoulder. The trunk space would be pitifully inadequate, except that the rear seats fold down to provide a huge luggage platform. Even with the seat up, there's a shelf under the rear window, like the VW's. Vision all around is excellent, as is the heater (it was one of the few cars in recent memory that could warm up its occupants on a sub-zero morning within five minutes of starting), but there was one thing we worried about constantly: the NASCAR style hood pins. The normal latching mechanism is removed, so the pins are the only thing holding the lid down. When some little brat steals the pins as souvenirs, you have to start looking for a 3/8-inch twig. Moral: When driving a GT 350 through treeless country, run over any kid who even looks at your automobile with envy. And for sure, they all will. **C/D**

PAXTON-BLOWN Novi Indianapolis engine, foreground, is a husky cousin to the Paxton-supercharged GT-350, rear.

MUSTANG GT-350-S

A Paxton Supercharger Provides an Added Kick

COMBINING THREE such unlikely elements as a highly-tuned and modified Mustang, a centrifugal type of supercharger and an automatic transmission is at best a chancy project. Any two of these elements would seem at the outset virtually incompatible. Yet, the subject of this road review proves that such ambitious assemblages, when undertaken by knowledgeable, competent craftsmen, can turn out remarkably well.

Carroll Shelby's Shelby-American Inc. took the chance when it began, last year, turning production-line Mustangs into GT-350s. It took a chance again when it started installing automatic transmissions in what is, for most intents and purposes, a pure sports car. Paxton Products took a chance, although a well-calculated one, when it topped this combination with its belt-driven, single-stage centrifugal compressor.

The outcome is a docile, 4-seat sports car that will out-thrust any 400-plus-cu. in. sedan on pure acceleration, or thrash it into submission on crooked roads. And, for only 289 cu. in. of displacement, this must be regarded as at least mildly sensational.

Yet, sheer performance isn't necessarily the most remarkable thing about the Paxton-supercharged Shelby-American GT-350. What is even more impressive is the car's ability to trickle through sticky traffic and perform mundane daily transportation chores in fuss-free fashion. The automatic transmission smooths out the highly-tuned engine's inherent lumpy low-speed characteristics with pleasant part-throttle operation. However, when the throttle is full down and maximum performance ordered, the response is instant and exhilarating.

The GT-350's virtues have been amply described in previous *CAR LIFE* articles (April and June, '65). Although an obtuse approach to the Mustang's basic problems, the GT-350 modification nonetheless makes it into as sporting a car as almost anyone could desire. Springing and steering may be a bit stiff for the ordinary motorist, but the enthusiastic driver generally seems to appreciate such attributes. GT-350 improvements for the '66 model year include a horizontally-barred grille, quarter-windows in place of the gill-slit ventilators and air scoops for the rear brakes. The rear seat, first removed to allow installation of the rear-end stabilizer arms, is now reinstalled as an optional item. The automatic transmission and the Paxton supercharger also are part of the option list.

The transmission is Ford's Cruise-O-Matic as strengthened for production-line attachment behind the 289-cu. in./271-bhp V-8 in Mustangs, Comets and Fairlanes. This option was announced by Ford in 1965, after augmenting clutches and bands and revising valving within the transmission to permit 6000-rpm upshifts. Full-throttle shifts now take place at 5500 rpm, first to second, and 6200 rpm, second to third. Gearing in this unit is the same as in the normal Cruise-O-Matic, i.e., 2.46:1 first and 1.46:1 second. The torque converter provides added multiplication of 2.1:1 at stall—about 1800 rpm.

Although not as ultimately controllable as Ford's GTA automatic transmission, this one does offer some independent control by the driver if he jockeys the lever between "D" and "L." For instance, *CL* produced its best acceleration by turning the engine nearly 7000 rpm in first gear, manually shifting out of first into second, then letting the transmission do its own 2–3 shift, which takes place at an adequate number (for this particular engine) of revolutions. Second could be re-selected from high at any speed over 20 mph, which made for thrilling, if noisy, passing and greatly more enthusiastic cornering. First in this automatic compares with first in 4-speed, while auto-second is fairly close to the 4-speed's third.

COMPARATIVE PERFORMANCES

	289/225*	HP-271	GT-350	Paxton	Unblown*
0–40 mph	4.5 sec.	4.5	3.6	3.6	(4.4)
0–60	8.5	8.3	6.8	6.2	(8.2)
0–80	15.8	14.2	11.2	10.5	(13.9)
0–100	26.8	22.5	19.0	16.8	(23.3)
Standing ¼-mile	16.8	15.9	14.7	14.0	(15.5)
Speed at end	84	85	90	92	(84)

(*Equipped with automatic transmissions, others, 4-speeds.)

To repeat, torque converter, supercharger and high-rev engine all are quite compatible. Where the engine needs its biggest help is at the very low end of its operational scale—below 2000 rpm. Here, the torque converter effectively multiplies what power there is into a useful quantity. Also, the supercharger at idling speed does not achieve an effective boost so the converter gives it breathing space by letting the engine turn up more rpm than would direct gearing. Where the converter phases out, the supercharger begins to pump—and the engine really starts to climb the power curve.

SUPERCHARGING, as a power and performance-improving process, has been around the automotive scene almost as long as the car itself. Renault of France took out patents on a rudimentary system in 1902; Chadwick of the U.S. fitted a Roots-type supercharger to a car which competed in a Pennsylvania hillclimb in 1913. Experimentation with aircraft supercharger systems began in the early 1920s. Both American and European racing cars made significant use of supercharging in the '20s and '30s, mainly because displacement restrictions tended to favor the supercharger engine.

Production-line automobile supercharging has been sporadic. Few applications have been made and those that were have been clouded by stories of superlative performance and dreadful unreliability. Production examples have been Duesenbergs, Auburns, Cords, Grahams in the between-wars era,

CHAN BUSH PHOTOS

GT-350-S

AIR CLEANER, plenum, compressor, pulleys, belt, tensioner, nuts and bolts are the necessaries for simple installation of a sizable boost.

Studebaker, Ford, Chevrolet Corvair and Olds F-85 in the more modern times. Of those, only the exhaust-driven turbo-supercharged Corvair still is produced. Most of these applications had something in common—design limitations prevented substantial power increases by the usual, and less costly, method of simply enlarging displacement.

They also had something else in common—they used the centrifugal type of supercharger for its greater flexibility of operation. Ford and Studebaker engineers turned to the McCulloch Motors Corp. for supercharger additions to their 1957 sedans. When Studebaker found it needed substantially more horsepower from its 289-cu. in. V-8s in the early Sixties, it returned to McCulloch. It seems somehow appropriate that the Shelby Mustang should also utilize Paxton (nee McCulloch) boosting for extra power.

The McCulloch supercharger has had a long and reasonably successful history. Company founder Robert Paxton McCulloch sold the first version out of a small shop in Milwaukee prior to World War II. Belt-driven to Ford flathead V-8s, it was modestly potent. After WW II, McCulloch moved his operation westward and started work on the Mark II version of his centrifugal blower. This model, actually designed by John Oehrili, utilized a planetary drive of king-size ball bearings to achieve the necessary step-up in impeller speed. This was the unit adopted by

SUPERCHARGING: THE BASICS

A SUPERCHARGER, reduced to its basic purpose, is an air pump. By simply pumping more air into an engine than can be supplied by atmospheric pressure, it improves the engine's volumetric efficiency, and therefore its power output. It is probably the simplest of all methods for increasing the power output of any given internal combustion engine.

Normally aspirated engines depend upon the pressure of the atmosphere (14.7 lb./sq. in.) to push in a fresh air and fuel charge when the intake valve is opened. And the power produced is in direct proportion to the density of the charge; combustion of the fuel/air charge results in greater power when greater amounts of oxygen are present. A practical example of this is the difference in engine power output between sea level and 10,000 ft. elevation where air is "thinner," i.e., atmospheric pressure is less.

The function of the air pump, then, is to compress air before it reaches the combustion chamber, thereby packing a greater number of oxygen molecules into a given volume. Then, when the intake valve is opened, this denser charge is forced into the cylinder where it burns hotter and longer, yielding greater energy to push down the piston.

There are three basic types of superchargers in common usage today on reciprocating, internal combustion engines. The most popular, used almost exclusively for competitive machinery, is the Roots type. The centrifugal supercharger has had perhaps the widest variety of usage, while the vane type has enjoyed modest popularity on small-displacement engines.

ROOTS—The best one extant is the GMC Diesel supercharger series which is extensively used in trucks and, after suitable modification, widely applied to drag- and straightaway racing engines where the supercharger's high delivery volume is important. In operation, gear-driven, counter-rotating, involuted impellers revolve on parallel shafts, trap and compress air within the housing. Close tolerances between impellers and housing must be maintained to reduce slippage of air past the blades, however, such slippage at high rpm cuts the Roots blower efficiency to as low as 50%. Primary drive by belts generally works better than gears because of pulsating air flow development. Matching of supercharger capacity to engine displacement is critical. Impeller gearing and air flow tend to create noise.

VANE—Literally a paddlewheel within a cylinder, the vane-type compresser is probably the least expensive to manufacture and the most efficient at low engine rpm. The paddlewheel in reality is a drum with sliding vanes mounted in radial slots. The axis of the drum is eccentric to that of the cylinder,

PAXTON SUPERCHARGER installation requires a minimum of plumbing. The plenum chamber simply bolts over the 289's standard 4-barrel carburetor.

GIL EDELSTEIN

IMPELLER, left, is driven by system of planetary balls.

Ford and Studebaker for their 1957 cars.

A third version, referred to simply as the Mark III, was developed just before McCulloch sold off the Paxton interests to John Thompson, who later sold to Andy Granatelli. The latter sold the small company to the Studebaker Corporation, although retaining direction of its operation. The Mark III was in principle similar to the Mark II, but smaller and more efficient and worked well on the Studebakers. Paxton-boosted Avantis and Larks established a whole handful of straightaway records at the Bonneville Salt Flats.

The Paxton Mark III also became the basis for a supercharging kit available

so that as the drum rotates, the vanes slide in and out to maintain contact with the cylinder bore. During rotation, the space between the blades first increases as it passes the suction port, then decreases to compress the air for the outlet. Usually belt-driven, this type of supercharger works best on small-displacement engines because of its low power requirement. However, the vanes have a sliding and rubbing friction which requires lubrication and the introduction of oil into the chamber causes a lowering of the fuel octane number. And, impeller speeds must be kept relatively low to minimize friction losses.

CENTRIFUGAL—An impeller rotating at high speed inside a casting increases the air pressure by increasing its velocity. Air enters at the center, over the crown of the impeller, and is thrown radially off the impeller tips. Routed through a diffuser, the rushing air's kinetic energy is then converted into pressure energy. Because pressure development is a function of impeller tip speed, size and weight are critical factors, and a large ratio step-up drive is required. Gears, belts and exhaust-driven turbines have been successfully used for power. Pressure varies as the square of the speed, so low engine and impeller speeds produce low pressures. However, the centrifugal type of supercharger offers the most performance per pound of weight and remains the most versatile in its operation. When combined with exhaust-driven turbines, they have proved particularly useful for supercharging Diesel truck engines and light aircraft gasoline engines. ■

ROOTS-type blower impellers rotate in opposite directions.

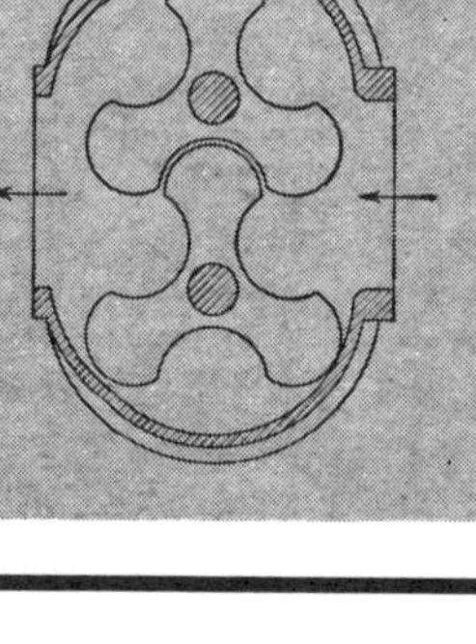

AIR trapped between vanes is forced into decreasing space.

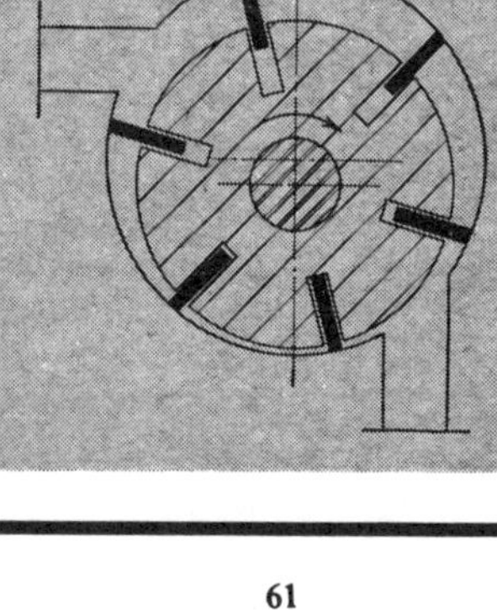

CENTRIFUGAL compressor changes kinetic energy into pressure.

GT-350-S

to automotive enthusiasts seeking bolt-on power increases. Other uses were found, too, particularly in the industrial field as air blowers and exhausters. One significant application is their usage in Polaris-class submarines' air-purification systems.

THE PAXTON Model SN is a fixed-ratio, single stage centrifugal type of supercharger. It has a 5.78-in. impeller which runs at 5.94 times engine speed, to give a maximum boost of 5 lb./sq. in. at 5000 rpm (engine). The drive pulley gives a 1.35:1 step-up and the ball planetary drive gives another 4.4:1 increase. Thus, at 5000 rpm crankshaft speed, the supercharger impeller is whirring away at 29,700 rpm!

Supercharger design theory suggests that impeller tip speed be kept just under the speed of sound for maximum efficiency. Thus the Paxton's 517 mph impeller tip speed (at 5000 rpm crankshaft) would be within the efficient area, and leave room for overspeed—7000 rpm would give 723 mph tip speed, or just under Mach 1 at earth surface level and standard temperature.

The supercharger drive consists of input and output shafts interconnected by a friction-operated planetary. The input shaft is belt-driven from the engine's crankshaft and its inner end forms the ball driver, or ring gear, of the planetary. Five hardened steel ball bearings act as the planet gears, friction-driving an inner race (sun gear) attached to the output, or impeller shaft. The driver is spring-loaded to maintain pressure against the balls and prevent their slippage on the race. The rotation of the outer ring thus is multiplied by the planet balls' diameter to drive the impeller 4.4 times faster.

The use of belt-and-pulley primary drive damps the surge-loading of acceleration and deceleration while the ball/race planetary is probably less expensive, and less vulnerable, than a gear system. However, the supercharger must have its own oiling system to cope with the high mechanical pressures on the balls and races. This utilizes automatic transmission fluid for its admirable load-bearing capabilities.

A well-engineered installation is at least half the battle in producing a satisfactory supercharger application and the Paxton kit is probably the neatest and cleanest on the market. Simple, sturdy aluminum castings mount the supercharger and its drive-belt tensioner, another neat set of castings encloses the carburetor. By pressurizing the carburetor, Paxton avoids some difficult mixture and linkage problems which would occur if it were mounted on the intake side of the supercharger. In the case of the test car, the only carburetor modification was one step richer main jets in the standard Ford 4-barrel carburetor.

The Paxton supercharger seems a natural for combination with the small Ford V-8 engine—just as it was with the Studebaker V-8. Of course, both are 289-cu. in. displacement, of short stroke/big bore design where in-head restrictions tend to limit breathing in the upper rpm range. Apparently, the Paxton's peak output of 400 cu. ft./min. airflow is more than enough to do the job on engines of less than 300 cu. in.

Ford's optional High Performance 289, the 271-bhp version with its mechanical valve lifters, hot valve

1966 FORD
MUSTANG GT-350-S

DIMENSIONS

Wheelbase, in.	108.0
Track, f/r, in.	57/57
Overall length, in.	181.6
width	68.2
height	55.0
Front seat hip room, in.	2 x 22
shoulder room	53.8
head room	37.4
pedal-seatback, max.	42.0
Rear seat hip room, in.	40.5
shoulder room	n.a.
leg room	n.a.
head room	n.a.
Door opening width, in.	42.3
Floor to ground height, in.	10.0
Ground clearance, in.	5.5

PRICES

List, fob factory	$4428
Equipped as tested	5365

Options included: Paxton supercharger, aluminum wheels, rear seat, automatic transmission.

CAPACITIES

No. of passengers	4
Luggage space, cu. ft.	5.0
Fuel tank, gal.	16.0
Crankcase, qt.	5.0
Transmission/diff., pt.	17.0
Radiator coolant, qt.	15.0

CHASSIS/SUSPENSION

Frame type	unitized
Front suspension type: Independent by short and long control arms, coil springs, telescopic shock absorbers, anti-roll bar.	
ride rate at wheel, lb./in.	105
anti-roll bar dia., in.	1.00
Rear suspension type: Hotchkiss drive with parallel, longitudinal, semi-elliptic leaf springs, telescopic shock absorbers, torque arms.	
ride rate at wheel, lb./in.	n.a.
Steering system: Recirculating ball and nut, parallelogram, linkage with 2 tie rods; ball-joint knuckles.	
gear ratio	n.a.
overall ratio	19.1
turns, lock to lock	3.25
turning circle, ft. curb-curb	38.0
Curb weight, lb.	2950
Test weight	3360
Weight distribution, % f/r	54.4/45.6

BRAKES

Type: Single-line hydraulic; caliper disc front, metallic duo-servo shoes in iron drums rear.	
Front disc, dia. in.	11.3
Rear drum, dia. x width	10 x 2.5
total swept area, sq. in.	381
Power assist	none
line psi @ 100 lb. pedal	700

WHEELS/TIRES

Wheel size	14 x 6.0
optional size available	15 x 6.0
bolt no./circle dia., in.	5/4.5
Tire make	Goodyear Blue Streak
size	6.95-14
recommended inflation, psi	28
capacity rating, total lb.	4560

ENGINE

Type, no. cyl.	supercharged V-8, ohv
Bore x stroke, in.	4.005 x 2.87
Displacement, cu. in.	289
Compression ratio	10.5
Rated bhp @ rpm	390 @ 6000
equivalent mph	109
Rated torque @ rpm	329 @ 4200
equivalent mph	76
Carburetion	Ford, 1x4
barrel dia., pri./sec.	1.562/1.562
Valve operation: Mechanical lifters, pushrods and adjustable rockers.	
valve dia., int./exh.	1.78/1.45
lift, int./exh.	0.457/0.457
timing, deg.	30-72, 78-24
duration, int./exh.	282/282
opening overlap	54
Exhaust system: Dual, reverse-flow mufflers.	
pipe dia., exh./tail	2.00/2.00
Lubrication pump type	rotor
normal press. @ rpm	50-60 @ 2000
Electrical supply	alternator
ampere rating	42
Battery, plates/amp. rating	54/45

DRIVE-TRAIN

Transmission type: Torque converter with 3-speed planetary gearbox.	
Gear ratio 4th ( ) overall	
3rd (1.00)	3.89
2nd (1.46)	5.68
1st (2.46)	9.56
1st x t.c. stall (2.02)	19.3
synchronous meshing?	planetary
Shift lever location	console
Differential type: Hypoid; straddle-mounted pinion.	
axle ratio	3.89

timing and higher compression thrives on Paxton supercharging. Although the nominal compression ratio is a bit higher, at 10:1, than most supercharging specialists would recommend, it proved no problem in the *CL* test car. At no time did pre-ignition or overheating occur, even after repeated dragstrip runs.

FOR THE sake of measurement of the supercharger's effectiveness, *CAR LIFE* ran acceleration tests with the supercharger both in normal operation and completely disconnected. The comparative tables show how it fared in these conditions. For the non-supercharged condition, the drive belt and the carburetor pressure box were removed. However, carburetion and ignition timing were not adjusted. Suffice it to say, the Shelby Mustang's performance with supercharger inoperative was virtually the equivalent of a standard Ford HP Mustang *CL* tested some months ago.

Belt on and blower going, the Paxton-charged Shelby Mustang is a potent pony. Performance, at least that measured on the straightaway, is superior to that of any Mustang *CL* has ever road-tested, including the 4-speed, street model of the GT-350. The supercharger makes the 289-cu. in. engine respond like 400 cu. in. In fact, Shelby-American and Paxton have established a 390 bhp rating (at 6000 rpm) for the supercharged GT-350, which is about what a high-performing 400-cu. in. engine would produce. Unsupercharged, Paxton calls it 271 bhp. Street model GT-350s are normally rated at 306 bhp; however, these are equipped with high-rise intake manifold and big Holley 4-barrel carburetor where the blown version retains the original Ford "low" riser and carburetor. Both versions have fabricated tubing exhaust headers.

Adding a supercharger to an already potent piece of machinery may seem an extra layer of icing on the cake to many, but, undoubtedly, there are those few who simply must have more power available than the next guy. The "basic" GT-350 goes out the door for $4428, mag-type wheels and automatic transmission extra; the Paxton supercharger puts $600 on top of that. *CL*'s test car thus represents a $5300 package. To those few all-out enthusiasts described above, it may all be worth it. But, might not that Paxton kit ($430 from Paxton Products, 929 Olympic Blvd., Santa Monica, Calif.), which can be installed in 2 hours, be even more effective at pepping up the lesser Mustangs? For little more than half the investment represented by the Shelby-Paxton combination the wise enthusiast could own a Mustang that performs like a GT-350. ■

SHELBY fillip is ducting which carries cool air to rear brakes.

CAR LIFE ROAD TEST

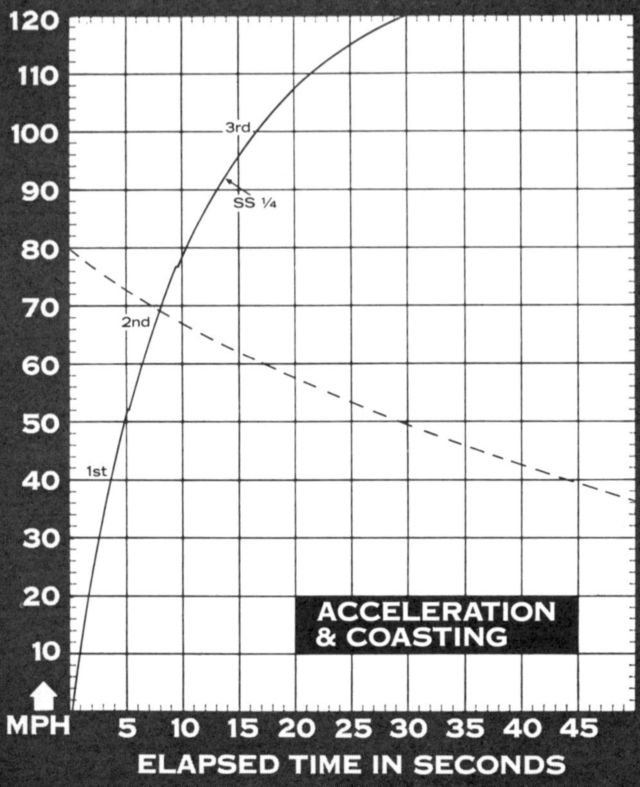

CALCULATED DATA

Lb./bhp (test weight)	8.6
Cu. ft./ton mile	165
Mph/1000 rpm (high gear)	18.1
Engine revs/mile (60 mph)	3320
Piston travel, ft./mile	1590
Car Life wear index	52.7
Frontal area, sq. ft.	20.8
Box volume, cu. ft.	394.5

SPEEDOMETER ERROR

30 mph, actual	27.8
40 mph	37.5
50 mph	44.6
60 mph	57.6
70 mph	63.4
80 mph	74.4
90 mph	81.7

MAINTENANCE INTERVALS

Oil change, engine, miles	6000
transmission/differential	as req.
Oil filter change	6000
Air cleaner service, mo	12
Chassis lubrication	36,000
Wheelbearing re-packing	30,000
Universal joint service	30,000
Coolant change, mo	24

TUNE-UP DATA

Spark plugs	Autolite BF-32
gap, in.	0.032-0.036
Spark setting, deg./idle rpm	12/1000
cent. max. adv., deg./rpm	21.5/5000
vac. max. adv., deg./in. Hg.	22/17
Breaker gap, in.	0.018-0.022
cam dwell angle	32-35
arm tension, oz.	27-32
Tappet clearance, int./exh.	0.016/0.018
Fuel pump pressure, psi	4.5-5.5
Radiator cap relief press., psi	12-15

PERFORMANCE

Top speed (7000), mph	127
Shifts (rpm) @ mph	
3rd to 4th ( )	
2nd to 3rd (6200)	77
1st to 2nd (7100)	52

ACCELERATION

0-30 mph, sec	2.6
0-40 mph	3.6
0-50 mph	4.8
0-60 mph	6.2
0-70 mph	8.0
0-80 mph	10.5
0-90 mph	13.4
0-100 mph	16.8
Standing ¼-mile, sec	14.0
speed at end, mph	92
Passing, 30-70 mph, sec	5.4

BRAKING

(Maximum deceleration rate achieved from 80 mph)

1st stop, ft./sec./sec	24
fade evident?	no
2nd stop, ft./sec./sec	24
fade evident?	no

FUEL CONSUMPTION

Test conditions, mpg	9.3
Normal cond., mpg	10-12
Cruising range, miles	160-192

GRADABILITY

4th, % grade @ mph	
3rd	22 @ 75
2nd	31 @ 60
1st	45 @ 40

DRAG FACTOR

Total drag @ 60 mph, lb	134

MUSCLE

FOR MUSTANGS

Tired of havin' other animals outrunnin' your horse? Well then we have just the item for you. A few little hop up tips that are easy to do and when completed, you'll be sittin' in the "Winners Corral" at the old drag strip.

BY BUD LANG

IF YOU'RE A MUSTANGER or have an inkling to be one, yet prefer just a few more "ponies" than Ford Motor Company is willing to give (though the potential is there), then take heart, for out in Glendora, California there is a whole stable of these machines being readied for the trail, drag that is. Les Ritchey of Performance Associates has made a habit of buying Shelby American GT 350 Mustang fastbacks with all the standard S-A equipment then adds much more of his own along with going through the complete engine. What P-A ends up with is a super-wild draggin's hoss known as a GT-350 "Performer". And perform these Mustangs do.

The stock GT 350 features a scooped fiberglass hood, fiberglass rear interior panel which replaces the back seat, 15" wheels, Cobra steering wheel, valve covers, pan and a few other refinements. Performance Associates make numerous changes to the body and chassis as shall be described as standard practice, but offer a few other treats as optional items. The engine in each car is blueprinted and modified slightly and again a few options are available regarding equipment.

Let's take a look at the car itself and see what alterations are made, then we'll delve into the engine itself. Starting from the ground up, rear rubber is exchanged for a set of 8.00 x 15 x 8" Casler Spraling cheater slicks. These tires have carried more than one winner to victory at local and national drag events, and like any good drag tire, are far superior to stockers when it comes to racing. Next a pair of adjustable lift bars are installed. First a cross-member is welded to the under body ahead of the rear end. In addition to providing adjustable mounting hangers for the lift bars, they add support to the body itself and will prevent the driveshaft from falling out should a U-joint fail. The bars are welded to the axle housings at rear, bolting to the new crossmember. Still at the rear of the car, the third member in the differential housing is replaced with one containing 4.86:1 ring and pinion gears. These low gears are essential to take advantage of the additonal power that will be available when the potent little 289 mill gets the works.

Mid-range in the Mustangs, we find that the Borg-Warner T-10 transmission has been re-worked. The synchro rings are gone over to allow faster, more positive shifts during racing. A Hurst Competition Plus 4-speed shifter replaces the stock linkage. Though the factory shift mechanism is good, it isn't considered "that good" when it comes to all-out racing where one missed shift can mean a race, and an engine. Front shocks have been replaced with Cure-Rides.

All of these changes are standard on the GT-350 Performers. As an option, P-A will also lighten the interior and/or chassis of the Mustang.

And now to the powerplant. The engine is pulled from the car and completely disassembled. The cylinder block is bored to allow a maximum of .006 inch piston-to-wall clearance, then honed to leave a specific cross-hatch pattern on the walls so the rings will seat quickly. The connecting rods are ground for exact center to center lengths though factory tolerances are limited to .003". Piston crowns are flycut for additional valve clearance. Next, the pistons and rods are aligned and pin fit.

At this point the crankshaft journals are micro-finished

and indexed off the woodruff key in the crank snout. Using the key groove as a starting point, each crank throw is checked to make sure it is at the correct 90 degree angle from that point. Should a journal be as much as .010 to .020 inch off either way, that set of rods and pistons would in turn be operating retarded or advanced, depending on which way the crank is off. This very important point is generally never considered though it is equally important as cam timing. The crankshaft dampener is also degreed 45 plus and minus from 0 degrees to allow quick alternation of ignition timing.

In addition to checking cylinder clearances, both main bearing and rod inserts are set to stock clearances on the journals. Connecting rod side clearance and crankshaft free end play are also checked again to make sure that they are right on. All pistons are matched to minimum deck height to get the most performance from this area. Once the entire assembly is clearanced and everything is ready to go together, the engine is disassembled and the crankshaft, rods, pistons, flywheel, clutch, etc., are sent out for balancing as a unit. After balancing each piece is carefully cleaned and lubed for reassembly. The standard camshaft used in the 289 block is the Ford 271 HP high performance shaft. A Cobra Drag camshaft, offering more duration and lift, is also available on an optional basis.

At the rear of the block, we find Performance Associates utilizes a Hays heavy-duty clutch assembly with the stock flywheel, though a Hays steel billet flywheel is available, again as an optional item. The clutch is rated at a reasonable 2400 pounds, sufficient for drag racing. Protecting the driver is a Cobra one-piece cast steel shatterproof bell housing.

Other engine modifications, outside of the cylinder heads include, installation of a Holley 715 cubic foot per minute (CFP) carburetor, dual-point distributor with reworked advance curve, and a choice of Doug's or Grossi headers. Doug's headers for the Mustang have individual pipes leading into short collectors. Gorssi's on the other hand are of Tri-Y design leading into long collectors. Another optional item is a clutch hub fan assembly that disengages the fan at speed.

And now to one of the most important areas of this engine, the cylinder heads. One of the top men at Les Ritchey's, Wally Cartwright, is in charge of building up most of the engines that go out of their shop. He personally performs or

(continued on following page)

Left — *The Mel Burns Ford 350 GT Mustang equipped with the Performance Associates refinements has already cracked the 12.40 e.t. mark at better than 113 mph. Car was one of the first setup with special "hoss" drag kit and what results!*

Right — *One of the first steps in getting your horse ready for the drags is installing a healthy set of lift bars which can be adjusted to fit any strip condition. Special crossmember is used with bars; will protect driveline should it break.*

Far right — *Along with some "hi-fi" internal engine work, the ignition system gets the full treatment also. Unit is checked out by removing the centrifical advance and installing a full mechanical advance mechanism for easier tuning.*

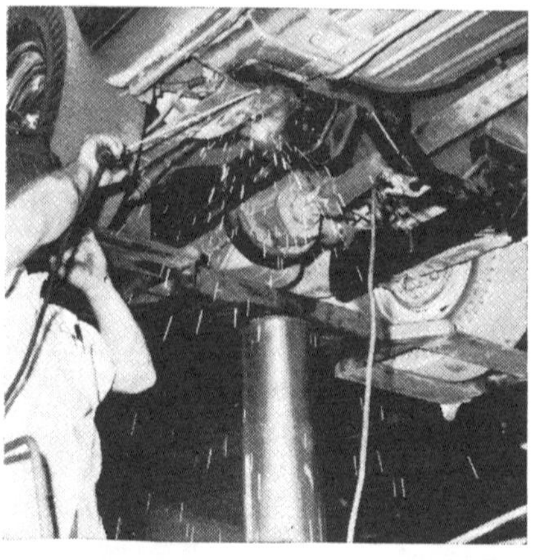

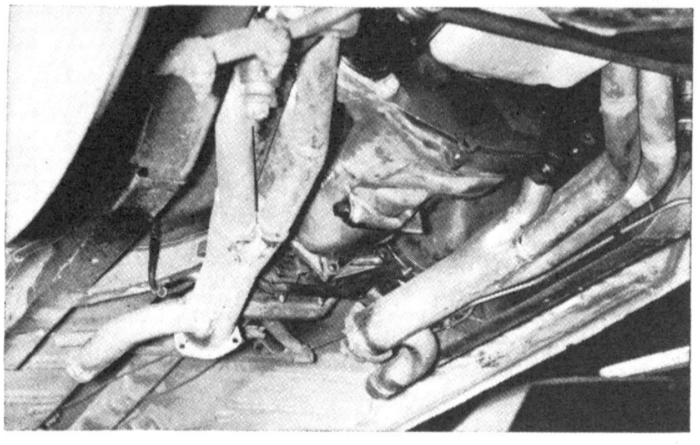

Upper left — *Custom headers are another important item if you want to be in the "winners corral." P.A uses either Doug's or Grossi's, headers. Both have worked out very well in test runs.*

Above — *Complete engine installed in Mustang looks just like the original item but most of the changes were on the inside. Atop engine is Holley 715 cubic foot per minute big performance carb.*

Left — *A dial indicator is used to make sure the valve pockets are perfectly concentric. Valve work is one of the most important phases of building up the high rpm 289 Ford engines.*

oversees all engine modifications. After receiving a "racing" valve job, where the seats and faces are precision ground to exacting dimensions, the heads are cc'd and milled so that total combustion chamber area is balanced. Deck height will have a bearing on head displacement as will gasket thickness. Therefore, one engine may have only 42cc's per cylinder, while another may have 50.

The combustion chambers are also machined about the oversize intake and exhaust valves. Both the machine process, which allows the heads to breathe better, and the oversize valves are to Cobra specs, making them legal for specific drag classes in the various associations. Likewise, the head ports are machined out to Cobra specs to make them that much more efficient. Valve guide bosses on the heads are ground to correct height and O. D. for use with Perfect Circle Valve Seals. If the guide height isn't correct, the valve retainer would stike the seal when the valve opens fully. The valves are also undercut to Cobra specs in an effort to lighten them and retard valve float. Iskenderian spring keepers are utilized with matched valve springs. Each valve is hand-lapped with Clover grade 4-A (extrememly fine) compound during seating after the valve seats are found to be concentric with a Kwik-Way dial indicator.

This wraps up the modifications to the Shelby GT 350 as performed by Les Ritchey. They are many as you can see, most are concerned with the engine. The little 289 inch Ford mill that powers the Mustangs packs plenty of power, but there is much more hidden beneath that skin. And this is what the men at Performance Associates are after, that "hidden hoss-power."

An example of what they have found can be seen in the Mel Burns Ford fastback Mustang in Long Beach, Calif. This swingin' machine is currently a record holder in Formula 4 XE class with AHRA. It romps through the traps at 113 mph in only 12.40; pretty good for a modified quarter-horse. If you already have a GT 350, Performance Associates would probably be willing to give it their "Performer" treatment, or you can buy one right across the board. Either way, your Mustang will be a first-class Performer all the way.

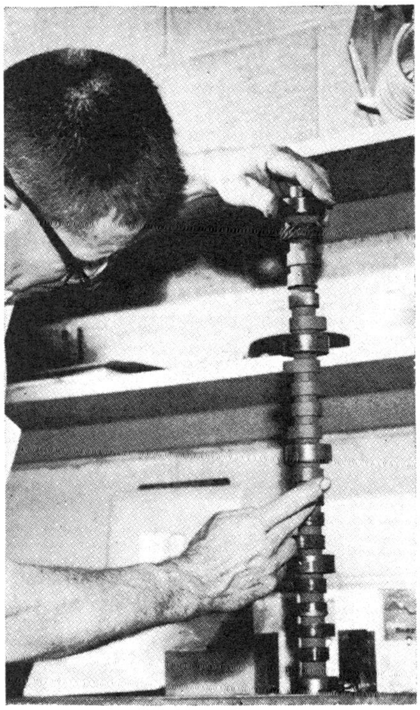

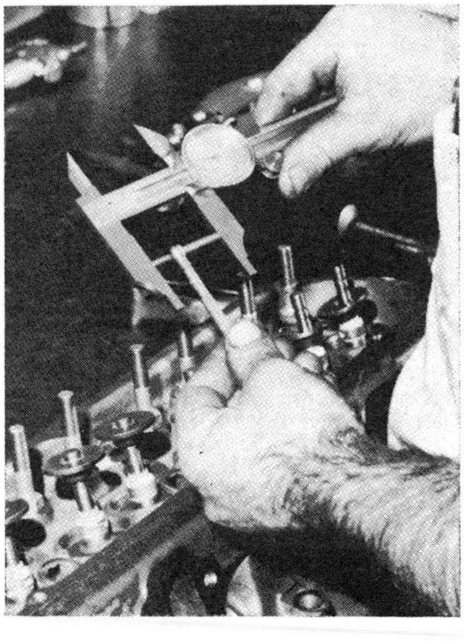

Above left — *Les Ritchey (left) and Ron Root of Performance Associates give a hot 289 "Performer" a little bit of "Super Tuning" prior to a day at the drags. Root drove Mel Burns Mustang to record marks while Les is known for A/FX exploits.*

Above — *A coating of Lubri-Plate covers the entire camshaft before it is installed. This will prevent galling when engine is first fired. Choice of 271 hp. camshaft of optional Cobra drag grind cam can be had. Type of driving will determine choice.*

Far left — *Vernier dial calipers are used to check the installed height of valve spring by measuring the distance between retainer and head when valve is closed. Measurement is made by using an adjustable mike which gives accurate readings.*

Left — *Each valve is given a careful hand laping job to make sure it is positively seated with a special air tool. Compound used is Clover grade 4-A (extremely fine). This is done after the seats have been closely checked with the dial indicator.*

Above — *Perfect Circle Valve Seals are installed on all valve stems to reduce chances of oil being drawn from the upper head through the guides and into the combustion chambers. Should this occur, the plugs will foul and deposits will cause the fuel to pre-ignite.*

Right — *A taper reamer is used to enlarge valves to proper size for use with special enlarged Cobra valves. Following this, the heads are given a thorough grinding to the exact angle required and then the valves are hand lapped for proper fit and clearance.*

Below right — *A cylinder head hone is used to achieve a good cross hatch pattern in the cylinder walls which will allow rings to seat very quickly. Following this, the block should be cleaned of any foreign particles with a good hand scrubbing soap job.*

Below — *Here are all the goodies that if used right, will make your engine not only a goer, but a winner to boot. Stock shifter is replaced with a Hurst unit while stock clutch is also tossed out in favor of a Hayes number. The ignition system s also rewired.*

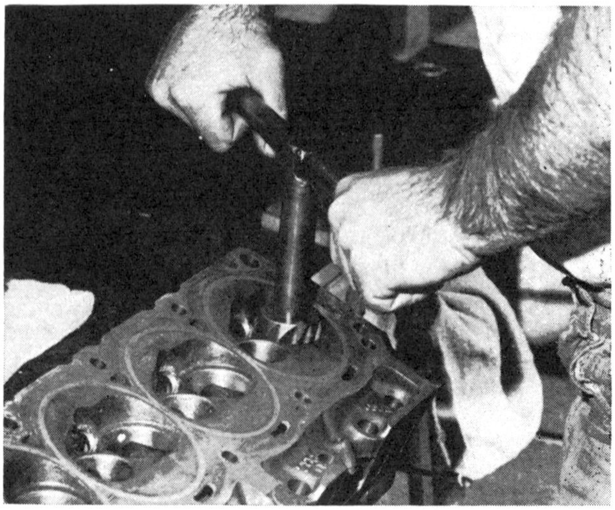

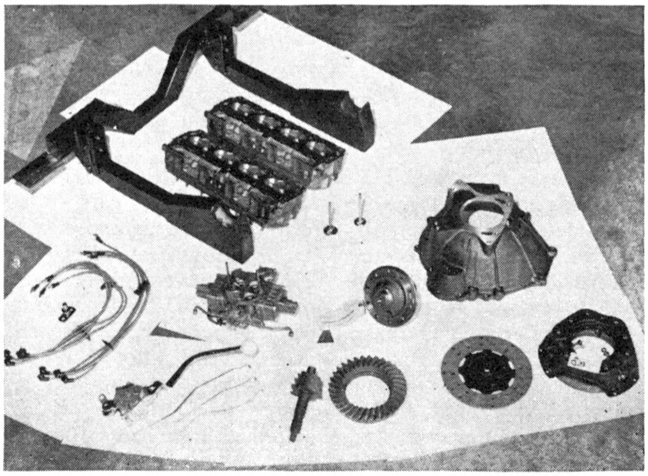

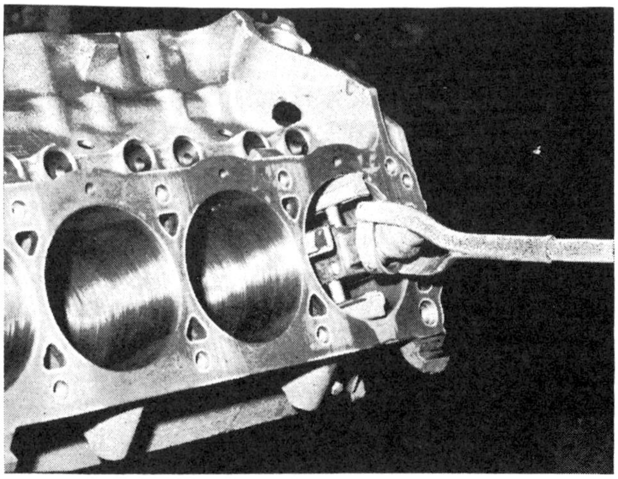

GT-350s

MT Road Test

STOCK & SUPERCHARGED: BOTH BOULEVARD BLISTERERS

THOUGH DIFFERENT IN TRANSMISSIONS AND INDUCTION, AS WELL AS IN PERFORMANCE, BOTH GT-350s HANDLED TURNS WITH EQUAL GRACE.

THE SHELBY-AMERICAN GT-350 Mustang can best be termed a "gentleman's sports car." It combines the flair, performance and handling qualities usually found in a true sports car with the practicality of a closed-body 4-seater model (with optional rear seat), plus the convenience of parts availability at most any Ford dealer.

In stock form – if a GT-350 can be termed stock – it will perform as well as any other car, American or European, in the same class and has proven it by being the Class B Production winner in SCCA national competition. Add a supercharger, though, to this already potent mixture, and you're in for a veritable explosion of power to the rear wheels of this lightweight machine.

Paxton superchargers are now an option on all GT-350 Mustangs, adding about $400 to the base price of $4428. For the resulting power and acceleration increase, a person would be hard pressed to equal it with $400 spent elsewhere in the engine.

Though there aren't a lot of options to be added to Shelby's version of the Mustang, our two test cars varied considerably. The dark ivy-green car came equipped with the standard 306-hp engine, 4-speed transmission, radio, and limited-slip differential. Our blown car was equipped with the optional high-performance Cruise-O-Matic, which according to Joe Granatelli, will out-perform any 4-speed around in acceleration times. Limited-slip differential isn't available with the automatic transmission, but our test car sure could've used it.

Otherwise the two cars were alike. Both had 130-mph-rated Goodyear Blue Streak tires, optional cast-aluminum

7.5-quart aluminum oil pan and tubular steel headers are added to 289 by Shelby.

Both cars were fitted with optional cast-aluminum wheels, which enhance appearance and reduce weight but present problems when tire is removed, due to small center.

Cross bar over engine is special fender stiffener from Shelby. After the alterations, 289-cubic-inch V-8 puts out 306 hp.

Neat installation of Paxton supercharger still allows access for periodic maintenance and boosts horsepower to over 400.

wheels, and fold-down rear seat, front disc brakes, metallic rear brake shoes, tubular steel headers, competition-type seat belts, functional scoops in the fenders for brake cooling, and a hood scoop for engine breathing.

Due to the 19-to-1 quick-ratio steering that is standard on every GT-350, drivers soon develop good arm muscles. No power assist is available for it, so out of necessity, GT-350 owners have become quite proficient at arm wrestling. Also, never challenge a GT-350 owner to a kicking contest. He has had opportunity to exercise his leg every time he touches the brake pedal. The pedal resists light touches and practically demands to be tromped on. When you do that, the car comes to rest – fast!

By the time arm and leg muscles have shaped up, and your kidneys have adjusted to the harsh ride given from competition-like suspension, truly enjoyable driving is at hand. We really didn't become impressed with the cars until we put them on the race course. In addition to the surprising performance registered in a straight line, especially with the Paxton-optioned car, GT-350s go around corners like they were a part of the road.

Joe Granatelli had told us that after a supercharger was added to a GT-350, it would go from 0 to 60 in five seconds flat. But by the time we had added two large-sized test drivers to the interior of the car (somebody has to read the clocks), plus all of our test equipment, the best we could manage was a flat six seconds.

With a little more experimenting with the engine, and a lighter load, we feel the 5-second figure would be attained. This, if done, would enable a time 1.3 seconds faster than the 4-speed version we tested, and .6-second quicker than the GT-350's competitor, the Corvette. The addition of the blower ups horsepower by 46% (estimated) giving the 289-cubic-inch Cobra a reading of around 440 hp. That's a big step in the right direction.

Both GT-350s handled the turns at Riverside Raceway with very little effort, but the 4-speed gave better control of the rear wheels on some of the shorter turns. The automatic didn't allow for much wheel slippage, sometimes needed to bring a car out of a turn properly set up. There is a noticeable lack of body lean, and the driver feels confident of control at all times.

As our spec chart indicates, we didn't take either of the cars to their maximum top speed. This was due to a logistic problem, but we'd safely say that they'd be pressing the rating of the tires if they were. Also, the 4-speed exhibited better passing times. The blower couldn't do much there, as it had to wait for the automatic transmission to kick down before it could go to work, while the stick was already in gear at the time the proper speed was reached.

For normal street operation, neither of the GT-350s presented any great problem. They didn't overheat or act sluggish at low speeds, but because of the hard brake pedal, we soon learned to keep extra distance from the car ahead. Though we knew we could stop as fast or faster than any other car on the road, we just didn't want to wear out the soles of our shoes.

The built-in scoop in the fiberglass hood draws in a lot of air at freeway speeds, and this air tends to force its way through any leaks in the firewall. Shelby would be ahead by taking extra pains to seal off any possible air leaks in this area to prevent driver complaints, as well as providing a tighter cockpit, isolated from noise and dirt.

We don't remember passing a stock Mustang owner while we were in the GT, without his doing an instant double-take. The car is a definite eye-catcher, but it's those side windows that draw everybody's attention. Owners of fastback Mustangs were quick to take note of how much better they looked than the louvers provided by Ford, and all wanted to know how to go about buying some or making their own.

Through a combination of experience in knowing what it takes to make a good sports car, and foresight into what people want in a car, Shelby-American has come up with an almost perfect result – the '66 GT-350. With the extra added grace of a Paxton supercharger, it becomes a car suited for duty in carrying its occupants through the commuters' rush every weekday morning, and over the finish line first on weekends. — ***Steve Kelly***

SLEEK LINES OF GT-350 ARE SHOWN TO BEST ADVANTAGE FROM SIDE. THROUGHOUT TEST WE AVERAGED 14 MPG WITH THE 4-SPEED VERSION.

Supercharged car had optional Paxton pressure gauges. Right dial indicates direct blower pressure, left one is for fuel.

Flat glass face on standard-equipment tachometer makes reading difficult at night due to light reflections from following cars.

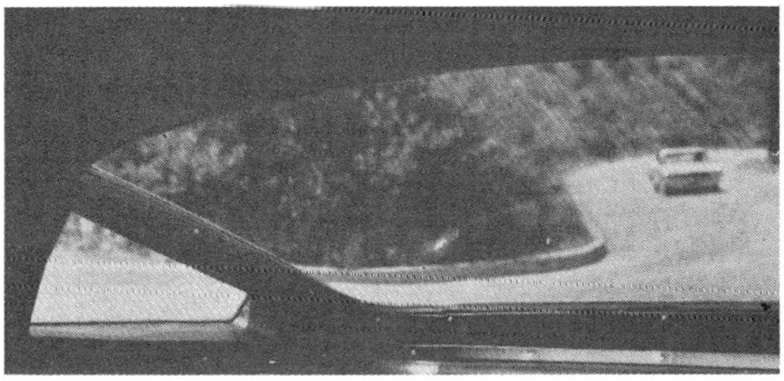

High window line of fastback and rear-quarter windows combine for good rear and side vision — handy when leaving driveways.

H-D rear housing is safety wired and has pair of torque arms going forward to body member. Koni shocks are part of pack.

PHOTOS BY DARRYL NORENBERG

STIFF STEERING MAKES CORNERS NOTICEABLE, AND STANDSTILL TURNS OF THE WHEELS ALL BUT IMPOSSIBLE. OVERALL RATIO IS 19 TO 1.

MT Road Test

GT-350s

SPECIFICATIONS FROM MANUFACTURER

ENGINE IN TEST CAR: Ohv V-8
Bore and stroke: 4.00 x 2.87-ins.
Displacement: 289 cu. ins.
Advertised horsepower: 306 @ 6000 rpm
Max torque: 329 lbs.-ft. @ 4200 rpm
Compression ratio: 10:1
Carburetion: 1 4-bbl.

TRANSMISSION TYPE & FINAL DRIVE RATIO: Fully synchronized 4-speed, or automatic torque-converter. Floor-mounted lever standard on either. 3.89:1 rear-axle ratio.

SUSPENSION: Independent front with coil springs above upper "A"-arms. Inner pivots of front control arms lowered 1 inch from stock Mustang location. Heavy-duty springs, sway bar and 1-piece rear-end housing, with H-D leaf springs and torque control arms running forward from top of housing to body member. Koni tubular shocks at each wheel.

STEERING: Recirculating ball and nut.
Turning diameter: 38.9 ft., curb to curb
Turns lock to lock: 3.7

WHEELS: Steel, 15-in. diameter, 6 ins. wide, or 14-in. diameter steel, standard. Optional: 15 x 6½ alloy.

TIRES: 7.75 x 15 or 6.95 x 14 high-speed, low-profile nylon

BRAKES: Caliper-disc (Kelsey Hayes) front with special racing pads. Drum rear with metallic linings.
Diameter of front disc, 11.3 ins; rear drum, 10 ins.

SERVICE:
Type of fuel recommended: Premium
Fuel capacity: 17 gals.
Oil capacity: 6.5 qts.; with filter: 7.5 qts.
Shortest lubrication interval: 6000 mi.
Oil- and filter-change interval: 6000 mi.

BODY & FRAME: Platform-type unitized construction
Wheelbase: 108.0 ins.
Track: front, 57.0 ins.; rear, 57.0 ins.
Overall: length, 181.6 ins.; width, 68.2 ins.; height, 55.0 ins.
Usable trunk capacity: 5.0 cu. ft.
Curb weight: 2940 lbs.

ACCESSORY PRICE LIST

Paxton supercharger kit $430.00
Paxton pressure gauges 35.00

Factory-installed Options

*Paxton supercharger and gauges 670.00
***Cast-alloy wheels 267.00
Chrome steel wheels 167.00
***Rear seat with fold-down shelf 50.00
Rallye Stripes (over the top) 64.00
**No-Spin rear axle (manual trans. only) .. 141.00
**AM radio 57.50
Automatic transmission. (beefed) ......no cost
*on supercharged car
**on stock car
***on both cars

MANUFACTURER'S SUGGESTED LIST PRICE: $4428 (plus license, taxes and freight, f.o.b. St. Louis, Mo.)

PRICE OF CARS TESTED: Stock, $4943.50; supercharged, $5415, (plus license, taxes, freight, & get-ready charges. f.o.b. St. Louis, Mo.)

MANUFACTURER'S WARRANTY: 4000 mi./3 mos.

PERFORMANCE
Stock

ACCELERATION (2 aboard)
0-30 mph .... 2.8 secs.
0-50 mph .... 5.4 secs.
0-60 mph .... 7.3 secs.
0-75 mph ....10.1 secs.

TIME & DISTANCE TO ATTAIN PASSING SPEEDS
40-60 mph ....3.1 secs., 227 ft.
50-70 mph ....3.4 secs., 299 ft.

STANDING-START QUARTER-MILE: 15.6 secs. and 94 mph

BEST SPEEDS IN GEARS @ SHIFT POINTS
1st .... 50 mph @ 6500 rpm
2nd .... 76 mph @ 6500 rpm
3rd ....107 mph @ 6500 rpm
4th ....(not maximum) 95 mph @ 5000 rpm

MPH PER 1000 RPM: 19

SPEEDOMETER ERROR AT 60 MPH: 1.5% fast

STOPPING DISTANCES: from 30 mph, 44 ft.; from 60 mph 155 ft.

PERFORMANCE
Supercharged

ACCELERATION (2 aboard)
0-30 mph ....2.65 secs.
0-50 mph ....4.6 secs.
0-60 mph ....6.0 secs.
0-75 mph ....8.1 secs.

TIME & DISTANCE TO ATTAIN PASSING SPEEDS
40-60 mph ....3.2 secs., 234 ft.
50-70 mph ....3.5 secs., 308 ft.

STANDING-START QUARTER-MILE: 14.0 secs. and 102 mph

BEST SPEEDS IN GEARS @ SHIFT POINTS
1st .... 45 mph @ 6000 rpm
2nd .... 75 mph @ 6000 rpm
3rd ....110 mph @ 6000 rpm

MPH PER 1000 RPM: 18.3

SPEEDOMETER ERROR AT 60 MPH: 1.5% slow

STOPPING DISTANCES: from 30 mph, 45 ft.; from 60 mph 119½ ft.

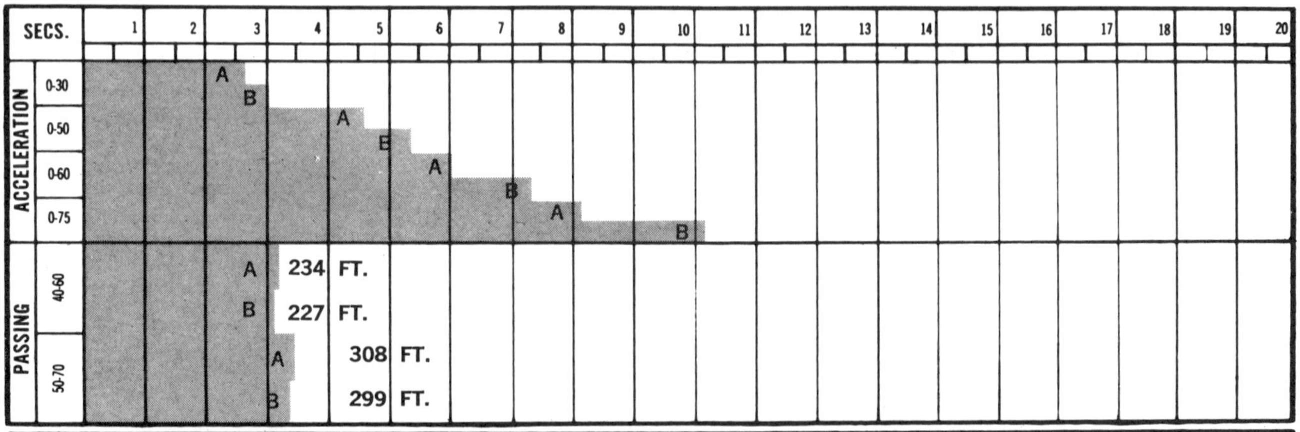

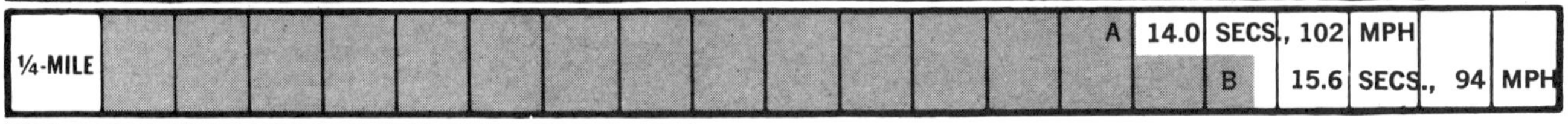

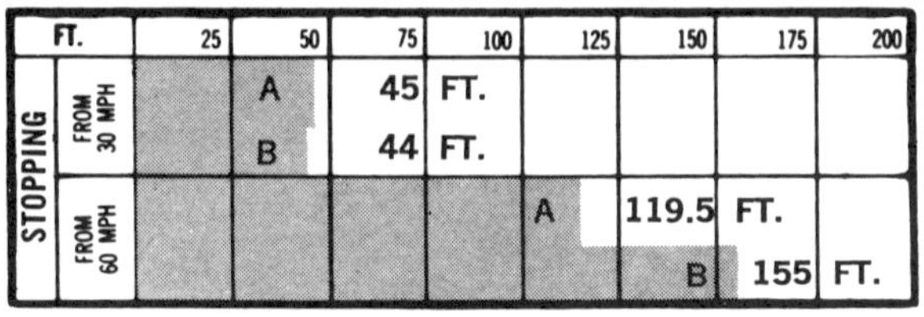

A = BLOWN B = STOCK

SHELBY-AMERICAN MUSTANG GT-350
2-door, 4-passenger fastback coupe

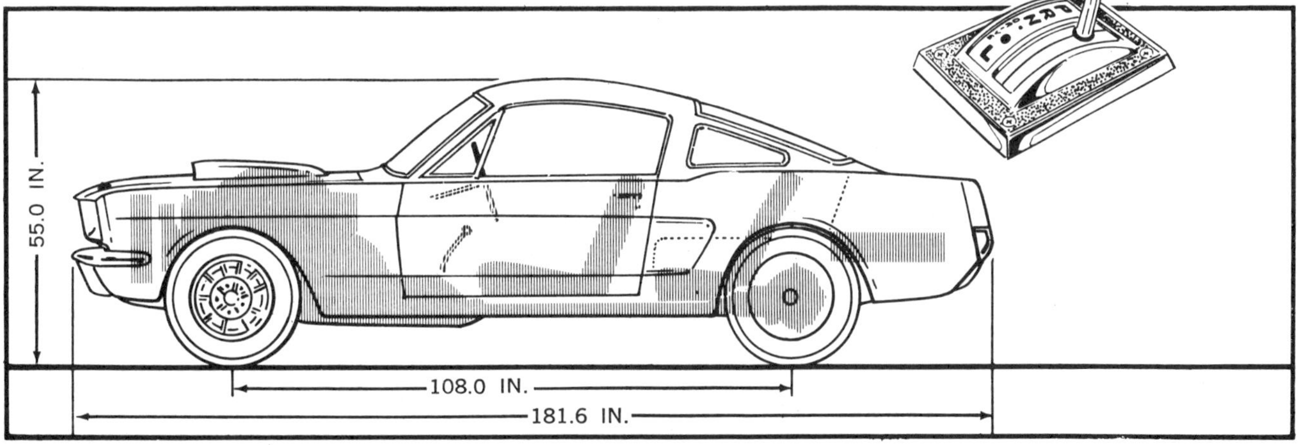

RENT-A-TIGER

WE TEST TWO VERSIONS OF THE GT-350 — THE SHELBY ORIGINAL AND THE HERTZ SPECIAL

BY JOHN LAWLOR

Rent a GT-350? The idea seemed preposterous. Shelby's modified Mustangs are highly specialized vehicles that not every driver can understand or appreciate. They're not at all difficult to operate but, to get the most out of them, something more than ordinary driving ability is required. It didn't seem reasonable, therefore, that they were now available to anyone with a Diners' Club card.

Yet, earlier this year, the Hertz Rent-A-Car system announced that it was purchasing a thousand GT-350's from Shelby and that they would be offered on an hourly, daily or weekly basis to qualified drivers.

We couldn't help but wonder – would Hertz supply the real thing? Would the rental firm's GT-350's be authentic examples of the Shelby product? Or would they be merely theatrical imitations with flashy paint jobs and noisy exhaust systems?

There was an easy way to find out. We would obtain a GT-350 from the Shelby factory and put it through its paces Then, without revealing our identity or purpose, we'd rent one from Hertz and see how it compared.

Our conclusion? The Hertz vehicle is absolutely authentic. In fact, the GT-350 we rented was better equipped than the one we borrowed from the factory!

Both Shelby and Hertz offered us a choice of cars with either four-speed manual or three-speed automatic transmission. We were particularly interested in the automatic because it's a new feature on the GT-350 this year and hasn't really received much attention. That meant we had to specify it on both the factory and rental vehicles, because we wanted them as similar as possible.

Beyond that, the Shelby car was virtually stripped. It had a couple of nominal options, a folding rear seat and racing stripes down the center of the body, but hardly a GT-350 rolls out of the factory without those features. However, the Hertz vehicle had some additional items, notably a radio and special alloy wheels, that weren't on the factory car.

The paint schemes differed. The Shelby vehicle was finished in the firm's traditional white with blue stripes, while the rental car borrowed its colors from the Hertz trademark, black with gold stripes.

Those purchasing a GT-350 can now get it in somewhat more subdued tones than these. The factory now offers the car in white, blue, red or green and, as we noted a moment ago, the stripes are optional. Frankly, we'd recommend more conservative coloration than our test cars had. The stripes, especially, attracted more atention than we wanted. We got tired of glancing in the rear view mirror and seeing other multi-colored cars trailing us, black and white ones with sirens and red lights. No, we were never stopped but we became awfully self-conscious about our driving!

Hertz adds a couple of ornamental touches to distinguish its version of the car from the factory model. It designates the car as the GT-350-*H* and it replaces the Shelby Cobra hubs on the alloy wheels with a design reading "Hertz Sports Car Club." However, those little details are the only differences. The engine, suspension and so forth are pure Shelby.

The modifications that separate the GT-350 – or GT-350-H – from the ordinary, run-of-the-mill Mustang are really nothing but minor refinements. Collectively, though, they transform the car from a satisfactory road machine to a superlative one.

Shelby starts with a high-performance Mustang fastback, with a 271-horse-

Light alloy wheel, left, is normal GT-350 option, carries Hertz Sports Car Club emblem on rental vehicle. Hertz version, above, costs $13 a day plus 13¢ a mile, a bargain in driving fun according to our test staff.

HERTZ GT-350 PROVED TO BE BETTER EQUIPPED THAN TEST CAR SUPPLIED BY SHELBY FACTORY! CAR HAS FINE HANDLING.

power, 289-cubic-inch engine and heavy-duty suspension. The stock intake manifold is replaced with an aluminum high-riser design and the exhaust system is rebuilt to incorporate tuned, steel headers and straight-through mufflers. On manually-geared cars, a special, enlarged Holley four-barrel carburetor is installed, though, with the automatic transmission, the standard Ford four-barrel is retained.

These simple breathing changes boost the engine's nominal output to 306 horsepower, well over one horsepower per cubic inch, without any real sacrifice in reliability or operating smoothness.

Shelby also equips the engine with Cobra rocker covers and an aluminum, baffled oil pan. The latter increases oil capacity, aids cooling and prevents oil surge during acceleration, hard cornering or braking.

The four-speed transmission is the old Warner T-10, rather than the unit Ford now produces. Shelby prefers the Warner because it's lighter. This gives the GT-350 owner an added benefit in that a wide variety of T-10 gearsets are available from parts suppliers. Swapping ratios for special driving purposes is a comparatively easy task. The set supplied in the car has gears of 2.36 in first, 1.62 in second, 1.20 in third and direct drive in fourth.

The automatic is a new, heavy-duty version of Cruise-O-Matic that Ford introduced last year. It has a 2.02 con-

Black paint with gold striping marks Hertz version of car. Those who don't want friends and neighbors to know they're driving rented vehicle will lose on this one! You can't miss it!

verter stall ratio, 2.46 first, 1.46 second and direct drive third and is engineered to withstand shifts at more than 6,000 rpm.

Unhappily, though, it is *not* the new Sport Shift version of Cruise-O-Matic offered in the Fairlane GT-A. It does allow some degree of manual control but not as much as we'd like to have in this type of vehicle.

Upshifts can be governed by starting in low for first gear, then shifting to the intermediate drive range and immediately back into low to engage and hold second and finally shifting back into drive for third. This technique works fine for straightaway acceleration and, with it, we were able to put the GT-350 through the standing quarter-mile in 15.6 seconds at 85 mph. That, incidentally, is just about a second slower than times we've recorded in the past with four-speed versions of the car.

But downshifts aren't as easy to control. Sure, the unit can be pulled back from drive into low to go from third to second. But, if speed drops to 20 mph or less, it's impossible to keep the gearbox from dropping into first.

Suppose you're heading into a tight turn, one you want to take in second. Fine. You pull the lever back into low and get the proper ratio. But, as you go into the turn, your speed drops slightly and pow! With a great, sudden roar, you've gone into first and have far more revs than you wanted.

With the new Sport Shift in the Fairlane, there's a "1" range for first gear and a "2" for second. Normally, there are no automatic upshifts or downshifts in either of these ranges. The unit allows positive control over both of the lower gears.

This is the answer to the problem we've described. We hope Ford expands production of this new version of Cruise-O-Matic so it can be offered in the GT-350. In fact, we'd like to see it available in all Mustangs.

Shelby reworks the undercarriage considerably but, interestingly, doesn't touch the springs themselves. The heavy-duty front coils and rear semi-elliptics supplied in the ordinary high-performance Mustang are considered satisfactory for the GT-350.

At the front, the inner pivots of the upper control arms are lowered one

Aside from black and gold paint job, GT-350-H designation and special club emblem on wheels, Hertz version of car meets full Shelby specs. Trunk space, left, is limited but, with only two passengers aboard, there's added space behind front seats. Photos of Hertz vehicle are work of Jim Gilbert.

Shelby test vehicle had traditional factory paint scheme of white with blue striping but more conservative paint is now available to GT-350 buyers. Stripes, in fact, are now extra-cost option. Wild setting for photos is appropriate. A wilder car — in the best sense — is hard to imagine!

inch. This keeps the wheels more nearly vertical during cornering maneuvers, improving bite and providing better slip angles to minimize the standard Mustang's understeering tendencies. In addition, a one-inch anti-sway bar is installed to increase roll stiffness.

There's only one basic change at the rear. Torque reaction arms are fitted to control axle wind-up during hard acceleration and braking. At all four wheels, adjustable heavy-duty shock absorbers are installed.

The steering is quickened with a lengthened pitman arm that cuts the overall steering ratio from 21- to 18-to-1, trimming half a turn from the lock-to-lock figure and bringing it down to 3.25 turns.

The brakes are based on the regular Mustang option of front discs and rear drums, but Shelby adds heavier pads to the discs and metallic linings to the drums. The result is a virtually fade-free if rather heavy-feeling system. In fact, the Hertz car carries a dash plaque alerting the driver that stronger than normal brake pedal effort is required.

It all adds up to a firm-riding vehicle and, at lower speeds, a stiff-handling one. But the reward comes out on the road in the form of absolute stability. Whether bombing along an interstate highway or snaking through a mountain pass, the car inspires feelings of confidence and security.

Shelby also alters the appearance — beyond the gaudy paint, that is. The grille is simplified, with the Mustang emblem offset to one side, and all side trim is removed. New this year are rear quarter windows to replace those silly louvers on the standard Mustang fastback, plus scoops to direct air to the rear brakes.

The hood was fiberglass on the original GT-350 but now it's sheet steel. It contains a functional scoop that feeds the four-barrel carburetor.

Shelby asks $4428 for the basic GT-350 package, plus $50 for the folding rear seat and $64 for the paint stripes. Alloy wheels, like those on our

ROAD TEST DATA: SHELBY GT-350

Base Price: $4428

Price as Tested: $4542

CHASSIS

Wheelbase: 108 inches

Tread: 57 inches, front and rear

Length: 181.6 inches

Width: 68.2 inches

Height: 55 inches

Tire Size: 7.75 x 15

Suspension: Independent coil springs front, semi-elliptics rear

Steering, Lock-to-Lock: 3.25 turns

Curb Weight: 2800 pounds

PERFORMANCE

Standing Quarter-Mile: 15.6 seconds @ 85 mph

ENGINE

Type: OHV V-8

Bore and Stroke: 4.0 x 2.87 inches

Displacement 289 cubic inches

Horsepower: 306 @ 6000 rpm

Torque: 329 @ 4200 rpm

Compression Ratio: 10-to-1

Carburetion: 1 x 4-bbl

DRIVE TRAIN

Transmission: 3-speed automatic

Rear Axle Ratio: 3.89-to-1

ROAD TEST: SHELBY GT-350
VALUE RATING CONSENSUS

	EXCELLENT	GOOD	FAIR	POOR
RIDE		XXXX	X	
HANDLING	XXXXX			
ACCELERATION	XXXX	X		
PASSING ABILITY	XXXX	X		
BRAKING	XXXXX			
SEATING COMFORT Front	XXXX	X		
Rear		X	XX	XX
NOISE LEVEL: Wind		XXX	XX	
Road		XX	XXX	
INSTRUMENTATION	XXXXX			
CONTROL ACCESSIBILITY	XXXX	X		
PAINT QUALITY		XXXX	X	
PANEL AND TRIM FIT: Exterior	XXX	XX		
Interior	XXXX	X		
FUEL ECONOMY		X	XXXX	
CARGO SPACE		XXX	XX	
STYLING	XX	XXX		

TEST AT A GLANCE

	EXCELLENT	GOOD	FAIR	POOR
PERFORMANCE	XXXX	X		
ECONOMY		X	XXXX	
VALUE	XX	XX	X	

rental vehicle, are another $267. The only other option on the regular factory list is a limited slip differential, which is actually installed by the dealer for $189. Shelby doesn't mention a radio but the dealer could undoubtedly put a standard Mustang unit in for approximately $60.

Hertz' rates for the rental version of the car are $2.50 per hour, $13 per day or $65 per week, plus 13 cents a mile. These figures include $100 deductible insurance but even that $100 liability can be waived for another $1 per day or $5 per week. We had our Hertz GT-350 just one day and drove it 129 miles. Our tab included the $13 for the day, $16.77 for mileage, $1 for the insurance waiver and $1.23 for tax, adding up to an even $32.

Hertz, incidentally, pays for gas and oil. If the customer has to buy any on the road, he need only present a receipt for it when he returns the car and the cost will be deducted from his bill.

Actually, the firm won't rent a GT-350 to just anybody. The driver has to be over 25 and he must sign a statement that he will be the only one operating the vehicle. If two or more people are taking a cross-country trip and want to spell each other at the wheel, they all have to sign the statement. This is an extremely important procedure because if an unauthorized driver is involved in an accident, the insurance is invalid and the person actually renting the car becomes liable for any damage incurred.

With the four-speed version, the customer has to be checked out by a Hertz driver to be sure he knows how to operate the vehicle. With the automatic, this isn't considered necessary. Once the customer has gone through this routine, he becomes a member of the Hertz Sports Car Club and, by presenting his club card, he can bypass the check out on future GT-350 rentals.

We were pleasantly surprised to find that the Hertz GT-350 meets full Shelby specs. Frankly, we wouldn't have blamed the rental firm if it had fudged a bit on some of the car's hidden details. Few people would have ever known the difference.

But Hertz has played it straight — even straighter, in fact, than we'd expected.

We ran an odometer check on the rental car and found its mileage reading to be dead accurate. The distance we covered — and had to pay for — was right on the money.

So the Hertz version of the GT-350 isn't just a good car. It's an honest one, too!

Details separate GT-350 from standard Mustang 2+2 fastback include special, functional rear brake scoops and quarter windows in place of louvers, top left; special steel hood with functional carb scoop and pin-type fasteners, top right; high-performance, 306-horsepower engine with high-riser manifold and special headers, center left; wood-rimmed wheel, dash-mounted tach and aircraft-type seat belts, center right; rocker panel striping, right. Modified suspension is also featured, with pivot point of upper front A-arms lowered, heavy-duty shocks and anti-sway bars front and rear. Front brakes are discs rear are drums, both with special heavy-duty linings.

G.T.500

SUPER STANGS FOR THE STREET

'Old Shel' ushers in the new year with a restyled and re-engineered GT-350 and a spanking new 428 Fang-mobile

BY HAL KEMPER

CARROLL SHELBY, better known in competition circles as "Old Shel the Little Old Snake Maker," is not one to rest on his laurels. Not satisfied with just a dressed-up and re-engineered GT-350 Mustang, the ex-championship sports car driver and Ford LeMans team leader, has been burning the midnight oil on a couple of new *fanged* creations. The results are two new GT cars, one with a 289 Cobra-type engine and the other powered by a big 428 wedge. Both have been fully re-engineered, restyled and re-powered to keep ahead of the times.

The GT-350, powered by the tried and true 306-hp 289 commands respect even though the factory offers a 390-cuber right off the showroom floor. The super-hot GT-500 packs a very conservatively rated 355-hp 428-cube punch and is truly an all-guts machine. Because of new merchandising policies at Shelby-American, the GT-350 actually tags out lower than last year's model. The GT-500 is priced right in there, lower than any other genuine GT car in its performance bracket.

For 1967 the GT-350 is once again based on the popular Mustang fastback. A competition-inspired grille, nose and hood assembly give the new GT an unmistakable look of authority and uniqueness. An exclusive rear deck spoiler patterned after those installed on high-speed race machines to keep the body from lifting, competition-proven fresh air intake ports and scoops which ventilate the interior, and the rear brakes are new this year.

The differences between the GT-350 and the showroom Mustang continue to be more than just skin deep. The 1967 model is again powered by the high-efficiency, high-rev Cobra 289 rated at 306 hp. Four-on-the-floor is standard, while a heavy-duty dial-a-win setup is optional.

Shelby has placed special emphasis on safety this year. In addition to its super-stop front disc brakes, steering has also been improved upon by the use of a new box with a 16-to-1 overall ratio. Nylon safety tires, track rated at over 140 mph and patterned after those proved at LeMans, are standard equipment. So is a fully-padded instrument panel, integral safety (roll) bar and an exclusive new shoulder harness for front seat passengers. All this, and the GT-350 is the lowest priced genuine GT car in the world!

The real kicker, however, in the Shelby camp is the GT-500. Here is a true GT machine with acceleration

Restyled fiberglass front end with air intake scoops, new side scoops and a rear spoiler set GT's off from production-line Mustangs. GT-500 packs 428 wedge punch.

and top speed usually only associated with out-and-out drag machines. This companion model to the GT-350 sports 428 cubic inches, dual quads and a questionable horsepower rating of 355 hp. Although road test figures are not yet available, early tests of a prototype "works" car produced a 0-to 60-mph clocking of 5.4 seconds. The standing quarter-mile netted the prototype driver with a 106-mph, 13.5-second time slip, which is mighty respectable for an off-the-floor machine. Just for the records, the 428 rig was able to leap from 40 to 60 miles-per-hour in just 2.4 seconds, or 162 feet. The boost from 50 to 70 miles-per-hour was attained in 3.1 seconds, 266 feet. Talk about Total Performance!

Super-wide 6½-inch cast aluminum hoops optional at extra cost on all GT's.

The top-line GT also features the safety goodies outlined in the GT-350 report. They're all standard at no extra cost. Because of the bigger engine the 500 model tips the scales at 486 pounds more than GT-350 and is fitted with more suitable gear ratios. New colors plus options galore are are available for both GT models, so that either can be set up for road racing, drag racing, or the supermarket sweepstakes.

SUPER STANGS FOR THE STREET

GENERAL DIMENSIONS: GT-350

Tread: Front, 58.0″ Rear 58.0″	
Wheelbase:	108.0″
Length:	186.6″
Width:	70.9″
Curb Weight:	2800 lbs.

Construction: Platform type unitized construction with reinforced floor members and export front and reinforcement.

ENGINE:

Type: Cobra OHV 289 cu in in 90° V-8
Bore & Stroke: 4.4005″ x 2.87″
Horsepower: 306 @ 6000 RPM
Carburetion: Cobra hi-rise intake manifold with Holly 4-bbl carburetor (715 cfm flow rate)
Solid valve lifters; die cast, polished aluminum "Cobra" rocker arm covers and air cleaner; dual exhaust system.

TRANSMISSION:

Fully synchronized four-speed manual (28 spline) standard.
Gear Ratio: First—2.32:1 Second—1.69:1
Third—1.29:1 Top —1.00:1
Heavy duty Cruise-O-Matic transmission optional.
First—2.46:1 Second—1.46:1
Top —1.00:1

FINAL DRIVE:

Heavy duty rear axle with straddle-mounted deep offset drive pinion.
Standard ratios: Manual transmission 3.89:1
Automatic transmission 3.50:1

SUSPENSION:

Front: Independent, with roll spring and ball joints, Shelby modified for flatter cornering. .94″ diameter front stabilizer bar.
Rear: 4-leaf springs with special rebound dampers to control rear spring windup.

BRAKES:

Front—disc, 11.3″, with high speed linings.
Rear—10″ x 2.5″ cast iron drum, self adjusting.
191.0 square inch effective lining area.
Independent service brake operating rear drum brakes.

WHEELS—TIRES:

Shelby 15″ steel wheel with 6.5″ rim width. "Speedway 350″ low profile 4-ply nylon E70-15 tires designed especially for Shelby GT cars.
Specifications subject to change without notice.

GENERAL DIMENSIONS: GT-500

Tread: Front 58.0″ Rear 58.0″	
Wheelbase:	108.0″
Length:	186.6″
Width:	70.9″
Curb Weight:	3286 lbs.

Construction: Platform type unitized construction with reinforced floor members and export front and reinforcement.

ENGINE:

Type: Cobra OHV 428 cu in 90° V-8
Bore & Stroke: 4.13″ x 3.984″
Horsepower: 355 advertised
Carburetion: Two 4-bbl Holley carburetors (600 cfm flow rate, each)
Special high-rev hydraulic valve train and camshaft; die cast, polished aluminum "Cobra" rocker arm covers and air cleaner; dual exhaust system.

TRANSMISSION:

Fully synchronized four-speed manual (31 spline) standard.
Gear ratio: First—2.32:1 Second—1.69:1
Third—1.29:1 Top —1.00:1
Heavy duty Cruise-O-Matic transmission optional.
Gear ratio: First—2.46:1 Second—1.46:1
Top —1.00:1

FINAL DRIVE:

Heavy duty rear axle with straddle-mounted deep offset drive pinion.
Standard ratios: Manual transmission 3.50:1
Automatic transmission 3.25:1

SUSPENSION:

Front: Independent, with roll coil spring and ball joints, Shelby modified for flatter cornering. .94″ diameter front stabilizer bar.
Rear: 4-leaf springs with special rebound dampers to control rear spring windup.

BRAKES:

Front—disc, 11.3″, with high speed linings.
Rear—10″ x 2.5″ cast iron drum, self adjusting.
191.0 square inch effective lining area.
Independent service brake operating rear drum brakes.

WHEELS—TIRES:

Shelby 15″ steel wheel with 6.5″ rim width. "Speedway 350″ low profile 4-ply nylon E70-15 tires designed especially for Shelby GT cars.

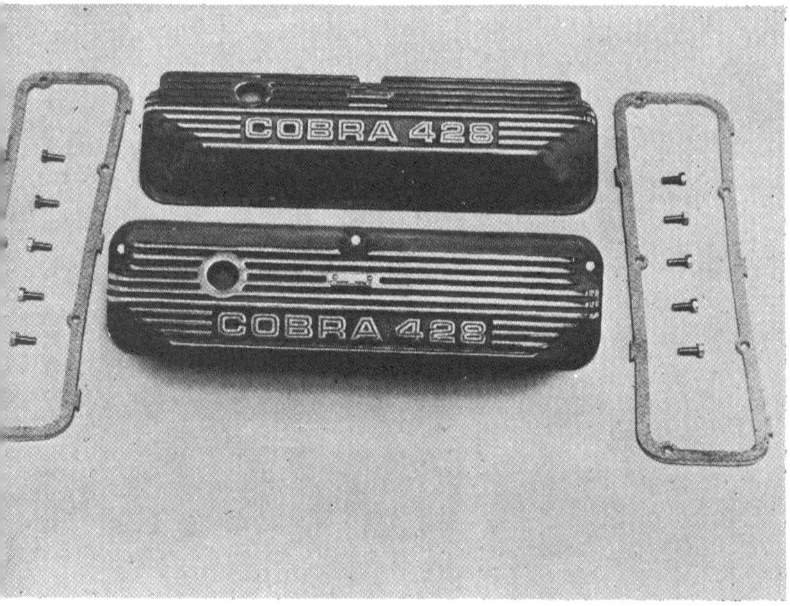

Above, air scoops cool interior and rear brake drums. Left, black crackle-finished aluminum valve covers are standard on 428 LeMans wedge. Below, rear spoiler and wall-to-wall tail blinker treatment are exclusive Shelby GT features.

SHELBY GT 500

A styled-up fastback that should appeal to a wider audience than previous Shelby Mustangs

ONE THING ABOUT Carroll Shelby and his cars, they aren't subtle. Sneaky, maybe. Sly, perhaps. Deceptive, undoubtedly. But not subtle. His latest entries in the automotive lists—the Shelby Mustang GT 350 and GT 500—are typical. They started with a Mustang fastback, re-styled the extremities, added a rollbar, put in shoulder straps, stuck on some trim and it became a different car. It looks something like a racing car, yet it isn't. It also has something of the flavor of a luxurious Grand Touring machine, but it isn't quite that either. It's different from anything Shelby has offered before—less brutal, less purposefully ugly, less stark performance—and yet it offers an abundance of those virtues for which Shelby's products have become famous. It goes, it handles and it stops.

The appearance is distinctive. At the front the nose has been given a revised shell that does away with the chrome grille above the bumper and adds a scoop below. Where the grille used to be there's now a finished-off snout that is backed by an expanded-metal bug strainer and encases a pair of close-together headlights mounted in the manner made popular on international rally cars. The hood, also fiberglass, has a big bulge molded in behind a functional airscoop and there are a pair of post-and-peg hold-downs.

On the sides of the car, the alterations consist of two pairs of scoops. Two fit over the cockpit air extractors in the rear quarter area; the others offer fresh air to the rear brakes.

At the rear, there's a moderate ducktail on the deck lid. This effect has been achieved by using a fiberglass trunk lid and replacing the standard rear fender caps with ones that have a matching upsweep. Across the back there are two wide tail-lights in place of the triplicated smaller ones on the standard Mustang.

These changes in appearance, plus shiny 15-in. steel wheels and Goodyear E70-15s, hang together well, in our opinion. It looks like what it is, a styled-up version of the Mustang fastback. Shelby would like to have you believe that this sort of thing "just happens" at Shelby American but the sure hand of a thoroughly professional stylist obviously had more than just a little to do with these changes. The original GT 350 looked appropriately purposeful but lacked the class of the current model.

The interior has the same sort of distinctive-but-similar flavor as the exterior. The Mustang seats, instrument panel, controls, etc., are retained but there is the addition of a proper rollbar and there are shoulder straps to supplement the standard lap belts. These shoulder straps are attached to the rollbar through an inertia reel that allows the wearer to lean forward as long as it is done slowly but locks up solid on being yanked.

The basic instrumentation of the Mustang is supplemented by an ammeter and oil pressure gauge awkwardly located on the bottom edge of the middle of the dash. There's also a wood-rimmed steering wheel with the Shelby emblem and this, being less deeply dished than the standard Ford wheel, is consequently a bit farther from the driver's chest. Which is good.

The back seat of the fastback is retained this year and though the seating is minimal, it does make a practical 2+2 where the original GT 350 was strictly a 2-seater. This back seat will also fold flat to make an attractive and practical flat deck and, borrowing an idea from the Plymouth Barracuda, there is a drop-down door between the small trunk and the back seat to make even more space.

The suspension of the GT 500 is stiff and the ride could be described as extra-firm. Until now, Shelby American has lowered the pivot point of the front upper A-arm but since Ford incorporated that change in the 1967 Mustang, this is no longer necessary. Shelby still uses stiffer springs front and rear, a larger anti-roll bar front (0.94 in., not as large as last year's 1.00) and Gabriel adjustable shocks all around.

At the rear, the trailing arms previously added to the GT 350 are gone, replaced by rubber snubbers mounted 8 in. be-

PHOTOS BY HENRY N. MANNEY & GORDON CHITTENDEN

hind the front eyes of the leaf springs. These still provide some resistance to rear axle windup and hop but are not as effective—or as harsh—or as expensive—as the arms. They did the job in our test car (the GT 500 with automatic transmission and without limited slip) but we don't know if they'd do equally well with the manual gearbox, hard clutch and limited-slip differential. The milder suspension alterations this year represent part of Shelby's effort to tailor the cars to a wider market—and at a lower cost.

The brakes are the disc front/drum rear combination available as an option from Ford but with a more fade-resistant organic friction material. These were power assisted on our test car and though the touch is a bit lighter than we prefer, they are comparatively easy to control. Because of the weight of the GT 500, the swept area per ton is not impressive (175 sq in/ton) and by the time we'd completed our sixth stop from 60 during our fade tests, the pedal effort had increased by 48%.

This year's Shelby Mustang GT can be had with the 289-cu-in. V-8 (the GT 350), a supercharged 289, or with the big-big 428 (the GT 500). This 428 (4.13 x 3.98 bore and stroke) is Ford's big cheap cooking engine used in the Thunderbird and the Police Interceptor variants, you should understand, not the celebrated 427 (4.24 x 3.78) developed for NASCAR stock car racing and used in the Le Mans-winning GT Mark II prototypes. As fitted to the GT 500, the 428 has hydraulic valve lifters, is equipped with two Holley 4-throat carburetors and is rated at 355 bhp at 5400 rpm.

In pure physical bulk, the 428 is bigger than king size. There is barely room for it in the Mustang's hull and though there have to be sparkplugs down there someplace (you can see the wires disappear under the rocker covers), changing them doesn't even bear thinking about.

That the 428 is also full of weight as well as bulk is demonstrated by the 3520-lb curb weight of the GT 500. The standard GT 350 we tested earlier had a curb weight of 2800 lb.

Our GT 500 had not only the 428 engine and automatic transmission, power steering and power brakes, but also air conditioning. All these things, with the exception of the air conditioning, come standard at no extra cost on the GT 500. Although these power assists may seem inappropriate to such a car at first glance, they blend into the car's personality without obtrusion. And unless you have biceps like Freddie Lorenzen, you'll find the power steering almost mandatory what with those big tires and all that weight pressing them against the ground.

The car is extremely easy to drive. The engine lights off with a whump, there's a clunk-jump when the shift lever is moved into gear, and if you mash on the gas pedal, you'll GO. The steering is easy, it tends to go where it is pointed, and on the open highway it rolls along without fuss or fury. There is an unmistakable detent in the throttle operation so

SHELBY GT 500
AT A GLANCE

Price as tested	$5114
Engine	V-8, ohv, 7014 cc, 355 bhp
Curb weight, lb.	3520
Top speed, mph.	132
Acceleration, 0–¼ mi, sec.	15.5
Average fuel consumption, mpg	9.8

Summary: More refinement and styling distinction than past GT 350, power assists standard; performance of big 7-liter disappointing. The ultimate attention-getter.

SHELBY GT 500

it requires a conscious effort to get through to the carburetors' secondaries. When you do push down hard enough to bring in the additional barrels there's a great hollow gasping, gouts of smoke pour out the rear and the car hunches forward with a bellow.

In pure acceleration, however, the GT 500 simply doesn't have anything sensational to offer. As we tested the car (two up, plus test gear), it would do 15.5-sec standing quarters consistently but that was about all. And that simply isn't very fast as dragstrip times go. A Ford Mustang with the 390-cu-in. engine option does as well and based on our experience, it would take about 400 bhp and a stick shift to get the GT 500 down to the 13.5-sec quarter claimed for it in Shelby American publicity.

As for handling, the GT 500 is something less than we've come to expect from Shelby's cars but still very good in comparison to the typical American sedan. On the other hand, considering the weight distribution, it's better than we would have thought possible only a couple years ago. With 58% of the total weight on the front wheels, we'd expect it to have understeer akin to that of the USS United States, but it doesn't. As we said, it goes where it's pointed. With that amount of weight at the front, there's no doubt that the front tires are operating at greater slip angles than the rears when the car is turning but we have a theory that modern wide-tread tires such as those on the GT 500 are so generous in their grip that non-break away understeer is almost undetectable. And especially when you're even further insulated from what's happening by power steering. If you push the car hard enough, the front end will finally slide but the grip is so good that cornering limits are higher than even normally vigorous driving will ever bring out.

Some of the things we didn't like about the GT 500 included the amount of attention given to it by traffic officers.

We also found our particular example needed the air conditioning working most of the time. Not only was there an uncomfortable amount of heat coming through the firewall, there were strong gasoline vapors that became even more objectionable when the throttle was depressed. With the air conditioner running, the heat was counteracted and the odor of raw gasoline diminished. The test car also deposited a pool of oil every time it was parked.

The GT 500 also recorded the highest fuel consumption of any street version car we've tested in years. In over 800 miles, we averaged 9.8 miles per gallon.

The noise level of the GT 500 was reasonably low and though there seemed to be a lot of valve clatter for an engine with hydraulic lifters, it blended in with the general rumble and was soon forgotten. There was a distinctive noise from the back end, though, that on a slightly rough surface sounded like the trunk was full of roller skates.

All in all, though, the GT 500 is a more civilized vehicle than the original GT 350 from which it descended. It rides better, it has more amenities and it is far more attractive. It isn't so closely related to a racing car, perhaps, but we have the feeling that it will appeal to a larger number of buyers than any previous Shelby American automobile.

Big 428 V-8 engine leaves little extra space under hood.

Trunk is minimal size but there's more space behind rear seat.

ROAD TEST

SHELBY GT 500

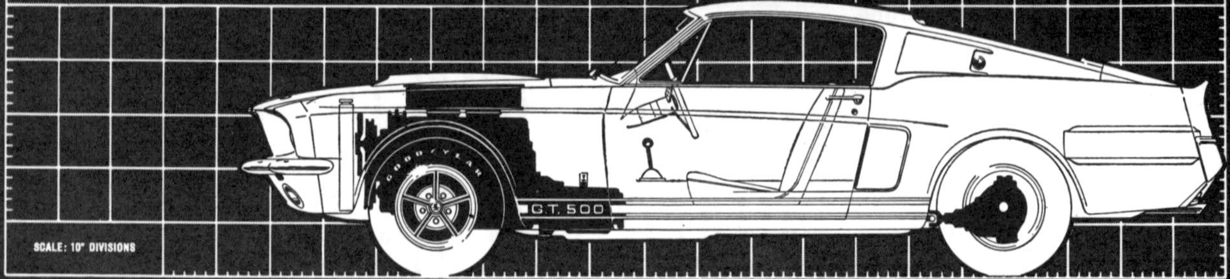

PRICE

Basic list.................$4395
As tested.................$5114

ENGINE

Type..................V-8, ohv
Bore x stroke, mm......115 x 101
Equivalent in........4.13 x 3.98
Displacement, cc/cu in...7014/428
Compression ratio..........10.5:1
Bhp @ rpm.........355 @ 5400
Equivalent mph...........127
Torque @ rpm, lb-ft..420 @ 3200
Equivalent mph.............75
Carburetion..........2 Holley 4-V
Type fuel required.......premium

DRIVE TRAIN

Automatic transmission: Ford Cruise-O-Matic (torque converter with 3-speed planetary gearbox).
Gear ratios: 3rd (1.00)......3.25:1
2nd (1.46)..............4.75:1
1st (2.46)..............8.00:1
1st (2.46 x 2.1)..........16.8:1
Final drive ratio...........3.25:1
Optional ratio............3.50:1

CHASSIS & BODY

Body/frame unitized with reinforced floor members.
Brake type: 11.3-in. discs front, 10x2.5-in duo-servo drums rear.
Swept area, sq in..........330
Wheel type & size, in.....15 x 6.5
Tires..Goodyear Speedway E70-15
Steering type: recirculating ball, power assisted.
Overall ratio............20.3:1
Turns, lock-to-lock..........3.6
Turning circle, ft..........39.4
Front suspension: independent with short & long arms, coil springs, tube shocks, anti-roll bar.
Rear suspension: live axle with semi-elliptic leaf springs, anti-windup snubber, tube shocks.

OPTIONAL EQUIPMENT

Included in "as tested" price: air conditioning, automatic transmission, radio, crankcase emission control, fold-down rear seat, styled wheels.
Other: limited-slip differential, exhaust emission control (Calif.), 12/12,000 warranty on powertrain

ACCOMMODATION

Seating capacity, persons...4 + 1
Seat width, front/rear...2 x 20/40
Head room, front/rear.....39/35.5
Seat back adjustment, deg......0
Driver comfort rating (scale of 100):
Driver 69 in. tall............95
Driver 72 in. tall............90
Driver 75 in. tall............80

INSTRUMENTATION

Instruments: 140-mph speedometer, 8000-rpm tachometer, fuel, water temperature, oil pressure, ammeter, clock.
Warning lights: parking brake, seatbelt, high beam, directionals, alternator.

MAINTENANCE

Crankcase capacity, qt..........5
Change interval, mi........6000
Filter change interval, mi.....6000
Chassis lube interval, mi.....6000
Tire pressures, psi.........28/28

MISCELLANEOUS

Body styles available: fastback as tested.
Warranty period, mo/mi: 3/4000 on powertrain; 12/12,000 on car.

GENERAL

Curb weight, lb............3520
Test weight................3900
Weight distribution (with driver), front/rear, %....58/42
Wheelbase, in..............108.0
Track, front/rear........58.0/58.0
Overall length..............186.6
Width.....................70.9
Height....................49.0
Frontal area, sq ft...........19.3
Ground clearance, in..........5.1
Overhang, front/rear....33.2/40.5
Usable trunk space, cu ft......6.0
Fuel tank capacity, gal.........17

CALCULATED DATA

Lb/hp (test wt).............11.0
Mph/1000 rpm (3rd gear).....23.8
Engine revs/mi (60 mph).....2520
Piston travel, ft/mi..........1670
Rpm @ 2500 ft/min.........3770
Equivalent mph.............89
Cu ft/ton mi.................160
R&T wear index.............42.1
Brake swept area sq in/ton.................170

ROAD TEST RESULTS

ACCELERATION

Time to distance, sec:
0–100 ft...................3.4
0–250 ft...................5.8
0–500 ft...................8.6
0–750 ft..................11.0
0–1000 ft.................13.2
0–1320 ft (¼ mi)..........15.5
Speed at end of ¼ mi, mph....95
Time to speed, sec:
0–30 mph..................2.8
0–40 mph..................4.1
0–50 mph..................5.7
0–60 mph..................7.2
0–70 mph..................8.8
0–80 mph.................11.0
0–100 mph................17.5
Passing exposure time, sec:
To pass car going 50 mph....4.5

FUEL CONSUMPTION

Normal driving, mph........8–12
Cruising range, mi........135–205

SPEEDS IN GEARS

3rd gear (5600 rpm), mph.....132
2nd (5600).................90
1st (5600)..................54

BRAKES

Panic stop from 80 mph:
Deceleration, % g...........84
Control..................good
Fade test: percent of increase in pedal effort required to maintain 50%-g deceleration rate in six stops from 60 mph..........48
Parking brake: 30% grade....yes
Overall brake rating.........good

SPEEDOMETER ERROR

30 mph indicated......actual 29.0
40 mph....................38.5
60 mph....................57.5
80 mph....................77.0
100 mph...................97.0
Odometer, 10.0 mi....actual 10.16

ACCELERATION & COASTING

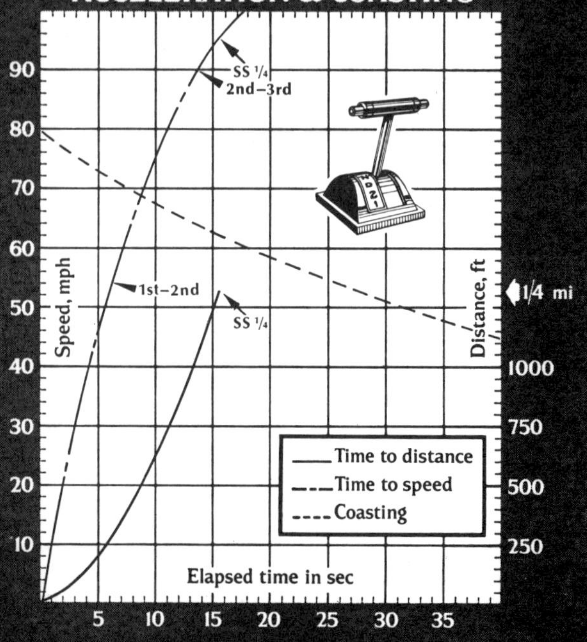

ROAD TEST: SHELBY GT 500 & 427 STING RAY

A contender for the throne meets the renowned king of speed, and the result is a clash between America's only pair of sports cars.

BY STEVE KELLY

PHOTOS BY RANDY HOLT, JR., GERRY STILES

67' GT handling isn't as good as forerunners', but above average "performance" cars'.

For the first time in its decade and a half of production, Chevrolet's Corvette – billed as "America's only true sports car" – is being challenged for the title. Shelby American's GT 500 Mustang-based sports fastback has all the earmarks of a "purist" type sporty-car with a just-right combination of comfort and convenience features.

Following Shelby American's announcement of a 428-cu.-in. V-8 engined GT, we were skeptical if this car would be suited for street duties. We've driven many small cars with huge engine transplants, and somehow they've always seemed to lack tameness during regular operation – unless regular operation was a full-throttle run down a strip. We found out, though, just how wrong we were after depositing ourselves behind the custom wood steering wheel of an air-conditioned GT 500 with Cruise-O-Matic transmission.

We could hardly believe its smoothness in bumper-to-tail-light traffic and had difficulty realizing it was built by the same people who used to peddle a rough-riding, hard-steering Mustang fastback with Bunyanesque brake pedal and an engine which would outshout a John Deere tractor.

In view of this, we arranged for an even hairier GT 500, and also set up an equally equipped Corvette to see if there really was a challenge.

The basic Shelby GT 500 is a true sophisticate compared to earlier cars built by the famous Texan. There was a brief period when the Ford Cobra, also built by Shelby, was thought to be a fair and equal competitor to the 'Vette, but a too-high price and rather impractical design for a street-destined car put it out of contention. Chevy has cause for some worry from the GT 500 though. There's more room inside, and it's easier to drive in traffic. There's a civilized luggage compartment that's accessible from the outside, and there's at least the same measure of racy styling. The GT 500 runs quieter, and in general is a more practical car for everyday use.

The Corvette can't be sold short though. Even in stock condition, it outruns the GT 500 by a solid 0.7 second in the quarter-mile, stops in less distance from any speed, is easier on gas, and exhibits noticeably better workmanship throughout.

The 'Vette started life back in 1953 – before the world or Ford Motor Co. had ever heard of a car called the Mustang – and was then equipped with an impotent 6-cylinder engine and fiberglass body. The 'glass body has remained, but the 6s were dropped two years later and the car since then has steadily climbed higher and higher in the esteem of performance addicts. Equipping the Corvette with a 427-cu.-in. engine was a natural move for Chevrolet, and for the past two years this version has been the image-maker of the line. Previous editions of the 'glass-bodied car were offered with as much luxury built in as speed and performance, and we remember a statement concerning Corvette sales cutting into

Stop-off at Irwindale Drags quickly revealed GT 500's popularity with ¼-mile fans. Most said they preferred it to 'Vette.

'Vette makes tight turns a simple operation; standard tires hinder serious contests.

Cadillac. The luxury is still there, but a hard ride comes with it, so we doubt if Cadillac is still worried.

There's more potential for making the Corvette a hard runner than there is the GT 500. The 428-cu.-in. V-8 in the Shelby car is too heavy for serious work, and it has a rather restricted breathing system compared to the 427-cu.-in. Chevy V-8. Called "porcupine" by Chevy engineers, the valve layout of the 'Vette engine is more ideally suited to getting big charges of fuel/air mixes into the combustion chamber, and the head configuration above the chamber is a "semi-hemi" type, similar to early Chryslers.

Weight distribution is better on the GT 500, with 56.4% being carried on the front wheels. The two total out almost equally, but so much of the Corvette *is* the front, that it is distinctly nose-heavy. This characteristic is abruptly apparent when one *attempts* to leave the line under hard throttle, or when you bring it around a sharp bend while applying power. The rear end becomes frighteningly "light."

Starting with two basic cars, more is standard equipment on the GT 500 than on the 'Vette, but there's some chicanery involved. For the base price of $4195, a GT 500 buyer gets the 428 engine, front-disc/rear-drum brakes, full instrumentation, a 4-speed transmission, and all the modified body work that goes into making a GT 500 the distinctive looking car it is. He also gets the beefed-up undercarriage which includes stiffer springs, adjustable shocks, and modifications to the front suspension to make it corner more flatly. On top of this though, he *must* buy power brakes, power steering, integral roll bar with inertia-reel shoulder harnesses (the best we've ever seen) and a fold-down rear seat. This comes to a total of $264.77 before he ever starts ordering a radio, air conditioning or whatever else he may choose. This is sort of like selling a car without seats and a steering wheel, but putting them on the car at the "buyer's request." All GT 500s are built with this batch of options which are called "standard equipment," but which really are not.

For the base convertible price of $4327.50, a Corvette buyer has all options still in front of him, with nothing mandatory except making the payments. The standard Corvette includes a 300-hp, 327-cu.-in. V-8, either a soft folding top or fiberglass lift-off type, and all instruments and gauges along with a 3-speed all-synchro gearbox. From there the owner can choose any one of four optional engines and two 4-speed manuals or Powerglide transmission. Our price and accessory list on page 29 shows the other options available to outfit the Corvette for street or track duties.

The primary function of an automobile, no matter what type it is, is to carry passengers wherever they may want to go. The GT 500 does this better than the Corvette. It has more room inside for people and packages, and will carry four adults for a short time or two children for a long time without complaints. The fold-down rear seat can become a parcel counter when only two are aboard, and the trunk bulkhead swings up for stowage of skis and the like.

The 500 is much easier to drive in traffic as it's not as low as the Corvette, and it's not as ticklish to keep running at slow speeds. That big 435-hp 'Vette engine likes to work hard and when it's in bumper-to-bumper traffic, it objects.

The ride on city streets is much better in the GT 500. There's very little bumping around, whereas the position of the driver in the Corvette is very close to the rear wheels, and any rebound action from them is strongly noticed by the man at the wheel.

Vision in the two cars is almost equal, but still not excellent. We liked the '66 Shelby GTs for their rear-quarter windows, but construction differences on the '67 Mustang prevented their continuance on Shelby's version. Corvette convertibles have blind rear quarters, too, and the lift-off 'glass top isn't much better. The fastback model allows full vision.

We liked the interior layout of the GT 500, but it doesn't have the Maserati look of the Corvette. The 'Vette has *all* necessary instruments right in front of the driver and all controls within a few inches reach. The Shelby car has the speedometer and tach right in front

Nose and hood of GT are fiberglass items. Scoops deposit air on top of carburetors.

Spoiler is integral part of 'glass deck lid; a la T-Bird tail lamps really light up rear.

Initial test GT had air conditioning and Cruise-O-Matic, making it "sporty-plush."

Second GT lacked optional comforts but ran faster. Shelby adds center two gauges.

"Speedway 350" tires break loose easily, but need to for the best acceleration times.

of the pilot, but the amp and oil pressure gauges (Shelby additions) are positioned centrally below the radio. It's not hard to see them, but they're not as readable as in the Corvette, nor are they in a direct line of sight with the road.

The really impressive points of the GT 500's insides are its great-feeling wood steering wheel and the integral roll bar with inertia-reel shoulder harnesses. The wheel is one of the most comfortable we've ever had our paws on, with a smooth lacquered finish and genuine "sporty" look. Shoulder harnesses can be cumbersome to attach, and restrict the normal movements of the driver, but not so in the GT 500. They fit around you like suspenders. The inertia retractor in the roll bar allows slow movement, but quickly holds you against any sudden jerk or action. You learn to be leisurely when reaching for the cigarette lighter.

The 'Vette spare is in a panel below the rear underside which, in the rain, may give the Triple A reason to upgrade its rates. There is room for perhaps two suitcases behind the seat.

It takes a while to get used to the stiff-riding Corvette, and the low driver position. Once we became oriented, however, we went along with the "true sports car" claim. The 'Vette is a lot of fun, but discretion must be used in driving around town. It's awful easy to bound past speed limits unknowingly.

Less of an all-out sports car, the GT 500 is more at home on the street than on the track. We received many more comments on the 500's styling than we

Clean 'Vette front end is product of "hiding" lights. Hood scoop can be opened.

Standard, horizontal back-up light throws excellent beam, distinguishes '67 Sting Ray.

Standard tires resist high-rpm starts by producing slow times and plenty of smoke.

Corvette gauge layout is nothing short of great, as is close-gate 4-speed shifter.

Entry to neat interior is easier than exit. Hand/parking brake is between seats.

did on the Chevy sportster's, but this can be attributed as much to its being new as anything.

Our forecast of how the cars would compare on the track turned out to be accurate. We were highly impressed with the solid-as-a-rock cornering of the GT 500, and its agility in quick turns. It ran impressively in acceleration tests, staying straight when the tires were spun, and it stopped quick and straight. But even with this good performance record, it fell short of outrunning and outmaneuvering the Corvette.

Unfortunately, Chevrolet couldn't find enough room under the fender wells for a tire comparable to the E70-15 nylons standard on the GT 500, so it suffers in this department. Standard 7.75 x 15 rayon tires are fitted, and this makes cornering and starting difficult. Gobs of wheelspin is all that results from a high-rpm start, but proper feathering of the gas pedal will help. The independent rear suspension gets an unfair shake when these regular tires are installed, and the rear end tends to leave the course when the car is pushed. Hooking up tires equivalent to the Speedway 350 Goodyears on the test GT 500 solved the Corvette's problems quickly. The potential is built in, but anyone desiring to put a 'Vette in proper handling form will have to start with a set of tires and the realization that lock-to-lock cornering will be slightly restricted.

It's quite a foe that Shelby's GT 500 has taken on, but not an unconquerable one. The Corvette is designed as a sports car, and that's what it is. It suffers somewhat as a street machine, but in no way is it reminiscent of early English sports cars with fold-down windshields and side curtains.

The 500 has more passenger car than sports car feel, but this could hardly be helped as the Mustang from which it stems has this quality. With a bigger engine (horsepower, not displacement), it would be close to the Corvette's acceleration times.

The two cars are apart yet fairly close. Shelby-American built the GT 500 with the idea of getting more customers for street-type vehicles than they could with the race-oriented GT 350s of the past. Oddly enough, this is what Chevrolet's theory has been too. They've been building a "hot" car for the street that would qualify for competitive use. Shelby's just reversed the game and taken a competition-type car and turned it into a street machine. Both gain in some respects and suffer in others from the compromise. /MT

Big 428 engine fills out GT compartment, leaves little space for working. Progressive linkage is used on dual 4-barrels.

Center carb of 427 has choke and is linked to other two by progressive hookup. Working room around V-8 is very good.

GT's shoulder harnesses are mounted to well-padded and fitted integral roll bar.

Inertia retractors stop sudden motion of GT wearer; allow slow, easy movement.

Hood pins are now wired to GT to prevent loss, or scavenger hunters from scoring.

427 STING RAY: REIGNING KING

Five-year-old Sting Ray body style still looks good, and all-around quality of fiberglass body is of a very high standard.

SHELBY GT 500: CANDIDATE FOR TOP BILLING

Well executed GT design results from extended nose, rear spoiler and scoops. 'Glass add-ons could stand quality check.

MT Road Test

sting ray at a glance . . .

Easily the most powerful production car made . . . designed and built as an all-out sports car, yet fairly suitable for everyday use . . . stiff riding suspension makes around-town jaunts and long trips uncomfortable on the torso, but free of handling problems . . . styling exhibits a "going" look even when standing still . . . practiced drivers can feel safe at all speeds and in all conditions, due to 4-wheel disc brakes, quick steering, and good stability.

how the car performed . . .

ACCELERATION (2 aboard)
0-30 mph . . . 2.5 secs. 0-60 mph . . . 5.5 secs.
0-45 mph . . . 3.8 secs. 0-75 mph . . . 7.5 secs.
TIME AND DISTANCE TO ATTAIN PASSING SPEEDS:
40-60 mph . . . 2.1 secs., 151 ft.
50-70 mph . . . 2.0 secs., 172 ft.
STANDING START QUARTER-MILE:
13.8 secs., 104 mph.

BEST SPEEDS IN GEARS @ SHIFT POINTS:
1st . . . 65 mph @ 6500 rpm
2nd . . . 87 mph @ 6500 rpm
3rd . . . 112 mph @ 6500 rpm
Top Speed . . . 143 mph @ 6500 rpm
MPH PER 1000 RPM: 22.0
SPEEDOMETER ERROR:

Calibrated Speedometer	**30**	**45**	**50**	**60**	**70**	**80**
Car's Speedometer	**28**	**44**	**49**	**58**	**67**	**77**

STOPPING DISTANCES:
From 30 mph, 31 ft.; from 60 mph, 135 ft.

specifications . . .

ENGINE: Ohv V-8
Bore and stroke (ins.): 4.251 x 3.76
Displacement (cu. ins.): 427
Max. torque (lbs.-ft.): 460 @ 4000 rpm
Horsepower: 435 @ 5800 rpm
Compression ratio: 11.0:1
Carburetion: 3 2-bbl. Holleys
TRANSMISSION: Optional 4-spd. manual, floor-mounted lever, all forward gears fully synchronous. Ratios of 2.20 in 1st, 1.64 in 2nd, 1.27 in 3rd, and 1.00:1 in 4th gear.
FINAL DRIVE RATIO: 3.55:1
SUSPENSION: Independent front, single lateral arm-type with coil spring, tube shock and spherically jointed steering knuckle at each wheel. Fully independent rear with fixed differential, transverse multi-leaf spring, lateral struts and universally jointed axle shafts. Tube shocks at each rear wheel.
STEERING: Optional linkage-type power assist, with semi-reversible recirculating ball nut. 17.6:1 overall gear ratio. Turning diameter: 39.9 ft., curb-to-curb. Number of turns lock-to-lock: 2.92
WHEELS: Short spoke disc steel, 15-in. dia. x 6-in. wide.
TIRES: 7.75 x 15 rayon, standard.
BRAKES: 4-wheel hydraulic, caliper disc actuated by dual system. Power optional. Dia. of disc front and rear: 11.75 ins. Effective lining area: 78.1 sq. ins.
FUEL CAPACITY: 20 gals.
MILEAGE RANGE: 9.0 – 12.0 mpg
BODY AND FRAME: Separate construction. All-welded, full-length, ladder-constructed frame with 5 cross-members. Fiberglass body.
DIMENSIONS: Wheelbase: 98.0 ins. Track: front 57.6 ins., rear 58.3 ins. Overall length 175.1 ins., width 69.6 ins., height 49.8 ins.
USABLE TRUNK CAPACITY: 8.1 cu. ft.
CURB WEIGHT: 3340 lbs.

prices and accessories . . .

MANUFACTURER'S SUGGESTED RETAIL: (includes federal excise tax but excludes state and local taxes, license, options, accessories and transportation) Soft top Corvette: $4327.50 (with 327, 300-hp V8).
OPTIONS & ACCESSORIES:

435-hp, 427-cu.-in. V-8	$437.10
4-speed transmission	184.35
Positraction axle (limited-slip)	42.15
Red Stripe, special nylon tires	46.65
Special cast aluminum wheels	263.30
Power steering	94.80
Power brakes	42.15
Heavy-duty brakes	342.30
Side-mounted, off-the-road exhausts	131.65
Full-transistor ignition	73.75
Special-purpose suspension	36.90
36-gal. fuel tank (coupe only)	198.05
Shoulder belts	26.35
Power windows	57.95
Removable hardtop (with soft top)	231.75
Air conditioning	412.90
AM-FM radio	172.75

---

gt 500 at a glance . . .

Eye catching styling is product of good judgment in restyling already good looking Mustang fastback . . . well suited for street duties, but refinements needed to make it an all-out champ on the track . . . one of the most comfortable cars for touring we've driven, but in-and-out from driver's seat is a tight fit . . . fiberglass additions would benefit from tighter quality control inspection . . . shoulder harness impressed us to the point of dissatisfaction with all others.

how the car performed . . .

ACCELERATION (2 aboard)
0-30 mph . . . 2.9 secs. 0-60 mph . . . 6.2 secs.
0-45 mph . . . 4.4 secs. 0-75 mph . . . 9.5 secs.
TIME AND DISTANCE TO ATTAIN PASSING SPEEDS:
40-60 mph . . . 2.6 secs., 190 ft.
50-70 mph . . . 3.0 secs., 264 ft.
STANDING START QUARTER-MILE:
14.52 secs., 101.35 mph
BEST SPEEDS IN GEARS @ SHIFT POINTS:
1st . . . 51 mph @ 5500 rpm
2nd . . . 68 mph @ 5500 rpm
3rd . . . 93 mph @ 5500 rpm
Top Speed . . . 120 mph @ 5500 rpm
MPH PER 1000 RPM: 21.1

SPEEDOMETER ERROR:

Calibrated Speedometer	**30**	**45**	**50**	**60**	**70**	**80**
Car's Speedometer	**27**	**40**	**45**	**55**	**64**	**72**

STOPPING DISTANCES:
From 30 mph, 40 ft.; from 60 mph, 144 ft.

specifications . . .

ENGINE: Ohv V-8
Bore and stroke (ins.): 4.13 x 3.984
Displacement (cu. ins.): 428
Horsepower: 355 @ 5400 rpm
Max. torque (lbs.-ft.): 420 @ 3200 rpm
Compression ratio: 10.5:1
Carburetion: 2 4-bbl. Holleys
TRANSMISSION: Fully synchronized (forward gears) 4-spd. manual. Floor-mounted lever. Gear ratios: 1st 2.32, 2nd 1.69, 3rd 1.29, 4th 1.00:1.
FINAL DRIVE RATIO: 3.50:1
SUSPENSION: Independent front with coil spring and ball joints, modified for flatter cornering, .94-in.-dia. front stabilizer bar. Straddle-mounted H-D rear axle, single unit, suspended with longitudinal 4-leaf springs with special rebound dampers to control rear spring wind-up. Preset, adjustable shock absorbers at each wheel.
STEERING: Recirculating ball and nut, linkage-type power assist standard. 16-to-1 overall gear ratio. Turning diameter, 37.16 ft. curb-to-curb. Turns lock-to-lock, 3.74.
WHEELS: Shelby steel wheel, 15-in. dia. x 6.5-in. wide.
TIRES: "Speedway 350" Goodyear low-profile 4-ply nylon E70-15 140-mph rated tires, standard.
BRAKES: Dual-system hydraulic. Front disc/rear drum with power assist and high-speed linings on disc caliper, standard. Dia. of front disc: 11.3 ins. Dia. of rear drum: 10 ins. Effective lining area: 191.0 sq. ins.
FUEL CAPACITY: 16 gals.
MILEAGE RANGE: 7.6 – 14.6 mpg
BODY AND FRAME: Platform-type unitized construction with reinforced floor side-members and export front end reinforcement.
DIMENSIONS: Wheelbase: 108.0 ins. Track: front 58.0 ins., rear 58.0 ins. Overall length 186.6 ins., width 70.9 ins., height 51.6 ins.
USABLE TRUNK CAPACITY: 5.1 cu. ft.
CURB WEIGHT: 3360 lbs.

prices and accessories . . .

MANUFACTURER'S SUGGESTED RETAIL: (includes federal excise tax, but excludes state and local taxes, license, options, accessories and transportation) 2-door fastback Shelby GT 500 Mustang – $4195.00 (f.o.b. St. Louis, Mo.)
OPTIONS & ACCESSORIES:

Mandatory options: power brakes, power steering, shoulder harnesses (2)	$200.00
Fold-down rear seat (mandatory)	64.77
C-6 Cruise-O-Matic	50.00
Air conditioning	356.09
AM radio	57.51
AM-FM radio	135.00
Styled steel wheels	185.00
Rally stripes (over the top)	24.95
Limited-slip differential	181.00

'68 COBRA GTs

Shelbys enter '68 with new flip-top and higher quality

BY STEVE KELLY

Big changes have occurred to both Shelby-American and the cars they produce. Auto production has been relocated to Ionia, Mich., nearer the Ford headquarters and a more centrally located shipping point. Administration, Parts and the Racing Division will stay in California, at a new Torrance facility.

The Cobra roadster will not be produced for '68. That name is now attached only to the GT 350 and GT 500. The roadster may reappear in '69, but with a new shape.

The new Cobra comes in two configurations: fastback and convertible. Both feature built-in roll bars and safety harnesses. The fastback roll bar is integral with the roof construction, attached securely to the sheetmetal, and supports suspender-type shoulder harnesses with inertia reel retractors mounted over both front seats. The convertible roll-bar is more prominent and is covered with resilient padding, and a form-fitting molded sleeve. A chest-type belt rather than shoulder harnesses is used on the flip-top. The belt pulls out from the rear quarter panel, across the driver/passenger, and secures to a center floor hook. An inertia retractor is used for take-up.

Body styling is changed for the better. All parts replacing stock Mustang components are smoother fitting, due to matched metal molds now used for the fiberglass pieces.

Front-end styling is tastefully executed, with added touches of chrome trim. The outboard fender headlights are a full $7^1/_4$-inch in diameter. These are augmented by rectangular high beam driving lights mounted inside the oval grille opening.

The twin-scoop and louvered "hatch" type fiberglass hood is Dzus fastened with turn-knob tie-downs, and electrostatic radio shielding incorporated.

The spoilerized deck lid fits much better than on earlier models and is adequately sealed to prevent dust and

Convertible body boasts same standard equipment as fastback: roll bar, shoulder harnesses, good performance. Shoulder belt extends from rear side panel, around driver's chest and fastens between front buckets. Hardtop retains "suspender" type units. Fiberglass hood has dual scoops and functional rear louvers. Center console is Shelby addition, contains armrest, glovebox, ammeter and oil gauge.

fumes from entering the interior. All models will have the rear panel below the deck lid finished in silver. Rectangular tail lamps housed in this panel have sequential turn signals.

Functional rear brake air scoops remain in use, and fastbacks have interior air extractors in the roof quarter panels.

Interiors are quite distinctive, having wood-grain applique on dash and door trim. A center console, blending into the dashboard, contains a padded glovebox/armrest combination, and carries auxiliary instruments and driving light switch. The steering wheel center is safety padded, with simulated wood rim, not the sporty all-wood type used on '67s. Shelby-American is hoping to add Ford's adjustable swing-away steering wheel to the option list, but this hasn't been confirmed.

The GT 350 fastback and convertibles use Ford's new 302-cu.-in. V-8 as the standard engine. It's a hydraulic lifter type, using a single 4-bbl. hi-riser intake manifold, and a 600 cfm (cubic-feet-per-minute flow) Holley carb. It's rated at 235 hp.

Optional for all GT 350s is a belt-driven, centrifugal-type supercharger for the 302 V-8. No hp ratings have been released on this optional engine yet, but it is expected to be over 300.

GT 500 Cobras have a 428-cu.-in., 390-hp Police Interceptor V-8 in standard form. It is Holley 4-bbl. carbed, using a 715 cfm unit, and hydraulic lifters that minimize engine noise. Optional for all GT 500s is a 425-hp, 427-cu.-in. V-8, with a medium riser intake manifold and 780 (approx.) cfm Hol'ey 4-bbl. carburetor.

All engines are mated to 4-speed manual gearboxes and can be optionally bolted to Cruise-O-Matic automatics. GT 350 automatics are C-4 series, while C-6 H-D units are used in GT 500s. GT 350s with 4-speed have a 3.89:1 rear ax'e ratio, and 3.50:1 with automatic. A 3.25:1 rear gear is used in all GT 500s.

Front disc/rear drum power brakes are standard, with Ford's new floating caliper operating on the 11.3-inch vented disc. Rear drums are 10 inches in diameter and $2^1/_4$ inches wide.

Air conditioning costs extra and is available only on the unblown GT 350 and 428-cu.-in. powered GT 500.

The fun of driving either Cobra model is as ever. We like the convertible for better all-round vision with the top down as well as up, and fresh air-in-the-face attributes. The maximum handling type suspensions are firm enough to allow vigorous cornering while not abusing kidneys in less-than-race-track operation. We stormed around Riverside Raceway in a GT 500 428-cu.-in. convert, really getting caught up in the spirit of the moment. We wheeled over a hill, down into a tight left turn, before we even had time to realize we may have over-stretched ourselves. But we encountered no difficulty whatsoever, and even our passenger was impressed with our great driving "prowess." The big E70 x 15 tires hang on beautifully, never letting out a squeal of objection. The .94-inch diameter front sway bar and Adjust-O-Matic shocks keep front and rear behaving properly, with only slight understeer.

Shelby-American's big year should be just around the corner. The new body styling, availability, and production-line quality offerings should appeal to a wide range of buyers. It is a long way from the first GT car, a hard-riding, heavy-steering, loud-sounding, all-out performance car with little of the comfort now offered. But it set the stage for the new breed; doing quite a job in the process. /MT

Roof scoops on coupe pull stale air from interior. Lower duct sends fresh air to rear brakes.

COBRA

THE NAME is the same, but the 1968 Cobra, essentially a Ford product, is a new kind of car. The new models are not all that's new at the ever-volatile Shelby American, Inc.

WHAT started the Ford Motor Company on its tremendous programme of international racing, the programme that led to two Le Mans victories, and Grand Prix and Indianapolis wins by Ford-powered cars? Many men and ideas contributed to the Dearborn decision to go racing. Not least among them stands Carroll Shelby. From past experience Ford knew whom to hire to go drag-racing and stock car racing in the United States, but sports car racing abroad was something entirely new to Ford. When Ford needed Shelby, and need him they did, he was ready.

As a retired race driver at the end of the 'fifties, Shelby had many schemes for making a production car. One was to be a Pontiac engine in an Austin-Healey chassis, for example. But when, in late 1961, he decided to marry the A.C. all-independent chassis to the brand-new Ford Fairlane V8, he came up with the ideal combination. He had no trouble naming it. Since 1956 Shelby had known that some day he'd build a car called Cobra.

Racing and special-bodied Cobras were gradually developed by Shelby's experts, among them Phil Remington, to meet the strains of international endurance racing. In 1964, before Ford had made appreciable headway with their GT40 project, Shelby defeated Ferrari to win the GT Championship. Cobras won it again in 1965, in spite of the diversion of Shelby time and talent to the Ford GT40 programme. Meanwhile, Shelby Cobras were being manufactured and sold. A total of 630 cars with the "small" 260/289 cu.in. engines were built, and 510 Cobras with the 427/428 cu.in. Ford V8s were produced.

The past tense can now be used, because the Shelby Cobras in their original form are no longer being made. A.C. are still making some very nice Ford-powered cars, but they're not Cobras, and they're not sold in the US. Several years ago Ford acquired the rights to the name Cobra, which had become so closely identified with their products, right down to dealer and customer level. Late in 1967 Shelby completed a prototype of a rear-engined two-seater, a possible successor to the Cobra called the Lone Star. It is not an attractive car or a practical one, especially when it is compared with the De Tomaso Mangusta, which offers far more for far less money ($9,000 against a projected $15,000 for the Lone Star). And the new US safety regulations pose serious problems for the small manufacturer.

Shelby isn't concerned about the Lone Star's future, because he has several other projects in the works. Backed by a recently-signed five-year personal contract with Ford, Shelby has set up headquarters in a four-storey office building in Playa del Rey, California, and has formed four separate companies to handle his activities. There's a parts company, for engine accessories for Fords; there's an accessory company, for miscellaneous items like T-shirts and Cobra badges, and there's Shelby Enterprises for some of Carroll's other projects, such as a proposed boy's ranch in Texas.

Closest to Shelby's heart is the fourth company, the Shelby Racing Company, which carry out development projects for other manufacturers. They've been modifying and testing Toyota's six-cylinder 2000GT sports car with an eye toward American sports car racing. "It will beat a Porsche 911 now," according to Carroll, who admitted that "we don't know what Porsche will do to improve their cars." Shelby Racing have also been active in Group 2 events, with Mustangs, and in Group 7 with the Len Terry-designed King Cobra.

One more responsibility of the Shelby Racing Company is the design, each year, of the new production versions of the Mustang which carry the Shelby name. The famous GT-350 and GT-500 Shelby Mustangs are "created" by Shelby and his crew, then engineered for production by Ford. Through 1967 the cars were built in California, partly at Ford's San Jose assembly plant and partly at the Shelby plant near Los Angeles International Airport. The

(*Left*) *Gaping maw and a bone in its teeth for the GT-500 Shelby Cobra—ant's eye view of the special glass-fibre front*

(*Right*) *Special high speed low profile Goodyear tyres are fitted; the roll-over bar is aerofoil shaped*

(*Below left*) *Sun glints off the impressive-looking coupé (pronounced henhouse in American)*

(*Below right*) *Safety-conscious interior with padded dished wheel and protection on facia edges*

latter 100,000-sq.ft plant has been given up in favour of an 84,000-sq.ft building at Torrance, California—but that's not where the Shelby Mustangs are now made.

Effective with 1968 production, the Shelby four-seaters are rolling out of a 965,000-sq.ft. plant in Ionia, Michigan, under the direct supervision of George Merwin of Ford Motor Company. The actual "rolling out" was delayed severely by the Ford strike, and the 1968-model Shelbys did not actually go on sale until 4 January. They had hoped to make and sell 4,000 of the 1968 cars, an increase of 500 over the 1967 output, but now they'll be happy to reach the '67 figure.

Two new concepts were introduced with the 1967 Shelby cars: the use of a 7-litre V8 engine as an option, and the design of a special bonnet and front-end of glass-fibre to give the car a special look. Both features are continued in 1968. GT-350 models in 1968 are fitted with Ford's new 5-litre V8 (4.00 × 3.00 in.) developing 250 bhp at 4,800 rpm. Earlier plans to equip this engine with a belt driven centrifugal supercharger, as used on some 1967 cars, ran foul of the new air pollution regulations for 1968.

When fitted with the big 7-litre V8 (4.13 × 3.98 in.) the Shelbys are designated GT-500. This engine was installed to provide potent performance without the noise and fuss that is necessary when you ask a small engine to do the same job. In the 1967 cars it propelled the coupé to 60 mph in 6.9 seconds and to 100 mph in 17.2, with a top speed of 132 mph, and the '68 model will be similar. Later in 1968, when Ford get caught up with all the post-strike engineering, they hope to be able to offer a more potent 7-litre V8 of the racing type (4.24 × 3.78 in.). This would give the GT-500 some 425-plus bhp at 5,600 rpm, against the "normal" 390 bhp at 5,400.

The GT-350 coupé weighs 3,146lb, the GT-500 3,445 lb, and most of the difference is over the front wheels. The bigger-engined car has 56.4 per cent of its weight on the front, against 53.0 per cent for the GT-350, which thus carries 276lb fewer on its front tyres. In compensation, the GT-500 is given the stiffer export-type front coil springs and adjustable shock absorbers at all four wheels. The GT-350 has them only at the rear. Light alloy wide-rim wheels are optional for the cars, which are fitted from the factory with special high-speed Goodyear low-profile tyres.

The sharp, professional styling of the Shelby cars shows that someone from Gene Bordinat's Ford Design Centre has taken a more than casual interest in the shape. Carried over from '67 is the coupé roof, with its impressive—and functional—air scoops and spoiler rear deck. The glass-fibre front end has been remodelled to look even more aggressive, adding upper "nostrils" and rectangular fog lamps mounted within the wide, squared grille opening. Pin-type fasteners, as used on stock (racing) cars, are fitted to hold the bonnet down.

Shelby and Ford considered producing a "notchback" coupé version of the cars in 1968, but then decided against it. They did add one new model to the Shelby line, a convertible, specifically dubbed Cobra. Its most distinctive feature is an exposed roll bar, a steel structure encased in airfoil-section, compressible foam plastic. The roll bar, built to meet racing specifications, is enclosed within the convertible top when it's up, making it somewhat different from the arrangement on the Porsche Targa. It will be interesting to see how the public reacts to this feature, which may predict the look of convertibles to come in the United States.

The Shelby-ized Mustangs have been very successful as an extremely exciting and specialized version of an already exciting and specialized car. They have been attractive enough to induce 111 Ford dealers to take Shelby franchises also, and Shelby hopes to have 150 dealers by the end of 1968. Still an impressive product, the Shelby Cobra may have changed its shape but it has by no means lost its sting.

By Eric Nielssen

SPORTS CAR TOURNAMENT: COBRA VS. CORVETTE

By Steve Kelly

Comparison shows Shelby car has much to do before overwhelming well-established 'Vette.

Just on a subjective basis, comparison between these two is scant. Appearance is vastly different, and on that basis alone, there'd be no stacking them against one another. But they do hit "head on" in the marketplace, and no matter how much either maker likes or dislikes it, they have to meet somewhere. If not in the showroom, then it might as well be here.

The Cobra costs less. Very few options are listed, so a full list Cobra adds up to fewer dollars than a full list Corvette with near-comparable equipment. Resale value may show a higher rate of depreciation, and the Corvette has more in the way of standard equipment goodies. Things like lamp-monitors, 4-wheel disc brakes, flow-through ventilation, forward opening hood, and independent rear suspension are standard Corvette items not even available for the Cobra.

Maintenance will show up as being less on a 428 Cobra than a 427 Corvette. This applies to only the biggest engined cars, and based on the smallest engined versions, it is about a draw. The Cobra, using the Mustang for a birthplace, is more readily acceptable in the average garage, but anyone who kicks out $4500 or so for *any* car isn't going to trust the "average" garage for service. He'll go to a specialist, where he will most likely find a Corvette. A draw here too.

Insurance rates are way over in the Cobra's favor. Being mostly steel, body repair costs are significantly lower than for the 'glass Corvette. There's also a reputation factor involved. In its 16-year life, 'Vettes have managed to become unwanted merchandise around many insurance offices. The feeling is—and justifiably so in lots of cases—that the Corvette driver is more likely to become engaged in an accident than if he were driving a more domesticated vehicle.

More people can fit in the Cobra, but judging from letters submitted in response to our '67 article comparing the two, not many people care. About the last thing either car is bought for is carrying the gang around town. Luggage though, is a more acceptable item in the Cobra. The small space behind the Corvette seats will take one or two suitcases, while the trunk and rear seat of the Cobra can hold many more.

Comfort-wise, the Corvette is nicer for drive. The seats don't wear you out on a long trip, nor give you bad aches where they most hurt. Entry and exit are easier in the Cobra since it's higher.

To park the Corvette takes some expertise in deciding where the nose ends. It drops down out of sight long before it ends. This problem exists to a much lesser degree in the Cobra.

True handling and acceleration are best handled by the Corvette. The wide-pattern tires (new, optionally for '68) make it superior to anything made in quantity production in the U.S. And that 427 "rat motor" is an engine which has left every other maker envious. Of course, a MoPar Hemi 426 seems to get the job done a bit better, but at more cost, and besides, you can't get it in the Cobra.

The comparison is here, and in a good number of instances it is straight across. Each has good and bad points, and each has distinctiveness all its own.

One of our office experts claims Shelby-American's last "good" Mustang based car was the '66. We go along with him — almost. The '66 was much more of a sports car than the '67 and '68, but finding a market for it was harder. Performance was there for the real hard-core rough-riding enthusiast, but their number is vastly overwhelmed by the buying group preferring both power and comfort. For Shelby-American and Ford's sake, the last two years have proved better cars, sales and acceptance being the criteria.

In a sense, Shelby has sold out to the add-on and chrome-it establishment. The new cars are more decorated than the old and show strains of having too much ornamentation. The convertible model seems especially so, with too many lines going too many ways.

Because of the now historic Ford assembly line strike, we were only able to obtain a GT 500 for instrumented testing. However, at the preview last summer of '68 Shelby products, we lived with the GT 350, 302-cu.-in. V-8 powered car for a brief session. It lacks substantially in power when compared to its earlier mates, but then has less horsepower. A hydraulic cam and cast iron exhaust manifolds are used, and they tend to tame the power and noise. It is many degrees quieter than the old 306-hp "street" combination on '67s, but is likely to please more than it will offend by this virtue.

Our GT 500 convertible came with the standard 428-cu.-in. V-8, and endeared itself to us by exhibiting calm behavior. It likes to eat gas, but otherwise seems no different from a 390-cu.-in. Mustang. Performance is substantially better than a 390, but not as much as you'd expect. This is more of an engine for "I want power" advocates than anything else. We doubt it'll beat any 427 Corvette for acceleration, and it falls in a bad class for drag racing, per NHRA specs. The size puts it out of contention for sedan racing, so the 428 is relegated to street duties primarily. At that, it is fine.

Slow speed operation isn't cause for trouble. Response from idle or sub 20-mph speeds is quite good. We had a Cruise-O-Matic transmission in our test GT, and prefer it for normal use. Shifting is crisp, and happens at reasonable speeds. Kickdown is easy, and you can usually reach passing-gear at all maximum speed limits.

The gear selector is of the "sport-shift" type, making for manual control throughout. Second gear has a high-speed monitor preventing over-revving on downshifts. The driver is responsible for this on upshifts. Gear changing on full throttle runs requires a 200-300 rpm anticipation. In other words, if you want to shift at 5500 rpm, better move the lever at 5200.

The 428 will live at 6000 rpm, but better times will come about shifting at 5500. Torque is hefty enough to slip the tires between 1st and 2nd, as well as from standstill.

Handling, Steering & Stopping

Suspension is less of a "springless" proposition than exhibited in '65-'66 GTs, but still firm. Relocation of the upper front control arms as was done in the early models isn't needed in the new Cobras. Expense and ride harshness rose with this move, and few people understood its value. The overall suspension is very close to Ford's own performance-handling option, yet the Shelbys have an edge. High-speed straight-line driving isn't at all abusive, and the cornering work is a safe proposition. We like cars with firm control, and found this in the Shelby Cobras.

Steering is right out of Mustang land. At least the feeling of "plasticity" is. Response is not at all like the Corvette's, and there just seems to be unwanted lag in the unit. The steering wheel is not up to sports car standards. We've a good idea Shelby folks agree here, but the standard wheel meets Federal specs for impact absorption, and a wood rimmed steel-spoke flat disc doesn't (such as the '67 item). It protrudes closer to the driver's chest than is comfortable, causing us to "kink" our arms even with the seat at full travel. We're not positive of the ruling for or against owner installation of the old-style wheel, but we'll bet it will be done often. *continued*

Photos: Bob D'Olivo, Pat Brollier

Shelby-American has sold-out to the "chrome-it" and "add-on" establishment. Ungainly roll-bar is best kept covered by the convertible top. Mustang interior has been added to, but not improved. Imitation mag wheel covers add final, phony touch.

Big Ford 428-cu.-in. V-8 is overly generous with torque, and acceleration is a roller-coaster thrill with Cobra. Louvers are functional and useful, but placed directly over spark plugs.

COBRA POWERTEAMS

Displacement	Carburetion	Horsepower	Torque	Transmission
V-8 302[1]	4-bbl.	250 @ 4800 rpm	310 lbs.-ft. @ 3200 rpm	4-speed Cruise-O-Matic*
V-8 428[2]	4-bbl.	360 @ 5400 rpm	459 lbs.-ft. @ 3200 rpm	4-speed Cruise-O-Matic*
V-8 427*	4-bbl.	400 @ 5600 rpm	460 lbs.-ft. @ 3200 rpm	4-speed Cruise-O-Matic*

1. Std. 6T350 engine 2. Std. GT500 engine *optional

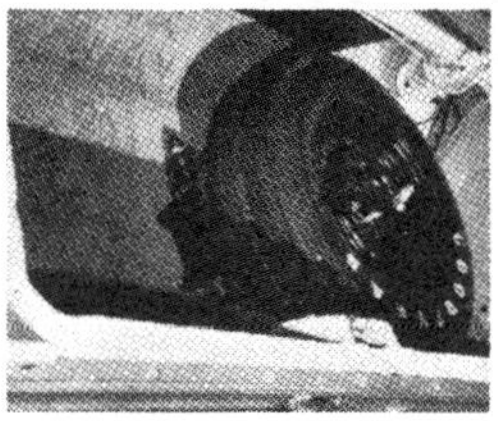

Stopping GTs is about like stopping Corvettes; no trouble at all. The front disc/rear drum setup is power assisted, using FoMoCo's "floating" caliper front brake. The rear drum is revised over the standard Mustang with a 10-inch diameter x 2¼-inch wide drum. High-performance lining material is used.

Braking is straight, quick and free from serious fade. Use of the Goodyear E70x15 high-speed tires contributes heavily by planting lots of rubber on the asphalt.

Comfort, Convenience & Ride

We'd like to see better front bucket seats in the Cobras. While those in use are fairly comfortable, they get tiresome after 90 minutes driving, or more. They are deluxe Mustang units, with good construction but a need for more seat padding for all-out comfort. The seatback is too straight, and could be neatly reclined a degree or two. Our test convertible had a crooked seatback, and though repairable, it is really a nuisance.

Addition of a center console to all '68 Cobras is a great idea, though decorating it with paper woodgrain applique is not. It's easily identifiable as paper and shows less class than vinyl covering. The just-right height armrest hides a huge storage bin. Because of the near-nothing capacity of Mustang gloveboxes, Cobra owners really make out. If the door armrest were as comfortable as the console pad, things would really be looking good.

Rear seat entry in either coupe or convertible is worthy of study. Some prefer stooping and diving under the roll bar in the convertible and through the shoulder harnesses on the fastback. Others manage to delicately thread their way in, turn around facing forward, then sit down contentedly. We never did set a pattern, but somehow managed the manuever. It's not all that bad, but a long way from having the convenience of a sedan.

Ride is, of necessity, stiff. Bounce is somewhat hard, but we didn't once incur loss of rear wheel traction over hard dips. The best description might be that you know it is stiff, but nothing more. The cars handle the job of plain riding in good style. This new ride-pattern is about halfway between stock Mustang and that of the "rough rider" '66 GT.

Plus & Minus Features

It might be a good idea to further separate the fog light switch from the convertible top button. They're right next to each other, and at night, without lights, it'd be easy to rip the top right off its anchors.

Another "better idea" would be to soften the trunk torsion bars. After removing the stock sheetmetal panel and replacing it with fiberglass, the hinges have much less weight to carry, and consequently spring the trunk up like a flag in a shooting gallery.

That great big "hatch" type hood interferes with forward vision, cutting off several feet of right-in-front sight.

The hood pins used to release the panel are neat, taking only a turn to undo them. They stay with the hood when raised so theft or loss is discounted. Functional louvers on each side allow air to escape, but we noticed the openings are right over the plug wires, with no deflectors for water being included. We louvered a hood once and put an aluminum panel below the louvers angled down and to the side of the engine compartment. If we can do it, anyone can!

Shelby's integrated roll bar idea deserves high praise. It also deserves to be copied. Any convertible is potentially more dangerous than a hardtop, and occupants lives are worth more than an additional few dollars of tubing.

Cobra shoulder belts are the best yet. The hardtop uses suspender type restraints, stemming from a single strap into an inertia reel harness, allowing slow body movement but locking up on sharp impact. The convertible restraint is housed in the rear quarter panel, pulling out and fitting against the front occupant's chest, and fastening to a center tunnel mounted buckle strap. The inertia reel is featured here also.

The "establishment" has had its impact on Shelby-American, and they've succumbed and resisted. Styling reflects the adoption of the "great" philosophy, but performance and safety still are Shelby's own exclusives.

Getting emotionally "hung up" on the Corvette styling takes somewhat longer than becoming enthused over its great driving characteristics—but not much. Getting in and out is more of a chore than on earlier models, but like styling and ride, is only a matter of accommodating one's particular personality.

From the looks of things, it appears that Corvette assembly line workers are taking time getting used to putting the new model together. Rough panels and ill-fitting sections, highly evident on early-run cars, indicate some practice is needed before perfection could be neared.

Our test cars were both convertibles. One came with a 327-cu.-in., 350-hp V-8, the other with a 427-cu.-in., 435-hp "semi-hemi" head engine. Compression on both is 11.0:1, and only 4-speed transmissions are offered here.

The 327 engine uses a single 4-bbl. Rochester carb and quiet running hydraulic cam and lifters. Lift is .4472-inch on intake and exhaust, with 306-degree opening duration on each. Valve opening overlap is 78 degrees, and valve head diameter is 2.02 inches for intake, and 1.60 inches on the exhaust.

Noise out of the engine room doesn't herald an "over-one-hp-per-cubic-inch" engine lurking there. A 3.70:1 final drive gear kept our top speed down to an exacted 117.15 mph at 5500 rpm. The 0-60 mph times and quarter-mile runs could be greatly improved by a stronger clutch and drag tires, but we weren't all that disappointed with the cars' recordings. City traffic operation was so near-perfect that we accepted the semi-compromise between "rump-rump" idling and reliability with praise. Some experimenting with tires and tuning will drop the 350-hp 'Vette into the mid 13-second bracket without difficulty.

The 435-hp test machine had Chevy's L88 aluminum head option, and frankly we thought it'd go quicker than it did. Tuning time was shortened by inclement weather, so we had to settle for 14-second quarter-mile times. But our earlier 427 cars ('66 & '67) both hit the 13s right off the bat. The '68 in proper tune should be high 12s in proper tune. The majority of drag-strip disciplined Corvettes hit the low 12-second mark, with a good many in the 11-second bracket. The potential is undoubtedly there.

We encountered trouble getting the triple carburetion to work in unison. Several times, the front and rear Holley 2-bbls. refused to work. Some jerry-rigging on the vacuum linkage helps, but there's no substitute for mechanical actuation—available at most speed or specialty shops.

This big engine objects to long idling periods, but cleans out quickly once on the move, with only a small amount of plug fouling. Clutch action is stiff, allowing the driver to object to idling time also. The throw is much shorter than on the 350-hp engine, negating much worry about the pedal hanging-up between power shifts.

Solid lifters are used on all 435-hp engines. Valve diameter is 2.190-inch intake, and 1.720-inch exhaust. Lift is .5197-inch on both, intake opening duration is 316 degrees and exhaust is 302 degrees with overlap being 80 degrees.

Handling, Steering & Stopping

Handling, both on twisting roads and straight-arrow

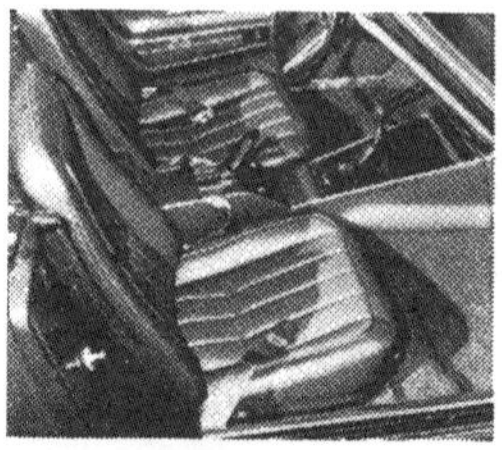

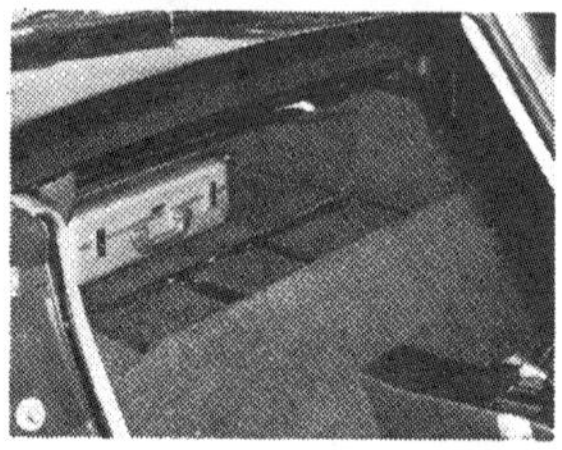

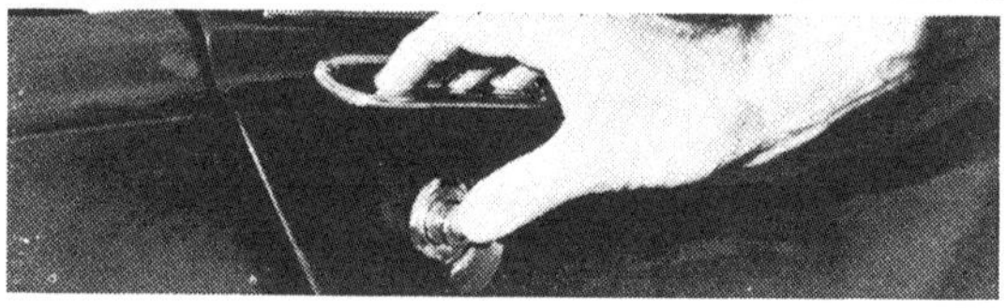

Corvetter interior is well laid out; all gauges are easily visible. Seats are extremely comfortable, but when driver seat is in full rear position, safety latch is impossible to reach. Handling and steering are excellent.

CORVETTE POWERTEAMS

1. Std. GT350 engine 2. Std. GT500 engine *optional

V-8 327 4-bbl.	300 @ 5000 rpm	360 lbs.-ft. @ 3400 rpm	3-speed 4-speed* Turbo-Hydra Matic
V-8 327* 4-bbl.	350 @ 5800 rpm	360 lbs.-ft. @ 3600 rpm	4-speed* (close or wide ratio)
V-8 427* 4-bbl.	390 @ 5400 rpm	460 lbs.-ft. @ 3600 rpm	4-speed* (close or wide ratio) Turbo-Hydra Matic
V-8 427* 3 2-bbl.	400 @ 5400 rpm	460 lbs.-ft. @ 3600 rpm	4-speed* (close or wide ratio) Turbo-Hydra Matic

stretches is improved over '67 by at least 100%. Introduction of wide-pattern tires — 7-inch wide wheels — as standard gets most of the credit. Also making big contributions are the reduced height (approx. 2 inches), wider rear track (.7-inch). Terming it a truly stable road car is far from exaggeration.

Manual and power steering units are quick to react. Power can't be had with the 435-hp engine, but we didn't mind. Only 3.4 turns of the 16-inch diameter wheel takes it from lock-to-lock, and only 2.92 are needed with power. Overall ratio on manual is 20.2:1, and 17.6:1 with power. Suspension feedback and/or sloppiness is not a problem here. It just doesn't exist.

Understeer is fairly inherent, but constant adjusting of the wheel isn't required while rounding curves. Once set — it stays. Oversteer is only a problem, or condition, encountered with the rear end braking loose.

Since '65, poor stopping Corvettes haven't been produced. With disc brakes at all 4-corners, they just plant themselves to the ground and come to rest straight and quick. Less than 120 feet from 60 to 0 mph (a distance we repeatedly accomplished) is something to write home about.

The 'Vette's 4-wheel disc brakes remind us of the auto company spokesman who recently defended his company's non-use of them by explaining difficulites in adapting a suitable parking brake to the system. Hmm? Wonder how Mercedes, Porsche, Volvo, Chevrolet, etc. did it? Gee!

Comfort, Convenience & Ride

Stetson wearers will find ample space for their headgear — behind the front seat. Rooflines are low, negating hat wearing.

Otherwise, headroom is not a problem. Nor is legroom a limiting factor. We did find that a 6-footer, or thereabouts, develops a tired leg on all but full throttle runs. Moderate gas pedal application causes the driver's knee to bend without having adequate thigh support. A large, softly upholstered roll, à la hot-rodders technique on channeled cars, would do away with the annoying non-support.

The semi-reclining seatbacks are enjoyable. We found ourselves hunching forward at first, but then learned to settle back, giving us plenty of comfort and also a just-right distance to the wheel.

Ride qualities of all but the biggest engined car are surprising. Bounce and rebound are relatively calm, especially for a sports car. The inside quietness is something we never expected in a 2-placer, though a good many imported sporty cars have this. The flow-through ventilation works, making for windows-up driving in all but the hottest weather.

Difficulty in entry and exit — for the average person — is just something sports car owners should expect. Considering this car is short of four feet in height, we'd have to rate it good.

Plus & Minus Features

"Rain, Rain, go away." We sang that song for a solid week. The Corvette handles admirably in the wet, but passengers didn't fare too well in our test convertible. The 350-hp car came with the removable hardtop, and oddly enough it leaked, and the soft top didn't. We found the major source of water-fill at the rear corners of the side windows, but weren't able to fix it on our own. A couple of comments on this have drifted in from '68 owners, and though they've cured the ill, it was a nuisance.

The hidden wipers should help prolong blade life, and they undoubtedly benefit appearance. We had a difficult time explaining this to the gas station attendant who gashed his finger on the non-flush fitting door edge corner. Seems something was keeping the cover from going flush to the cowl, and he found out about it the hard way.

The glove compartment is behind the seat, along with the other identical-looking hatches. One holds the jack, and the other — right behind the driver contains the battery. Be careful which one you reach for.

Using the press-type door handles is easy, but a couple of drawbacks were illustrated to us. Ladies are prone to breaking fingernails in the spring-acted door-top grip, and closing the door and locking it GM-style takes some deft hand movement.

Regardless of its few distinct minus points, the Corvette still holds the position of one of the world's all-out "class" vehicles, and justifiably so. Kinda wish we had one.

Raw power is Corvette symbolism, and five different engines are available for '68 in 327- and 427-cu.-in. sizes. Road handling is much improved over '67 models. Suspension got stiffer.

PERFORMANCE

	Cobra 428	Corvette 327	Corvette 427
Acceleraton (2 aboard)			
0-30 mph	3.0 secs.	3.2 secs.	2.7 secs.
0-45 mph	4.3 secs.	4.8 secs.	4.0 secs.
0-60 mph	6.5 secs.	7.1 secs.	6.3 secs.
0-75 mph	9.4 secs.	10.0 secs.	8.3 secs.
Passing Speeds			
40-60 mph	2.5 secs. 183.0 ft.	3.1 secs. 227 ft.	2.7 secs. 197 ft.
50-70 mph	3.0 secs. 264 ft.	3.6 secs. 316 ft.	2.9 secs. 255 ft.
Standing Start Quarter-mile	14.75 secs. 98 mph	15.0 secs. 92 mph	14.1 secs. 103 mph
Speeds in Gears			
1st . . . MPH @ RPM	47 @ 5500	53 @ 5500	66 @ 6500
2nd . . .	79 @ 5500	71 @ 5500	88 @ 6500
3rd . . .	116 @ 5500	92 @ 5500	116 @ 6500
4th . . .		117 @ 5500	160 @ 6500
MPH per 1000 RPM	21.1	21.2	24.7
Stopping Distances			
from 30 mph	39 ft.	27 ft.	29 ft.
from 60 mph	151 ft.	117 ft.	119 ft.
Speedometer Error			
at 60 mph	5% fast	6% fast	none
Mileage Range	9.0-15.0 mpg	10.6-20.0 mpg	7.8-17.5 mpg
Average Mileage	12.8 mpg	15.7 mpg	13.6 mpg

TEST CAR SPECIFICATIONS

	Cobra 428	Corvette 327	Corvette 427
Engine	OHV V-8	OHV V-8	OHV V-8
Bore & Stroke	4.13 x 3.984 ins.	4.0 x 3.25 ins.	4.25 x 3.76 ins.
Displacement-cu. in.	428	327	427
HP @ RPM	360 @ 5400	350 @ 5800	435 @ 5800
Torque: lbs.-ft. @ RPM	459 @ 3200	360 @ 3600	460 @ 4000
Compression Ratio	10.5:1	11.0:1	11.0:1
Carburetion	1 4-bbl	1 4-bbl	3 2-bbl
Transmission	Automatic; 3-speed	Manual; 4-speed	Manual; 4-speed
Final Drive Ratio	3.50:1	3.70:1	3.55:1
Steering	Recirculating ball & nut w/power assist	Recirculating ball & nut w/power	Recirculating ball & nut. Manual
Overall Steering Ratio	2.03	17.6:1	20.2:1
Turning Dia. — curb-to-curb	39.4 ft.	39.9 ft.	39.9 ft.
Wheel turns — lock-to-lock	3.6	2.92	3.4
Tires	E70 x 15 hi-speed	F70 x 15	F70 x 15
Brakes	front disc/ rear drum with power assist	4-wheel disc with integral power	4-wheel disc with integral power
Fuel Capacity	17 gal.	20 gals.	20 gals.
Curb Weight — lbs.	3665	3445	3425
Body Frame Construction	Steel unit	Ladder frame	Ladder frame
Wheelbase — Ins.	108	98	98
Front Track — Ins.	58.1	58.3	58.3
Rear Track — Ins.	58.1	59.0	59.0
Overall Length — Ins.	186.81	182.1	182.1
Width — Ins.	70.9	69.2	69.2
Height — Ins.	51.8	47.8	47.8
Front Suspension	Independent with coil spring	Independent with coil spring	Independent with coil spring
Rear Suspension	Hotchkiss type w/semi-elliptic multi leaf spring	Independent with fixed differential and multi-leaf transverse spring	

COMMENTS

	Cobra	Corvette
Handling	Better than most domestic cars – firm with plenty of control – just slightly below that of all-out sports car.	Best of all U.S. cars, and one of the upper echelon of sports "handlers."
Ride	Stiff but not abusive.	Good on all but the big-engined. 435-hp cars are abusive to passengers.
Comfort	Not bad, but seats could stand refinement. Rear seat room is very marginal.	Very good on long trips as well as in-town. Interior access is difficult for all but the very agile.
Quality	Surprising. Vast improvement over '67s.	Needs upgrading, but is progressively getting better.
Performance	Adequate for car's intended use.	Good, and better than its competitor. Has the best potential.
We like	Most of the styling – ease of operation – shoulder harnesses and roll bar.	Style – instrumentation – lamp monitors – 4-wheel disc brakes – good tractability.
We don't like	Phony "mag" hubcaps that ruin looks – too light trunk lid and too-open hood louvers – obstructed forward vision – lack of greater distinction from Mustang.	Leaky bodies – front ends that end past line of vision – unreachable left seat-back release – absence of control lighting and good interior illumination.

OPTIONS & PRICES

	428 Cobra	427 Corvette
Mfg's Suggested Price	$4317.39 Coupe $4438.91 Converitble	$4636.00 Coupe $4320.00 Convertible
Engine Options		105.35 350 hp 200.15 390 hp 305.50 400 hp 437.00 435 hp
4-Speed Trans.	Standard	184.35
Auto Trans.	50.08	237.00
Limited-Slip Diff.		46.35
Hi-Perf. Tires	Standard	31.30
Special Susp.	Standard	36.90
HD Ignition	N.A.	73.75
Adj. Steer. Col.	66.14	42.15
Power Steering	84.47	94.80
Power Disc Brakes	64.77	42.15
Power Windows	N.A.	57.95
Air Conditioning	356.10	412.90
AM Radio	57.59	N.A.
AM/FM Radio	181.36 (Stereo)	172.75
Fold-Down Rear Seat	64.78	N.A.

HOT ROD ROAD TEST

Late arrivals in '68 wrappings could prove to be closest threat yet to 'Vette from Ford. But then again, they might not

Shelby GT500-KR
CORVETTE THREAT?

Two versions tested are shown above. Though lighter-weight fastback out-accelerated convertible, both displayed admirable high- and low-speed handling traits. Thanks to automatic trans and air conditioning, flip-top machine became our choice for highway touring. BELOW — Formed cover conceals convert rollbar. Console addition provides extra storage and includes oil and amp gauges.

photography: Eric Rickman

by Steve Kelly

Had it not been for Carroll Shelby's 1965-'66 series GT350, with its worthwhile features as a sports and performance machine, Ford probably wouldn't be marketing the current breed of GT350's and GT500's. Nor is it likely that such a car would've developed. Who knows, FoMoCo may not even *like* perpetuating the brand, but they are, and with greater success than when control was in Shelby's hands. Now the cars are made in Ionia, Michigan, feature bigger powerplants and more closely approach the deluxe appointments of Chevy's Corvette, although lacking the almost complete chassis revision given to the '65-'66 machines. No longer are the upper A-arms repositioned for greater cornering ability and lower roll center; no longer are rear axle housings replaced by big car units and safety wire-looped. They are now much quieter, feature power disc brakes optionally (power was absent on the early model, although discs were standard), can be had with engines more powerful and much heavier than Shelby's original and have more distinctive bodies than before. Although this last item doesn't contribute a great deal to the GT's sports car value, it does separate it appearancewise from the regular Mustang. When you're marketing a car, this sometimes is more of a sales tool than ultra-strong and competitionlike underpinnings.

Enough on commentary. As a singular piece of merchandise, the car we tested comes off good and bad. This latest Shelbyized Mustang, the GT500-KR, was late in making the scene for a number of reasons. Included in them are Ford's first-of-the-model-year labor strike, relocation of assembly from the West Coast, and finally certifying the cars for smog emissions. The latter was the biggest hang-up, and had it not been for this factor, there might've been a supercharged big-block Shelby for '68. The KR arrived as the '69 models were beginning production, and it's a safe bet there'll be little change other than the body on '69 KR models. The KR designation, besides signifying a Cobra Jet-equipped car, reportedly means "King of the Road." We decided against further telling of this during our use of either test KR after the first reply to the "What's the KR mean?" question evoked incredulous laughter. Evidently the young set isn't ready for that title yet.

The KR's we tested consisted of a fully equipped convertible with automatic transmission and air conditioning, and a lighter-weight fastback machine with fewer options and quicker performance. Both carried Shelby-instituted integral rollbars, inertia-reel shoulder harnesses, full instrumentation, center console and larger-than-standard tires. The convert had a 3.25:1 rear axle, while the fastback was 3.50:1-geared. Limited-slip differential, as it was on the four-speed car, is standard on all but refrigeration-equipped models. Offsetting that to a degree is inclusion of an oil cooler on cars with "air."

We admire the ease of street operation with current GT's. Power-assisted brakes make stopping easier on your foot, since a great deal of pedal pressure was required for the discs without assist on the '65-'66 versions. Short stops from low speeds are now easier to attain. The steering is also power-helped, adding to street-use pleasure. The "originals" were about as noisy inside as out. The latest GT's, though, are well-padded with sound deadeners which, while making them heavier, do wipe out lots of road noise. But we can't say the '68's were totally quiet inside. We discovered a number of squeaks and rattles in each but, ironically, more in the fastback. Convertibles usually tend to loosen up more — and quicker.

With the GT500-KR, there isn't the "inch-off-the-ground" feel experienced with a 'Vette or other sports car. That's one part of the car Ford can't do much about, short of making a new body. And the over-the-hood view is filled with a large dome which we likened at times to the foredeck of a seagoing runabout, especially when it blocked our view. Being a Cobra Jet car, there are fresh-air inlets via twin hood scoops. And the Shelby cars also feature functional hood louvers at the rear of each side of the hood. They do a good ventilating job for the crowded engine compartment. They also contribute a lot of "refuse" to the windshield area in front of the driver and passenger. The final blow is that air coming from the hood louvers feeds into the cowl vent grilles for interior venting. The cowl feed is a high-pressure area, and that's why it's used for this purpose. With GT Shelbys, though, you get more than "fresh" air.

The added center console not only helps interior appearance; it gives the Mustang a storage compartment as well. The regular glove box needs help just to hold a warranty book. The deep opening is capped with a padded top that's just right for an armrest. Good thinkin', guys. It can hinder quick shifts with a four-speed, but for the sake of an armrest as good as this one, we'd buy an automatic.

Don't get us wrong; we like the Mustang dash. It's just not a Corvette dash, and that's a hard one to fight. Once you're cradled behind the heavily padded steering wheel that juts too far into your chest, you've still got a Mustang dash facing you in a car that cost nearly as much as a 'Vette. But then there might be some who'd say the Shelby car makes more sense 'cause it'll hold four people. Well, okay. But for how long and how far? Try running between Los Angeles and San Francisco with two good friends in the back of that fastback Mustang. That'll decide true friendship.

After we'd made the rounds of street usage, collected our thoughts from these ventures and recorded mileage figures, it was time for thrashing through the quarter at Orange County International Raceway with each KR.

Running the automatic takes practice before getting a system developed for best times. Stall speed is relatively low, so we found that starting at 800-900 rpm and easing into the power from there was best. Very little wheelspin developed and better e.t.'s resulted. Leaving the selector in Drive is fruitless. The C-J engine hooked to the three-speed automatic needs all of 6000 rpm before going to the next gear, and the "D" position barely reaches 5500 before upshifting. After detaching belts and air-cleaner element, plus raising front tire pressure to 45 psi and rear to 32, we achieved a 97.71 mph top end and e.t. of 14.58 seconds. This was the best. It had dropped to a poor 14.94 e.t. earlier.

Our four-speed car never dropped below 100 mph in all its runs and never traversed the quarter in more than 14.20 seconds. Much better and more respectable than the convert. Also, we found six grand too much for this one, recording consistently better e.t.'s by shortshifting at 5500 rpm. While both engines (automatic and stick) will take 6000 without objection, they seem "all done" past that point in stock condition. The stick shift linkage leaves much to be desired, and the stock 11.5-inch-diameter clutch unit is hurting when it comes to long life. Five or six close runs produce unreal fade and slip. We did feel content — relatively speaking — when the four-speed car ticked off a 14.01-second quarter and top end of 102.73 mph. This is a long way from the 13.56 e.t. and 106 mph Pure Stock run made in a Mustang Cobra Jet tested for our March issue, so we know the KR can use improvement. Richer jetting is the first step to take, and one we didn't do, for reasons of no carb jets and no time.

The KR series Shelby Mustangs are built to attract a wide variety of tastes and interests, which accounts for their stable sales picture. Some sports car customers don't care to sit an inch off the ground, and this is why they don't buy the 'Vette. There must be a reason. The Shelby GT500-KR may be a better sports vehicle than the production Mustang, but this Ford appears to fall short of the 'Vette. We're bound to get a stack of mail on this one; better warn John in the mailroom what to expect. ■ ■

VEHICLE

Shelby GT500-KR	Fastback	Convertible
PRICE		
Base	$4472.57	$4594.09
As Tested	$4857.13	$5350.39
ENGINE		
Type	OHV V8	
Cylinders	8	
Bore and Stroke	4.13 x 3.984 in.	
Displacement	428 cu. in.	
Compression ratio	10.6:1	
Horsepower	335 @ 5200 rpm	
Torque	440 lbs.-ft. @ 3400 rpm	
Valves: Intake	2.06 in. dia.	
Exhaust	1.625 in. dia.	
Camshaft:		
Lift	.481 in. intake; .490 in. exhaust	
Duration	290°	
Tappets	Hydraulic	
Carburetion	Single Holley 4-bbl	
Exhaust	Dual	
TRANSMISSION		
Type	Manual 4-speed. Synchro all forward gears	Selecto-O-Matic 3-speed automatic
Ratios: 1st	2.32:1	2.46:1
2nd	1.69:1	1.46:1
3rd	1.29:1	1.00:1
4th	1.00:1	
Clutch	Long; 11.5 inch dia., single disc, dry plate	
DIFFERENTIAL		
Type	1-piece housing, live axle. Straddle-mounted ring gear and offset pinion. Locking-rear std. except with a/c	
Final drive ratio	3.50:1	3.25:1
BRAKES		
Type	Power-assisted front disc/rear drum. Floating-caliper design	
Dimensions: Front	Disc, 11.3 in. dia.	
Rear	Drum, 10 in. dia.	
Total effective area	180 sq. in.	
SUSPENSION		
Front	Independent with coil spring over upper control arm. Direct-acting Gabriel adjustable shocks	
Rear	Hotchkiss type. Four-leaf semi-elliptic leaf springs. Gabriel adjustable shocks	
Stabilizer	Front only. .94-inch diameter	
Tires	E70-15 Goodyear Polyglas wide pattern	
Wheel rim width	6.0 in.	
Steering:		
Type	Recirculating ball and nut with linkage-type hydraulic power assist	
Gear ratio	16.0:1	
Overall ratio	20.32:1	
Turning circle	37.16 feet, curb to curb	
Wheel diameter	16.0 in.	
Turns lock to lock	3.74	
PERFORMANCE		
Standing start quarter-mile (best)	14.01 sec. and 102.73 mph	14.58 sec. and 97.71 mph
FUEL CONSUMPTION		
Best reading	11.47 mpg	13.58 mpg
Poorest	10.6 mpg	10.21 mpg
Average	11.02 mpg	11.96 mpg
Recommended fuel	Premium	Premium
DIMENSIONS		
Wheelbase	108.0 in.	
Front track	58.1 in.	
Rear track	58.1 in.	
Overall height	51.8 in.	
Overall width	70.9 in.	
Overall length	186.81 in.	
Test weight	3570 lb.	3740 lb.
Body/frame construction	Unitized	
Fuel tank capacity	17 gal.	

PHOTOS BY SCOTT MALCOLM

SHELBY'S COBRA GT 500-KR

"I'M EMBARRASSED," said the man from Ford one Monday morning, "but we've got a problem. The Shelby GT 500-KR you asked for arrived in Los Angeles Friday, and we put it on our parking lot. Someone stole it. The police found it yesterday, but it's, well, it's in no condition to be tested."

Oh, goodness. We're too big to cry, and it hurt too much to laugh. The Shelby KR test was scheduled for the September issue, and we're tight on a deadline. "Stolen, eh?" "Yes," said the Ford man. "It's the first time it ever happened."

Not for one moment would we condone theft, but in this case, we understand it. The penalty for stealing a Shelby KR is the same as for stealing a Falcon 6, and the Shelby might be worth it.

The KR stands for King of the Road. Like the king of beasts, the Shelby KR gets half the job done just sitting there, looking invincible. The KR starts life as a Mustang, either fastback or convertible, and receives a fiberglass hood and front panels, with a huge radiator opening, functional cool-air scoops and hot-air extractors; scoops for interior ventilation (they work) and scoops in a position to direct cooling air to the rear brakes (they don't work); and a fiberglass trunk lid with an integral spoiler, which looks keen, and may even have a helpful effect on air flow.

Inside, there's a built-in roll bar, with inertia-reel shoulder harness attached. Too bad all passenger cars don't have attachment points for such harnesses. There's a single mounting point, and the harness divides, one strap for each shoulder, just behind the driver's head. The division is padded and reinforced. In front impact, the force of the crash would be divided between the two straps, not taken diagonally, as with the single strap harness. If the KR was rammed from behind, the driver's head would be rocked back into the harness, which should lock up and reduce, if not prevent, whiplash.

The factory Mustang's array of lights and little gauges is replaced in the KR by big, legible speedometer and tachometer, in front of the driver for easy reading, and smaller but useful gauges on the panel and console, where they can be referred to in moments of calm. The semi-bucket seats are covered with vinyl, and thick carpeting is standard. The non-padded surfaces of

—Was It Worth Stealing?

the dashboard have walnut-grain applique, and the gearshift knob is real wood. Everything matched, looked good, worked, and could be reached.

With the seat pushed back from the steering wheel, the gearshift was within reach in first and third by only one inch. The shift is one of the better factory units, though. The new, stiff transmission added to shifting time, but the lever didn't.

Visibility was acceptable. The fastback KR has no rear quarter windows (that will be changed in 1969), and a narrow rear window. There are blind spots just behind the car on both sides. The outside mirror helped on the left, but right side vision was limited, a minor but continuing annoyance.

KR suspension is Mustang, beefed up. Front and rear springs are stiffer than even the Mustang handling package, the shock absorbers are bigger, and the front anti-roll bar is thicker. A limited-slip differential is standard equipment, and Shelby clamps anti-windup dampers on the rear springs.

The KR engine is Ford's 428-cid Cobra Jet, and it's some engine. The original 428, which made a wonderful boat anchor, has been improved almost beyond measurement with the cylinder heads and manifolds from the 427 racing engine. It's big and strong, and very highly tuned. At 6000 rpm, the Cobra Jet will pull a semi-trailer up Pikes Peak. At less than 2000, it wouldn't pull the petals off a daisy.

The Cobra Jet is obviously underrated at 335 bhp; educated guesswork and gossip puts the true output at something close to 500. This little gambit is confusing, and even the factories can't keep their stories straight. In the Mercury Cyclone, the Cobra Jet is rated at 335 at 5600 rpm. In the Shelby KR, with the same camshaft and a bigger carburetor, the Cobra Jet is rated at 335, but at 5200 rpm. Taken at face value, the bigger carburetor doesn't affect horsepower, it just moves the power peak down a few hundred revs. Moral: Don't take the rating at face value.

Even with 500-odd ponies, the King of the Road is vincible. But, beating the KR's quarter-mile e.t. of 14.57 will take a well-tuned Hemi, or a Corvette, a Ram-Air GTO, or a Buick GS 400 Stage One, or a Cyclone with the Cobra Jet engine.

But, there's more to life than the quarter-mile. The test KR had the standard 3.50:1 final drive gearing, while all but one of the quicker cars listed above were geared for acceleration, and engine speeds on the highway ranged from annoying to impossible. The KR cruises at half its permissible engine speed. The drag-prepared cars top out just past the timing lights, and the KR is just getting into fourth gear at the first light. Top speed is an estimate; there wasn't time to take the car to good ol' Nevada, and the KR was still accelerating at the end of the straight at the test track. The KR has the power, and the gears, to hit 130, and maybe more.

A cynic might question the utility of a 130-mph top speed. It's just as use-

BALANCED with throttle, KR drifts through turn.

GT 500

ful as a 14-sec. e.t., with a bonus of highway ease and even reasonable fuel mileage.

The KR also has traction problems. With all that power at high revs and no power at all at low revs, the KR driver can either ease off the line, with the engine bogged, or blast away full-bore, with tires spinning for the first 100 feet. Even with big, sticky Goodyear Polyglas tires, a limited-slip differential and a properly controlled rear axle, the KR couldn't transmit full power to the pavement.

This led to some experimentation, with unexpected results. The KR made some test starts at the first speed timing light, so the drivers could know, rather than guess, how to cover that vital first section quickest.

With a *banzai* start—4000 revs, right foot down and left foot up as fast as they'd move, the KR moved 132 feet in 3.296 sec.

Traffic trickling—engine just above idle, clutch eased up and accelerator eased down as the car moved off the line—brought the time down to 3.125 sec.

A brisk start—engine at 2500, clutch out quick enough to pop the tires, full power as soon as the tires caught—was the best way, with a 132-ft. time of 3.082 sec.

Very interesting. Finding the optimum combination of clutch and accelerator would take more time than we had, but the dedicated KR owner, provided he has some test space, a stop watch and a friend to operate it, may improve on our figures. What we did conclude is that smoking tires make impressive marks on the pavement, but don't fool the clocks.

Oddly, the KR's vincibility doesn't matter. The car is so impressive, so intimidating to challengers, that there are no challengers. The KR breeds confidence bordering on arrogance. The KR driver coasts along, mighty engine rumbling, and looks with a condescending smile on the driver of a lesser car who creeps away from the lights,

1968 SHELBY
GT 500-KR 2+2 FASTBACK

DIMENSIONS

Wheelbase, in. ... 108.0
Track, f/r, in. ... 58.5/58.5
Overall length, in. ... 183.6
width ... 70.9
height ... 51.4
Front seat hip room, in. ... 2 x 22
shoulder room ... 53.4
head room ... 37.1
pedal-seatback, max. ... 40.5
Rear seat hip room, in. ... 43.2
shoulder room ... 53.2
leg room ... 27.0
head room ... 34.0
Door opening width, in. ... 45.0
Trunk liftover height, in. ... 30.0

PRICES

List, FOB factory ... $4473
Equipped as tested ... $4857
Options included: Power disc brakes, power steering, inertia-reel shoulder harness (factory fitted, mandatory options); fold-down rear seat, AM radio, Tilt-Away steering wheel.

CAPACITIES

No. of passengers ... 4
Luggage space, cu. ft. ... 9.2
Fuel tank, gal. ... 16.0
Crankcase, qt. ... 4.0
Transmission/dif., pt. ... 15.0
Radiator coolant, qt. ... 20.5

CHASSIS/SUSPENSION

Frame type: Platform.
Front suspension type: Independent by s.l.a., coil springs, telescopic shock absorbers.
ride rate at wheel, lb./in. ... 123
antiroll bar dia., in. ... 0.94
Rear suspension type: Hotchkiss live axle, multileaf springs, telescopic shock absorbers, anti-wind up dampers.
ride rate at wheel, lb./in. ... 150
Steering system: Linkage assist, recirculating ball gear, parallelogram linkage behind front wheels.
overall ratio ... 16:1
turns, lock to lock ... 3.7
turning circle, ft. curb-curb ... 37.6
Curb weight, lb. ... 3500
Test weight ... 3780
distribution, (driver), % f/r ... 56.8/43.2

BRAKES

Type: Cast iron disc front; single leading shoe, cast iron drum, rear.
Front drum/rotor, dia. x width, in. ... 11.3 x 2.07
Rear drum, dia. x width ... 10 x 1.75
total swept area, sq. in. ... 180.0
Power assist: Linkage.
line psi at 100 lb. pedal ... 1050

WHEELS/TIRES

Wheel rim size ... 15 x 6JJ
optional size ... 15 x 7
bolt no./circle dia. in. ... 5/4.5
Tires: Goodyear Polyglas.
size ... E70 x 15
normal inflation, psi f/r ... 28/28
Capacity @ psi ... 5200 @ 28

ENGINE

Type, no. of cyl. ... 90° V-8 ohv
Bore x stroke, in. ... 4.13 x 3.98
Displacement, cu. in. ... 428
Compression ratio ... 10.6:1
Fuel required ... premium
Rated bhp @ rpm ... 335 @ 5200
equivalent mph ... 111
Rated torque @ rpm ... 440 @ 3400
equivalent mph ... 72.5
Carburetion: Holley 1x4.
throttle dia., pri./sec. ... 1.68/1.68
Valve train: Hydraulic lifters, pushrods and overhead valves.
cam timing deg., int./exh. ... 18-72/82-28
duration, int./exh. ... 270/290
Exhaust system: Dual with branched headers.
pipe dia., exh./tail ... 2.25/2.25
Normal oil press @ rpm ... 40 @ 2000
Electrical supply, V./amp ... 12/55
Battery, plates/amp. hr ... 66/80

DRIVE TRAIN

Clutch type: Single dry plate.
dia., in. ... 11.5
Transmission type: Four-speed manual.
Gear ratio 4th (1.00:1) overall ... 3.50:1
3rd (1.29:1) ... 4.52:1
2nd (1.69:1) ... 5.92:1
1st (2.32:1) ... 8.12:1
Shift lever location: Console.
Differential type: Hypoid.
axle ratio ... 3.50:1

then chuffs past in traffic, knuckles white on the steering wheel.

Carroll Shelby might not be a prisoner in the Ford works, but every year the Shelby Mustang is a little less Shelby, and a little more Mustang. The first Shelbys, still competitive in road racing and as sports cars at that, were thoroughly revised, with improved front suspension geometry and trailing arms to change and limit movement of the live rear axle. Maybe it wasn't necessary, maybe the customers wouldn't pay for something that didn't show. But now, as mentioned, the Shelby version is Mustang, stiffened.

There is a temptation to say that Ford learned nothing from its well-publicized triumphs in International racing, but in fairness, they probably learned lessons that merely haven't been applied yet to the production passenger cars. The factory Mustang has a soft ride, the handling package gives more control with some roughness, and Shelby's changes bring still better handling, and less comfort.

The KR is at its best on a good road, smooth and fast. Aided by the cross-ply belted tires, the KR is stable at any speed this side of revocation. It stays balanced over dips and camber changes. Rough pavement is bad news.

TOO MUCH power brings the front up, and the rear out.

The live rear axle skitters sideways, and bounces the tires off the road. Driving the KR around medium-speed curves at high speeds is driving at time and a half.

The usual technique for Supercar driving is to rush up to the corner at speed, flick the steering wheel to start the front end sliding, then apply power and balance the car, so it goes through the turn in a controlled drift. With the KR, the technique only works for the first half of the curve. The Shelby doesn't lean much, and the driver feels secure as he zooms into the corner, cranks on more steering lock, and—

The Cobra Jet is a big engine, with a double-big carburetor. Applying power isn't as easy as it sounds. It doesn't turn on in one smooth rush, more like a rocket sled, with power coming in a series of jolts, fire two, fire four, fire eight.

With the KR understeering toward the tulies, floor the gas pedal. The primary barrels open, and nothing hap-

CAR LIFE ROAD TEST

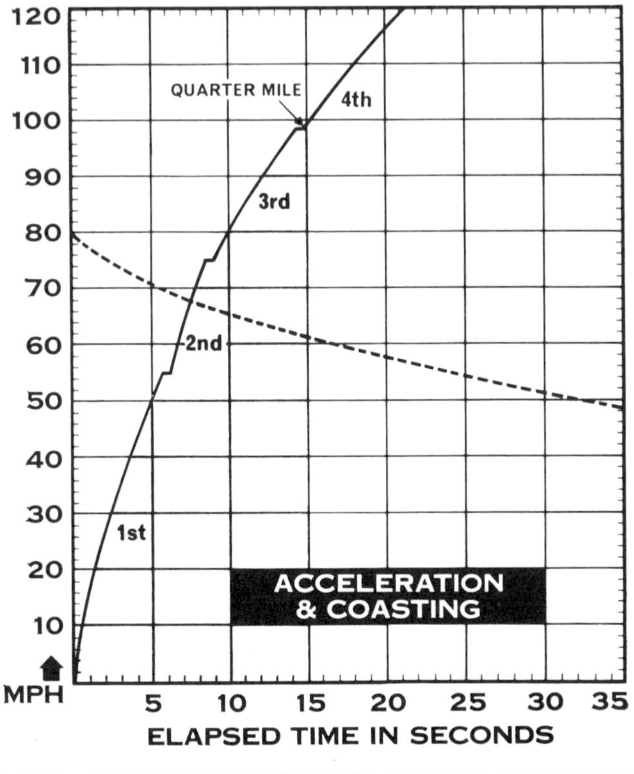

CALCULATED DATA

Lb./bhp (test weight)	11.3
Cu. ft./ton mile	184.6
Mph/1000 rpm (high gear)	21.34
Engine revs/mile (60 mph)	2825
Piston travel, ft./mile	1850
CAR LIFE wear index	52.2
Frontal area, sq. ft.	20.2

SPEEDOMETER ERROR

30 mph, actual	29.20
40 mph	38.22
50 mph	47.51
60 mph	56.46
70 mph	66.27
80 mph	75.44
90 mph	85.38
100 mph	95.13

MAINTENANCE

Engine oil, miles/days	6000/60
oil filter, miles/days	6000/60
Chassis lubrication, miles	24,000
Antismog servicing, type/miles	clean valve/12,000, replace hoses and spacers/24,000
Air cleaner, miles	12,000
Spark plugs: Autolite BF32. gap, (in.)	0.032
Basic timing, deg./rpm	4-8/600
max. cent. adv., deg./rpm	24-29/4000
max. vac. adv., deg./in. Hg.	19/15
Ignition point gap, in.	0.020
cam dwell angle, deg.	25-29
arm tension, oz.	17-21
Tappet clearance, int./exh.	0/0
Fuel pressure at idle, psi	8
Radiator cap relief press., psi	12-15

PERFORMANCE

Top speed (6100), mph	130
Test shift points (rpm) @ mph	
3rd to 4th (6000)	98.7
2nd to 3rd (6000)	75.6
1st to 2nd (6000)	55.2

ACCELERATION

0-30 mph, sec.	2.5
0-40 mph	3.6
0-50 mph	4.8
0-60 mph	6.9
0-70 mph	8.2
0-80 mph	10.2
0-90 mph	12.6
0-100 mph	14.6
Passing, 30-70 mph, sec.	5.7
Standing ¼-mile, sec.	14.57
speed at end, mph	99.55

BRAKING

Max. deceleration rate from 80 mph ft./sec./sec.	23
No. of stops from 80 mph (60-sec. intervals) before 20% loss in deceleration rate	4
Control loss?	Slight.
Overall brake performance	poor

FUEL CONSUMPTION

Test conditions, mpg	8.5
Normal cond., mpg	10.5-13.7
Cruising range, miles	168-219

GT 500

pens, then whump—the power comes on and the rear end slides out. Correct with the steering, and whump—more power, more slide, more correction. Then the secondary barrels, controlled by vacuum and not the driver, open and whump—still more power, slide and correction. It's about all the enjoyment the driver can stand.

The brakes should have been among the best available, with big front discs and rear drums and a power booster, but they were disappointing although they were carefully broken in, as per instructions from Shelby's crew, with several hard stops, separated by several minutes of cool-off time.

They faded, though, during the cornering session. The brake test was postponed, and the coasting runs and speedometer checks were done first, so the brakes would be cool. The first hard stop from 80 mph showed a maximum deceleration rate of 23 ft./sec.^{2}. The KR stayed in a straight line and the brakes didn't lock, but the big brakes and tires should have done better. Next time, the highest reading was 19 ft./sec.^{2}, then 18, then 15, then 14. The driver had both feet planted on the brake pedal, but the KR just kept rolling along, with smoke pouring from the front pads. The brake test ended there. The KR still had to be driven back to the office, via freeway. The testers are devoted to duty, but home and mother come first.

Several weeks before the test, Ford had a bone to pick with Car Life. We keep saying our road test Fords rattle, are assembled in haste, and inspected indifferently. No one else complains, said Ford, leading us to the conclusion that all the rattlers are set aside for testing by Car Life while the solid Fords go elsewhere.

Not so. Shelby doesn't assemble Shelbys. He designs the changes and specifies the equipment, but the cars are built by another firm. The test KR came straight from there. The Shelby crew in Los Angeles didn't have a chance to check the car out fully, and they worried about it.

With good reason. First the power steering hoses leaked. A Shelby mechanic fixed it. While he was at it, he removed a rattle from the rear seat. Another remained somewhere in the trunk. Then the transmission locked in second gear. When the driver eased into a driveway the KR went into reverse, while remaining in second. Even 500 horsepower cannot prevail against a transmission trying to turn simultaneously in opposite directions. The KR spent the night in the garage, then went back to Shelby for another transmission. It seemed some minor piece, needed only to keep the transmission in one gear at a time, had been left out during assembly.

Now we know. Ford's careless cars are reserved for Car Life *and* Shelby.

The Shelby is thoroughbred Mustang. We might wish for more: a new suspension, maybe; and rear bucket seats designed to be sat in, not climbed out of; and an adjustable steering column instead of that tilt-away thing that goes WHANG! at the driver's chin when the door is opened. But the King of the Road will wow the neighbors, cover ground and make drivers of ordinary Mustangs eat their hearts out. Well worth stealing. ■

INTERIOR is nicely done, with walnut-grain paneling, extra engine gauges and huge tachometer added to standard Mustang dashboard.

HUGE COBRA Jet 428-cid V-8 completely fills engine compartment. Factory claim of 335 bhp is more than 100 less than true output.

SHELBY GT 350/500

1969 Prices Not Available

Ford Mustang with modifications. 350 features 351 cu. in. Ford engine, the 500, the 428 cu. in. version. Most of the modifications favorably affect handling, such as the heavier duty stabilizer and shocks, others are more for esthetics or for lightening or cooling. Formerly remanufactured in Los Angeles, Shelby GT's are now assembled in Ionia, Michigan.

The factory notes that the 1969 models are the fifth generation of Shelby custom road cars. It doesn't seem that long, but in retrospect they have come a long way since the first relatively crude Shelby GT-350 was rolled out in 1965. Back then the cars were no-nonsense performers with little compromise in terms of creature comforts. Today's units, convertible or SportsRoof, are still no-nonsense performers but they offer just about any automotive creature comfort one might require.

There are significant design changes — engineering and appearance — over last year's cars. Front and rear treatments are completely new. Chromed lower bumper impact bar blends smoothly with chrome fender and hood trim. Of interest is the fact that hood and front fenders are made of fiberglass. The hood has racing-type hood locks and a trio of functional scoops. Scoops are nothing new, but these are supposed to be based on NASA designs for Gemini space craft. What they do is reduce turbulence and skin friction, provide extra engine compartment cooling, add a supercharging effect at high speed, and operate more quietly.

A big side stripe differs from '68, while the air scoops for the rear brakes are about the same. At the rear it's all Shelby — a unique high kick-up deck lid spoiler. Back here deck lid, rear fender extensions and lower back panel appliques are also of fiberglass.

On the inside the Shelby GT is a combination of Detroit luxury high style with racing touches. It comes off reasonably well: high back bucket seats, vinyl covered wheel with 'rim blow' horn feature, floor console, simulated teakwood grain trim, electric clock and bright trimmed pedal pads. One feature which doesn't work very well is the roll bar on the convertible; it simply doesn't permit honest rear seat occupancy. Yet we're not going to knock a factory-stock roll bar because it is a step in the right direction.

There now are sufficient power train options that a Shelby GT fancier can tailor his car to suit his tastes. Start with engines: For the 350 it is the all-new 350-cubic-inch Ram-Air V-8, rated 290 bhp at 4800 rpm, 385 lbs.-ft. of torque at 3200. Carburetion is a special Autolite four-barrel on a high-rise intake manifold. There are special cast rocker covers (for show) and a new low restriction dual exhaust system. Power for the 500 is the 428-inch Cobra Jet Ram-Air V-8, 335 hp at 5200 rpm, torque 440 at 3400. It, too, has a four-barrel carb (Holley), a medium-riser intake manifold, high-capacity fuel pump, die cast rocker covers and low restriction dual exhausts.

Standard transmission for both models is a full synchro four-on-the-floor. Close ratio is standard on the 500, optional for the 350. Clutch for the 350 is 11 inches in diameter; an 11.5-inch clutch is standard on the GT 500. Clutch housings are high strength malleable cast iron, a helpful safety idea.

Available for both models is an optional heavy-duty automatic transmission featuring a three-speed planetary gear arrangement with an automatic, manually operated kick-down. This transmission can be manually shifted through all gears and offers a shift handle detent to minimize the possibility of skipped shifts.

Optional at the rear end is a new Ford-designed Traction-Lok differential in a variety of ratios. It is less violent than previous limited-slip rear ends on Shelby GTs, yet does the job quite well, indeed, as the degree of lock-up varies directly with the torque applied. This offers distinct advantages in hard acceleration or in extreme handling situations. Very

The driver's 'office' . . . ventilated, leather-padded steering wheel, wood finished dash, T-handled shifter.

Shelby scoops really work. The gauze-meshed intake cools the floating caliper, 11.3 in. front disc brakes.

Cooling for the rear brakes is through a deep, high-intake scoop. The 10 inch drum brakes at the rear help give the GT a total effective brake area of no less than 324.8 inches.

good on the drag strip or road race course.

Chassis design overall is routine, but in detail there are some keen features which add up to make the car a handler. Front coils are special high rate, high performance. GT 350's and GT 500's with automatic trans utilize 1-3/16-inch extra-heavy-duty Adjust-O-Matic shocks with export-type mountings. GT 500 four-speed models have a unique 'staggered' rear shock absorber mounting which locates the left shock absorber to the rear of the axle and the right shock absorber in front of the axle. This effectively reduces spring wind-up and rear wheel hop under hard acceleration.

Brakes on Shelby's GTs are sensational, stopping from 60 mph in the less-than-150-foot category. Front brakes on all models are 11.3-inch diameter, floating caliper, ventilated discs, with low-fade linings. Rear brakes are heavy-duty drums, also low-fade linings. Total effective brake area is a most ample 324.8 square inches.

These days it is difficult to tell the difference between a hub cap which simulates a cast racing wheel and the real thing. The Shelby GT has the real thing — 15-inch x 7 wheels with alloy cast spider and chrome-plated full safety-type steel rim. Standard tires are F60 x 15 super-wide oval, low profile, bias cord with fiberglass belts — the latest in wide oval design for improved traction, ride and wear. Major differences from previous rubber are in the tread which has extra width and a molded-in convex shape. When inflated, this results in a flatter cross-section and more uniform contact with the road throughout the tire's footprint.

We mentioned the roll bar in the convertible. The SportsRoof model also incorporates a heavily-padded vinyl-covered bar. It is semi-concealed which provides increased headroom as well as easier entrance to and exit from the rear seat.

Shelby pioneered the use of inertia reel shoulder harnesses and continues them this year. Convertibles have the inertia reels mounted on the body inner panels, concealed inside the rear quarter trim panels. SportsRoof

The aircraft-type intakes do two jobs. They keep down the under-hood operating temperatures and the center one is ducted into the carburetor air cleaner to give better breathing.

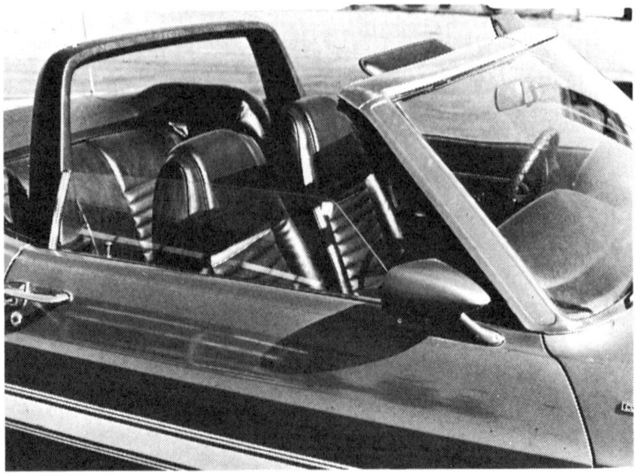

With safety in mind the Shelby GT has contoured, high-backed bucket seats and standard built-in roll bar that is heavily-padded and vinyl-covered.

For impressive, read Shelby GT. The frontal aspect of the Mustang-based sports car is big, brutal and beautiful.

shoulder harnesses are designed with inertia reels equipped with padded covers mounted on the roll bar to provide ease of entrance for rear seat passengers.

Special cockpit instrumentation is in keeping with the Shelby GT's performance character. Speedo records to 140 mph, while the tach tops at 8000 rpm. The new molded plastic center console also houses matching oil pressure and alternator output gauges, as well as special finger tip switches for the Lucas road lamps and the interior courtesy lamps.

The optional equipment list makes it fairly clear that while the Shelby GT is a super high performance car, there is little intent for it to be raced. Rather it is designed for the enthusiast who wants something distinctive which also carries his favorite motoring necessities — air conditioning (automatic trans required on the GT 500), AM/FM stereo radio, stereo tape, Tilt-Away steering wheel, fold down rear seat and others. We don't say this emphasis on luxury is a bad thing; those first GT's were ground thumpers that required muscle and skill to drive while 1969's crop can be driven by mama to the supermarket without the issue of her safe arrival being in doubt. In fact, with standard power steering and optional automatic trans, we can't imagine a more pleasant way to go motoring.

Some of the numbers which affect acceleration and handling are weight and weight distribution. The GT 350 looks like this: The SportsRoof has a curb weight of 3600 pounds and front/rear weight distribution of 55/45. The convertible weighs 3689 and spreads it out 54/46. In the GT 500 series weight of the hardtop is up to 3850 and distributed 57/43, while the convertible is 3939 and 58/42.

As a general rule the heavier the front end the less desirable the handling. However, the new tires have eliminated much of the sweat here, and the GT 500 gets through the corners just about as smoothly as does the smaller-engined, lighter-weight 350.

Last year we said that the GT's may have improved themselves right out of racing competition. This year it appears even more true. But we still like the car and consider it one of the neatest handfuls of luxury gran turismo that this country has ever produced. ♠

Shelby T 350/500

Data in Brief

DIMENSIONS	
Overall length (in.)	176
Height (in.)	53
Turning diameter (ft.)	37.2
WEIGHT, TIRES, BRAKES	
Weight (lbs.)	3286
Tires	E70 x 5
Brakes, front	11.3 in. disc
Brakes, rear	10 in. drum
ENGINE	
Type	OHV pushrod V8
Displacement (cu. in.)	GT350 351 GT500 428
Horsepower	GT350 290 GT500 335
SUSPENSION	
Front	wishbones, coil springs
Rear	leaf springs

RAM AIR, SHELBY-STYLE!

The old Snakemaker comes on strong for '69 with fresh air fed 351 and 428-powered GT Mustangs

SHELBY Automotive Company, Inc., builders of the Shelby GT-350 and GT-500 Mustangs, has announced a totally new Ram-Air induction system for their '69 models.

The new system utilizes an advanced design, centrally located air intake scoop molded into the precision molded fiberglass-reinforced plastic hood. Hood scoop contours are similiar to those designed by the scientists of the National Aeronautics and Space Administration for Gemini space craft.

Extensive wind tunnel testing of this type of air scoop setup has shown that the design produces maximum laminar air flow in a wide speed range with extremely low levels of turbulence and skin friction. These qualities were sought by Shelby engineers to provide a highly efficient air system for engine compartment cooling and Ram-Air induction. All '69 Shelby models incorporate this unique feature.

The two outboard scoops on each Shelby hood vent directly into the engine compartment, providing additional cooling air to reduce under-hood operating temperature. The center air scoop is ducted directly into the carburetor air cleaner. As vehicle speed increases, the air entering the center intake produces a supercharging effect, resulting in increased engine breathing for improved performance and economy.

Shelby engineers were pleased to discover an added benefit when testing the new design. The NASA-type intakes also are the quietest at high speeds, a direct result of maintaining smooth laminar air flow over the hood.

Shelby GT-350 (351-cube) and GT-500 (428-cube) models are available through Shelby—franchised Ford dealers throughout the United States and Canada.

Ram Air front end setup is standard on all GT-350 & 500 models. Scoops are NASA-type.

GT-350/COBRA II COMPARISON

Does the new Mustang stand up against the old Shelby versions?

by Jim Norris
PHOTOS BY ERIC RICKMAN

Carroll Shelby's 1965 Mustang GT-350 cannot be duplicated and yet the measure of its success is in a brand new baby brother. The Cobra II is an attempt to satisfy a growing segment of the car-buying public. That segment represents a holdout against the diet of anti-smogged pap that Detroit has been lately induced into offering. The Cobra II is an economy package, with at least some of the venom restored to its rightful place. And it doesn't come off a necessary second best, either. In a head-to-head comparison, the new snake is anything but a half-asp. The boys at FoMoCo should be proud of their latest efforts.

By taking a 1965 Mustang 2+2 fastback coupe, Shelby injected some Cobra venom, toned up the leg muscles, added lightness, and produced a hot number that racked up a decade of steady and impressive demonstrations of speed.

In contrast, the 1976 Cobra II fastback coupe was not produced by Shelby, but by Detroit penthouse brains, who, in their hair-pulling need for new buyer inducements, figured that they could capture past glories with a package that would satisfy all concerned—not the least of which are government agencies.

To trap the Cobra in its den was not an easy maneuver, but was accomplished with the aid of long-time Shelby buddy, Eric Rickman. With Rickman and a few conversations with "The Man," Shelby's personal GT-350 was made available on a serene day, near his home in Marina del Rey, California.

INTRODUCTION

The Cobra II came through the assistance of Ford public relations department, strictly a factory operation. The Mustang GT-350, in days past, left Ford assembly plants in semi-completed trim, and showed up in Shelby's Venice facility for reworking, before sale to the buyer. Anyone who intended to drive on the street wouldn't want the fat racing slicks, magnesium wheels, oil cooler, plastic rear and side windows, gutted interior or fiberglass seats. For them, Shelby arranged a trifle more civilized package. For the course competitor, there was a hot GT-350 with the above features. Both worlds were therefore serviced by the former U.S. Road Racing Champion—the creative genius of the super-hot Cobra roadsters.

The Cobra II—like Shelby's—is also white, a 2+2 coupe, and like Shelby's, has a full package of dark blue racing stripes. Both cars have chin spoilers, a hint of a rear deck spoiler, hood scoop, and general look-alike shapes.

Shelby's suspension has been a bit more tweaked, and does in fact have an edge in high speed cornering—but we mean high speed, in excess of 100 or so. The latest Cobra doesn't need it. Why sell something that puts every owner behind bars? Times have changed. Yet even so, one instinctively feels, with the Cobra II, that the Gods have been uncommonly kind by not forgetting a general feeling of high-speed stability.

A large measure of that sense of stability has to revolve with the Cobra II's BR-70x13-in. Firestone Steel Radial 500's. (That is how our test car was equipped.) In cars with nearly identical curb weights (Cobra II's 2835 lbs. vs. GT-350's 2800 lbs.), the smaller tires on the Cobra II only begin to lose their secure feeling at road speeds that are well past the 55-mph legal driving limits. While Shelby's 15-in. Goodyear Polyglas GT's are a bit wider and bigger, the tough rubber and mag wheel combination becomes strong at the top end, yet of only hair-splitting margin at the lower speeds that the law has established.

Both vehicles have the high-mounted coil springs in the front and solid axle with semi-elliptic leaf springs in the rear. Double-acting

1

tubular shocks are, at a quick glance, similar, with spring rates calibrated for high speed on the GT-350 and for lower speeds on the Cobra II. Be it advised, however: Shelby's car is of a mean hue in suspension; the Cobra II is not. In fact, it becomes immediately apparent, in the driving comparison, that the Cobra II is a small car with a luxurious ride, while the GT-350 doesn't really intend to make popularity points with Aunt Agatha's backside. The amazing thing is that the Cobra II, with all of its civility,

1. Ford's new Cobra II, when compared to Carroll Shelby's legendary GT-350, is a chip off the old block in the looks department.

2. Both father and son are white with blue racing stripes, thin-line bumpers, and fold-down rear seat. Front and rear spoilers and large rear window are common to both.

3. The GT-350—this one Shelby's personal car—is a mean machine. Oil cooler and radiator breathe well through gaping jaws. The car, though, is not out to gain Brownie points in the ride department—it has other things in mind.

4. But the Cobra II is anything but a half-asp. Its 302-cu.-in. V-8 is** highly **responsive all the way up to legal speed limit.

can hold as well as it does in the hard-running exercises.

Of hard-running exercises, one must note FoMoCo's hot-out-of-the-box Cobra II GT, which represents the first serious road racing challenge since the factory pulled out of racing in 1970. It's a similar body shape to the streetable Cobra II, but is a full-bore racing effort; not compatible with the Cobra II, or the long line of streetable GT-350's.

One could always feel, though, that Shelby's GT-350 would hang together in a tight spot, whereas the Cobra II, for all of its comfort and surprising roadability, was simply not designed to slog it out in the upper mach ranges.

Handling on the GT-350—over the original '65 Mustangs—was vastly improved. By cornering hard enough to remove all slack out of the suspension in one direction, the sensation was one of absolute superb stability. The tires really hung in there. In the GT-350, there was a trifle more CG force to contend with, which was only noticeable when the body was forced to roll from one extreme to another. In that aspect of our test situation, the tail was forced out of line briefly, but the way Shelby's car is set up, a brief steering compensation was all that was needed in what was still a basic understeer configuration.

One doesn't easily forget the excellent high-speed road manners of this vehicle, but it is well within the limits of propriety to dare to compare the Cobra II and the GT-350. It is not fair to call them an apples-and-oranges proposition. We are living now in an age of degree of strengths and weaknesses that are interchangeablemanias.

The GT-350, with its recirculating ball and nut steering and non-power, maneuvered its way in and around parking lots with the ease of a tank, but once on the open road, the 19:1 overall ratio was exceedingly precise for a car of the Mustang's bulk.

The Cobra II on the other hand, employed the TRW power unit, with an overall ratio of 19.03:1, on center, and its rack and pinion linkage matched up closely with the

2

3

4

beefier Mustang's unit. At the higher speeds, of course, where steering feel is crucial, the non-power unit had the slight edge. But remember, the Cobra II is a super go-to-the-market car, while the GT-350 needs more elbow room you must give a little to get a little. Again, we're getting into areas of mighty concern to current engineering from Detroit. To the people who must pay more for their transportation than ever before, it is almost a direct relationship that comfort needs loom larger than ever before. In 1965, there was no fuel crisis, and the euphoric national automotive scene was still in full swing. Today, there are a different set of variables. Even if these two cars *were* an apples-and-oranges proposition, it would still be possible to *think* that they are alike, that only the *times* had changed. Today, also, there are more people at work *trying* to change attitudes and beliefs than ever before. Never before, therefore, has a new automobile so much *captured* a ripe moment for introduction as the Cobra II.

THE CLOSER YOU LOOK

From any looks angle, the Cobra II comes across like a son, standing bravely before its father—the GT-350. But look closer. On the GT-350, Shelby relocated the inner pivot points for the upper suspension arms. As the roll center was slightly raised at the front, the GT-350 dropped its nose. On Shelby's car the top of the front fender wells slant lower over the tire tops. It is a hint of meanness that attracts hot cars and the men in blue—an unnecessary event if a GT-350 driver motivates his machine with care.

Yet the Cobra II, doesn't come off that badly, perched at an equally rakish angle over its 13-in. steel radials, a look that more than makes up for its stock Mustang II suspension. Heavy-duty springs and shocks can be ordered, but there is a definite hesitation on their use for the ultimate in relaxed driving. Part of the charm of the Cobra II is that it can *hint* of these hairier pursuits while still providing a luxury car ride. It's a hard illusion to pull off, make no mistake, but the Cobra II has done it.

Both cars are painted white. Both have identical dark blue trim stripes, the Cobra II's being thicker at the door line, and the name of the snake being advertised for all it's worth on the bottom of the extra-wide doors. A profile match-up, with the Cobra II in front, shows clearly that the Cobra II has lost none of the masculine appeal of the GT-350. "Where ignorance is bliss, 'tis folly to be wise," said Thomas Gray.

But he who buys a Cobra II is certainly not ignorant. Look at the figures. With an original status as a "limited production" vehicle, the Ford Motor Company has reported that the 5000-unit run for the first 6 months was completely sold out even *before* the next revised production schedule could take shape. It's obvious that FoMoCo now knows exactly what it's doing.

As you look closer at the father-and-son configuration parked in tandem, there is noticed a hood scoop on both, a white bumper on both, a relatively small chin scoop, likewise on both, and a triangle-shaped rear quarter window on both. Here, the Cobra II clearly has an advantage. Its quarter window is given the modish louvered, or venetian blind look, with four louvers blocking out excessive light rays. Moreover, the entire window *opens*—a ventilation asset that is welcome. The original Mustang 2+2, with a smaller set of abbreviated louvers, began looking the GT-350 part after Shelby substituted a fixed glass insert. To a 6-footer who has ridden in the back of a GT-350, the situation was one of reduced headroom and excessive temperatures. In the smaller Cobra II, with its improved ventilation, this is not likely to be the case—a comfort improvement plus in anybody's league. There is a partition behind the 2+2 rear seat in Shelby's car, which can be folded down for a restricted clear shot to the rear trunk lid, but this is unneeded in the Cobra II, with its open design; another clear advantage, considering the latter's plusher carpet.

From the slant of the front windshield to the end of its rear quarter window, the overall body shape of the two cars is awfully close. Shelby has added a fresh air scoop for his rear brakes, and, although total swept area is a bit bigger than on the Cobra II, it is pleasing to note that both have the front disc/rear drum combination.

As could be guessed, by the man behind its concept, the GT-350 has a no-nonsense, functional hood scoop. On the Cobra II, the hood scoop, which is somewhat flatter and shorter, is suffocated by a strip of sheet steel at its mouth. One wonders why the son could not look more like the father in this respect. Certainly the engine, with its retarded timing and higher operating temperatures wouldn't balk at

1

more fresh air sent its way.

On the GT-350 of Shelby's, there is a massive rubber compression-type bumper, added not only for its lightness, but for the obvious impact-absorbing qualities. On the producton GT-350 models, there was the traditional strip of chrome in this area, erstwhile known as a bumper. Shelby's car, however has

1. Hatchback design of the Cobra II is a real winner in the luggage department. Notice what it says on the bottom of Shelby's license plate. Interesting comment.

2. Cobra II has a plush interior—to challenge anything in the comfort department. High mileage drivers would like it.

3. But as the man said, "This one came to play." Scaled down and geared up for action would be more like it. Note huge tach and fire extinguisher.

4. It is a note of the changing times and consumer tastes that the Cobra II does away with its weapons of war, prefers luxurious appeal, gas mileage.

5. Cobra II V-8 is a 4x3-in. V-8 of 302 cu. ins. Unleaded gas is a must, but underhood maintenance should not be too difficult.

6. Shelby's GT-350 has been stripped of unneeded obstructions. Horizontal tube stiffener keeps things from going away at the higher mach ranges. It's a 289 with 11.5:1 compression.

had a large center chunk of bumper removed, in order to provide a clean gulp of fresh air to the radiator. There is no doubt that such a design move is the way to go for an engine that is asked to provide 130-plus-mph speeds.

In another advantage for the Cobra II, the lip of the hood on the GT-350 protrudes over its narrow, horizontal grille, while the Cobra II's hood is set well back from the edge of the cowl. Not counting the dummy hood scoop opening on the Mustang II, the whole front end is a bit more civilized than the competition tie-down hooks on the Shelby. In fact, the front end on the Cobra II is one heck of a clean, modern design. And always, of course, there's that coiled snake of chrome plate on a black background—front and center in the Cobra II's grille. From a distance it's not hard to imagine a cobra in a waft of dark jungle grass, sounding a warning—the unheeded kind of which claims thousands of those unschooled individuals who annually die from snake bites around the world.

Then there's that low-slung chin spoiler on both cars, which, in theory at least, is supposed to disrupt the high-speed airflow and keep the front end on the ground—to act in other words as a negative airfoil. It is a situation that is not needed on the Cobra II, for obvious reasons. But why fight it? Who knows? It can't hurt.

Unlike Shelby's bomb, the Cobra II has a plush interior. The tach and all other instruments are grouped underneath two rows of idiot lights, and the seats are among the most comfortable ever experienced. Our particular test car had the Cruise-O-Matic, but a 4-speed is available on special order. In the test car, an AM-FM stereo radio heightened the sense of controlled excitement in the civilized cockpit.

Regarding the GT-350, much has been written about its time-distilled accomplishments. Few cars of recent memory have evoked such inspiring prose. Do not confuse the GT-500, with the 350. The GT-350 was *the* pony with the pounding hoofs, the latter only another marketing activity.

Petersen's *Motor Trend* Magazine devoted numerous pages to the 350's exploits, and the car was a favorite of most of those involved with it. Other publications follow-

2

3

4

5

6

ed, with reams of copy extolling its virtues. One writer, when comparing a stock Mustang with a GT-350, could only mutter something about "A Butterfly and a Dragon." Perhaps his approach was the most logical of all. But in a world where appearances are crowding a grief-stricken definition of the word, "reality," Ford's new Cobra II is appealing transportation.

"WILL THE REAL ONE PLEASE . . . "

It was the same vinyl, the same carpets, the same T-Bar automatic trans shift lever—the looks concept anyway was the same. The Cobra II felt more luxurious, and hence more expensive (it was, $4745 base price, f.o.b. Detroit, vs. the GT-350's base of $4500). But even the price is deceiving. Time and inflation have seen to that.

"The man" had said that his car was not running right, when the writers eased into the eastward traffic near his garage. One drove the Cobra II, the other the GT-350, then doubled up, one car at a time for the tests.

Behind the wheel of the GT-350, the writer was afraid that he'd meet a Corvette, or some foreign exotica, as "the man's" *pukata-pukata* cam idled with care-free abandon. Some things the GT-350 cannot stand, and one is another fast car. A GT-350 has never refused a challenge. Anyway, looking around the cockpit, one was instantly aware of the following items. Item—Floor-mounted fire extinguisher, complete with pressure gauge. (The Cobra II is not aimed at the fire extinguisher set.) Item—Enormous driver-slanted tachometer, mounted high atop the dash, with its own housing. (The Cobra II uses a factory-mounted tach next to the speedo.) Item—Competition full racing harness, reminding one of green drag-strip lights, and a past history of more red lights than it is pleasant to remember.) Item—A flat, rear carpeted area of large size. (The Cobra II has one, too.)

Of the streetable versions, all GT-350's received the following Shelby treatment. At least 200 lbs. was chopped out of each car. This brought the curb weight to under 2800 lbs. The standard hood was replaced by one of fiberglass. The locking studs on the hood saved weight. There is a strap-metal safety loop at the forward end of the driveshaft to prevent the shaft from falling to the ground in case of U-Joint failure. Ratios in Shelby's 3-speed Cruise-O-Matic trans are beefed for max use by his 3.89 "NoSPIN" axle. Front discs use heavy-duty lining material, developed for Grand Prix racing. Rear drums are 10-in. by 2.5-in. cast-iron with metallic linings.

Under the hood of the GT-350, it was the same story: do the job, and no funny stuff. High riser intake manifold, finned aluminum oil pan. Thin wall tubing headers, the magnificent 289 V-8 pumped to 11.5:1 compression, and some 285 gross horses awaiting at five grand. Koni-equipped rear axle, control arms, and so on. The list is for a man's man. The GT-350 was (and is) all that it's supposed to be. The libraries are a testimony to its decade of deeds.

But all of a sudden, something enters this marvelous mix. Something that takes breath away, and brings honest tears to the eyes. What is it, smog? No. Any auto man has long since taken an optimum amount for granted. No, it's world's worse—Fuel. . .a lack of it.

A GT-350 had a 16-gal. tank. Its normal consumption rate was some 12 to 15 mpg. A Cobra II, with its 16.5-gal auxiliary tank setup, and the way the engine's timed, will *start* at close to 17 mpg around town, and go up from there—*with* its 302 V-8. You're burning lead-free gas; the engine will stay cleaner, the sky will stay cleaner. Increased temperatures may affect valve life, but the factory is using strong materials in an effort to compensate. The net result should be the same, as far as durability is concerned. If the purists are now chewing nails and adding this writer's name to their blacklists, nevertheless, it is a problem that must be faced. The horse has been slowed, the reins are tighter, but it's not

1. In the GT-350, the battery was placed in the trunk for better weight distribution at higher handling speeds. The Cobra II's is up front.

2. Cobra II interior is lush, with thick carpets—this one in a metallic blue color. Seat folds flat.

3. Quarter windows in the Cobra II open—not so in the GT-350. Somebody did a good job on Cobra emblem.

4. Production for GT-350 was 562 in 1965 and 2500 in 1966—the two finest years. Plenty of space with the rear seat down, but the Cobra II has a bigger, more comfortable seat.

5. Shelby's car has a racing safety harness with a military aircraft quick-release seat belt. The Cobra II has the pushbutton design with standard Detroit-designed harness.

1

2

necessarily a bad thing. Even a cobra will not strike if it's properly hypnotized.

A Cobra II is still a Cobra, there is still that magnificent swaying hood, only this time to the gas pump's well modulated flute. It will not strike. But it's still a cobra.

In a pair of 2-way speed runs, the Cobra II ran nearly neck-and-neck with Shelby's semi-starving horse that "wasn't running right." Allowing both cars to start in low, a 2-way average for both cars (using the full three speeds of the torque converter transmission), gave exceedingly close times, with *Motor Trend's* multiple stopwatch board.

The Cobra II hit the zero to 60-mph mark at 10.51 secs. This was with gas, oil, water, and more than 370 lbs. of writers. The GT-350, with the same load, ran it in 10 seconds flat.

Zero to 30 mph was 3.4 secs. in the Cobra II, and 4.5 secs. in the GT-350. Zero to 50 mph was 7.8 secs. in the Cobra II, and 8.6 secs. in the GT-350.

One must remember the hot cam setup in the Shelby, however. It was once said that a GT-350 will gag and choke, and that "anything less than 50 will cause the GT-350 to do an agonized crawl." But again, consider the overall picture on today's highways. Consider the interfering restrictions. Consider the speed limit. Consider the smog problem. Consider the fuel shortage. Consider everything. But consider that this is a Centennial year. Consider the red, white and blue . . . apple pie and the flag . . . an American emblem, the eagle, with powerful talons and fearless bearing. Yet how many are left? The thought causes strange emotions, as the GT-350 eases slowly onto a new stretch of deserted concrete freeway. How can one rate these two machines—a father and son by birth? Always there is the image of Henry Ford, when selling cars was an iffy proposition, taking to the wheel of "old 999," in his own words, "scared to death," but doing it, until he turned the tiller over to a half-crazed ex-bicycle racer named Oldfield.

But all of a sudden it was 70 mph and not yesterday any more. At 70 mph, the "man's car" finally realizes that someone has a foot in its head. That's when the cam comes on, and when all the painstaking goodies earn their money. What's it like, above 55 mph—on a deserted stretch of beautiful new concrete freeway? Well, first of all, it's against the law. Secondly, in a GT-350, things happen too fast to accurately describe.

From 40 to 60, the GT-350 runs dead even with the stopwatch times of the Cobra II. From 50 to 70, the horse responds to a silent command from the greatest snake charmer of them all—the tall Texan with the Lone Ranger's drawl. One can almost hear the radio words from long-ago:. . ."Come-on, big fella." There is a full second of clean advantage from 50 to 70 for the GT-350. The stopwatch gap suddenly widens. Above 70? No contest, period. At 80, there is no doubt what is happening. The tach goes wild, the rear-end scoots gently as it swallows torque and brute power; the whole chassis seems to leap with slow rhythm on its spring perches, and whoever is riding as shotgun, starts yelling, "Whoa, Nelli, there are cops in these parts!"

You know you can let off at the last minute because of the big discs, but you are now caught up in the ride of the Four Horsemen of the Apocalypse. Scenery is a blur. War, famine, pestilence, death. There it is, 100 . . . 110 . . . 115 . . . 117. The chassis is responding like a beautiful song. The pedal is down, but the tach keeps winding. The engine is going crazy. You are now past the point of feel. It's eerie to describe. The wild horse in full gallop- . . . . full defiance of anybody or anything. You must depend on basic logic . . . tach limits, how fast the distant mountain range is approaching, how much rubber you've got to the road. The speed is past 120; the wild thing is still coming on. You wonder if the other writer will do something rash. He sits there, stunned. But suddenly you remember, from the fogs of your mind: "This thing ain't running right." And, as suddenly as it started, the foot comes off, clear off. The brakes come on . . . gently, now. The Four Horsemen disappear. The end of the freeway looms too fast on the horizon, yet the GT-350 of Carroll Shelby comes to a nice-and-easy stop. There is no one in sight. Miraculously, there were no men in blue—or jails, either. You lucked out. But not with the other writer. "Never again," he says. "No company can stay in business this way."

The test is over. The machines are dutifully returned to their proper owners. Luck may have run out after all, for any thoughts of repeating this sort of thing. The sun comes down, as always it must. What we are dealing with is more than two cars with the same heritage. We are living at the transition—in two periods of history. There are new equations, as there are new concepts of automobiles.

But to find one, bearing the *name "Cobra II,"* is a lasting monument for as long as its owner rides behind its wheel; for as long as its owner remembers what it once was, and contemplates on what it now must be.

3

4

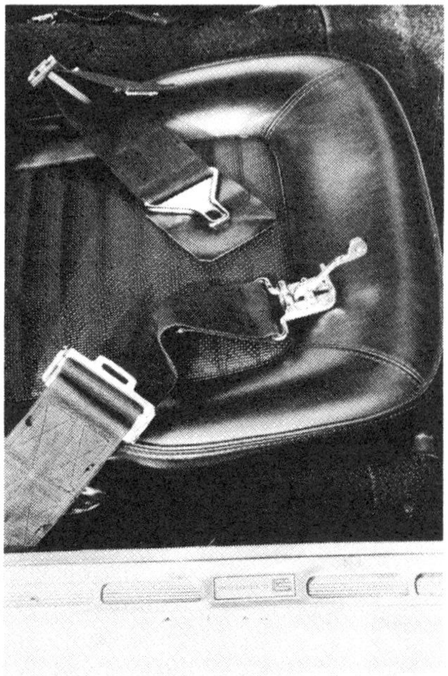

5

The CAR COLLECTOR'S Crystal Ball

In this column each month the CAR COLLECTOR will present a collector car that we feel is being neglected by collectors although very worthy and collectible, a car often referred to as a *sleeper*. Of course some of these cars may not be sleeping, they may be dead. But in our opinion the cars that you will find in this series may well be ones that five years from now you will be wishing that you had collected while the price was still right.

The Shelby Mustangs 1965 - 1970

by Rick Kopec

Black and White Photos by Ray Hofmann, Jr.

Before we gaze into the Crystal Ball this month, it would, perhaps, be helpful to define the term "collector car" which we throw about so often. Generally speaking, it is a car that some person or group of persons will care about in the future, due to one or more of the following factors:

A. Limited production
B. Historical significance
C. Innovative features
D. High performance
E. No longer in production

If enough people care about such a car, its value will increase as time goes by (called appreciation). Conversely, if no one cares about them their value will decrease (depreciation). Shelby Mustangs fit into all five of the above categories.

The Shelby Mustangs are awkward to categorize, and thus are often given their own niche. They share some general characteristics with their peers, but on the whole, they have more differences than similarities. True, they *are* basically Fords, but the parent company – Shelby American (and later, Shelby Automotive) – was separate and distinct; it even had its own network of dealers. Many of the "Shelby-unique" parts often found their way onto standard Mustangs, thus further clouding the differences, but the Shelby Mustang was much more than just another member of the Ford product line.

The Shelby Mustang story, in a nutshell, is this. Near the end of 1964, Ford's new Mustang had proven, very quickly, to be a sales bonanza. It broke all records for sales of a particular new model and had both GM and Chrysler scrambling to come

This article's author, Rick Kopec, has been intimately involved in Shelbys and Cobras for over ten years, having served as crew chief for a 289 competition Cobra in SCCA A/Production in 1966. He is also vice-president of the Shelby-American Automobile Club and writes a monthly column in its excellent magazine. Rick is an avid collector of Shelbyana and is deeply involved in the SAAC's many projects, such as the soon-to-be published Shelby American Guide. He and his wife, Colleen, reside in Norwalk, Connecticut, where Rick teaches in an alternative high school and only incidentally terrorizes the populace with a 1966 Shelby GT350 he has owned since 1970.

1965 – The first generation Shelby Mustangs resembled stock Mustangs most closely; the main difference was what lurked under the hood and in the driveline. The over-the-top "Le Mans Stripes" which originated on the 1965s have become synonymous with Shelbys ever since. Rear brake scoops and rear quarter windows appeared on the 1966 models. 1965 GT350, pictured here, belongs to Austin Craig of Reading, Pennsylvania. (Wheels are aftermarket, not original.)

Fiberglass rear deck lid and spoiler made the 1967's really stand out in a crowd. Tail lights, borrowed from 1967 Cougar, were sequential and became another Shelby trademark. Side brake scoops directed air to rear brakes while top scoops served to duct air out from inside the car.

up with a sales competitor. After a season of serious development, Shelby's Cobras began to get their act together and the cars were quickly becoming synonymous with winning. But in spite of the "Powered by Ford" tie-in, there just wasn't enough product identification to pull prospective buyers into Ford showrooms. Ford wanted to add some muscle and sparkle to the Mustang line and also to offer a showroom competitor to GM's Corvette.

Ford's choice of Carroll Shelby was obvious, but in order to get this special Mustang model certified as a bona fide sports car (according to the Sports Car Club of America), it would have to be a two-seater. It would also have to be competitive in heads-up road racing with the Corvette. Shelby assembled a development team, headed by Project Engineer Chuck Cantwell and Development Driver Ken Miles. They began modifying a pair of Mustangs and within a few weeks, had the initial parameters worked out. During this time, Ford was also thinking about building a special Mustang with an independent rear suspension. In fact, engineer Klaus Arning (famous for his later work on the Ford GT and 427 Cobra suspensions) oversaw the actual construction of a prototype, but side-by-side testing yielded very little actual difference, so the project was terminated.

By November, 1964, Shelby had a prototype built out of a Mustang fastback. It was lettered "COBRA MUSTANG Grand Sport", but later the official name was changed to GT350. That designation was actually an arbitrary number chosen by Shelby when no consensus on a name could be reached. Shelby reportedly asked engineer Phil Remington what the distance between the assembly line and the dyno shop was. When Remington responded "350 feet", Shelby said that that would be the name they would use. In point of fact, the first GT 350s didn't have 350 anything; later competition cars were reported to have 350 horsepower, but this was pure coincidence.

Shelby needed to produce a minimum of 100 cars in order to get the model homologated with the SCCA. He made arrangements to take over Ford's San Jose assembly plant for two days and had 100 white fastbacks produced with black interiors, high performance engines, 4-speed transmissions and a few other goodies culled from Ford's parts lists. These cars were delivered semi-complete to Shelby's Venice, California, plant, because certain items, such as hoods, exhaust systems, grille bars and rear seats, were deleted. When the SCCA Inspectors arrived, they were somewhat surprised to find the required 100 cars.

In the GT350 *racing* version, the SCCA dictated that such

Specific Data	1965 GT350	1965 GT350"R"	1966 GT350	1967 GT350	1967 GT500	1968 GT350	1968 GT500	1968 GT500KR	1969 GT350	1969 GT500	1970 GT350	1970 GT500
List Price (Base Car) *Convertible – Add $122 **Convertible – Add $318	$4547	$5995	$4428	$3995	$4195	$4116*	$4438*	$4472*	$4434**	$4650**	$4434**	$4650**
Engine Size (Cu. In.)	289	289	289	289	428	302	428	428CJ	351	428CJ	351	428CJ
Horsepower (Advertised)	306	350	306	306	355	250	360	335	290	335	290	335
Curb Weight (Advertised) †Convertibles – Add 250 lbs.	2800	2550	2800	2923	3286	3146†	3445†	3445†	3600†	3850†	3600†	3850†
Wheelbase	108"	108"	108"	108"	108"	108"	108"	108"	108"	108"	108"	108"
0-60 time	5.7	N/A	5.7	7.1	6.7	6.3	6.5	6.8	6.9	6.6	6.9	6.6
Top Speed (Est.) Δ	133	165	133	129	132	119	130	132	120	130	120	130

Δ Note: All figures shown are as advertised. Varying test conditions and other factors such as rear end ratios, state of tune, etc., may be responsible for differing figures.

1967 – GT500 means 428 cubic inches of Ford big block. Front end (except for bumper and hood are all fiberglass. Roll bar was standard and wheels are optional 7" x 15" aluminum alloy type. This beautiful example belongs to Henry and Linda Busch of Berryville, New York.

cars could have either different suspensions or different engine configurations from the street version – but not both. Shelby chose the latter for the competition cars, so all GT350s – street and competition – had identical suspensions. Race cars received fully balanced and blueprinted engines with extensive headwork and a race camshaft, as well as racing wheels and tires, gutted interiors, PLEXIGLAS windows (except the windshield) and an overall lightening. During the 1965 model year, Shelby built 562 GT350s. All were street cars except about 25 or 30, which were built as full-tilt race cars, or "R" models. These cars were so successful in SCCA B/Production racing that many people assumed that at least a hundred were built. It was hard to imagine that so few cars could win so often – but they did. The GT350 "R" models, built essentially to 1965 specifications, were National Champions in 1965, 1966 and 1967.

By 1966, Ford was providing more of an influence on Shelby. The Cobras were about to be phased out and Shelby was deep into the Ford GT program. Ford wanted Shelby to sell more cars (for better product identification) and this meant more compromises. The first generation GT350s, built in 1965, were loud, hard-riding two-seaters that were meant to be driven with a passion. Compromises were few and functionality was given priority. As such, they appealed to a small segment of the market and, thus, they were not produced or sold in what could be called large numbers. A few "R" models, built near the end of the 1965 production run, were left over and sold during the beginning of the 1966 model year, but no

1968 Shelby Mustangs acquired new front ends, (again, fiberglass) large hood scoops, functional exit louvers and real driving lights (for which special Federal exemptions had to be granted).

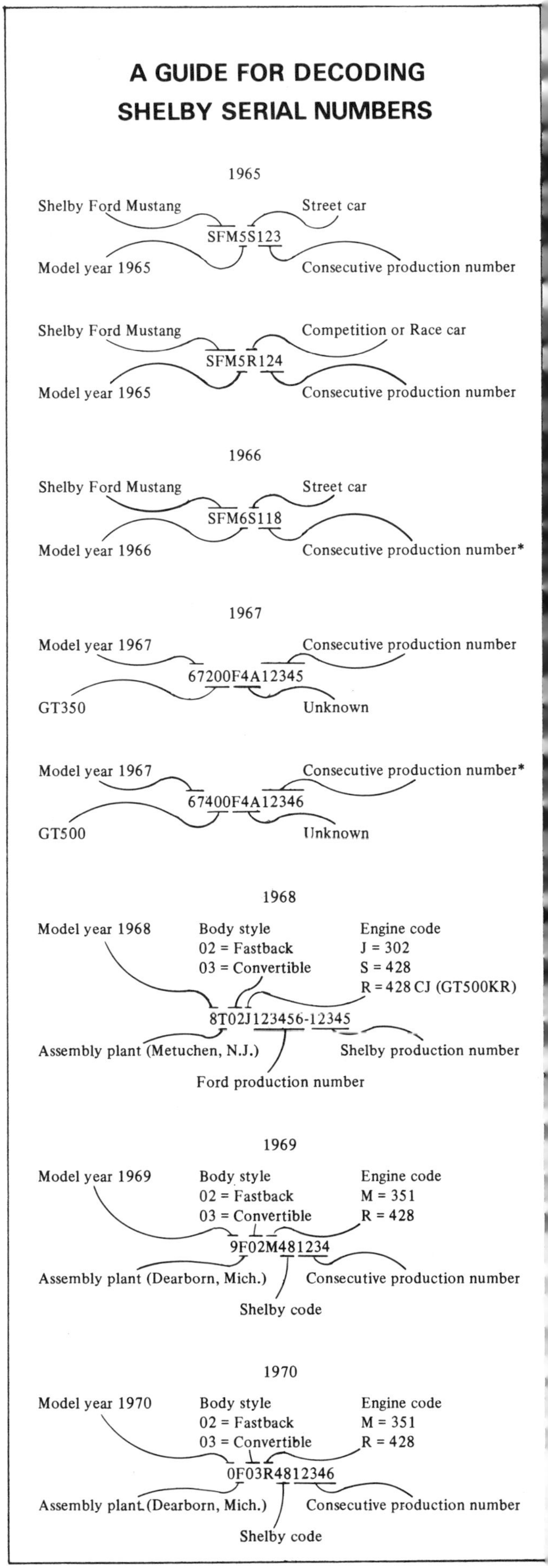

actual 1966 race cars were built by the factory. In 1966, Shelby's Mustangs began to change, slightly, in appearance, becoming more distinctive. All 1966 cars were built at Ford's San Jose plant and delivered, semi-complete, to Shelby's sprawling Los Angeles airport facility for completion. Here they received the aluminum engine dress-up parts, fiberglass hoods, side scoops, side stripes, rear quarter windows and interior parts. By the end of 1966 production, the GT350 had lost many of the suspension modifications with which they had gained their reputations (lowered front "A" frame mounting points, Koni shock absorbers and over-ride rear traction bars). Shelby and his marketing staff did manage to interest Hertz in purchasing a special model – the GT350H, which was painted black with gold stripes. These cars, usually fitted with automatic transmissions, were rented only to customers over age 25. Exactly 1000 Hertz cars were ordered, but by the close of the 1966 production, only 936 had been delivered. The last six 1966 GT350s to come down Shelby's production line were convertibles. These cars were not sold to the public; rather they were given by Shelby to his friends and associates. His wife received one, as did his personal secretary.

Each time that Ford changed the Mustang, Shelby was required to update his version. As more and more Shelby buyers ordered comfort and convenience options, the thrust of the total Shelby GT package changed from pure performance at all costs to performance with style and luxury. Each generation of Shelby Mustangs moved farther away from the original concept, but each year more were sold. So, in a sense, the buyers shaped what the cars were ultimately to become.

When Ford offered a 390 c.i.d. big block in the 1967 Mustang, Shelby went one better by offering a 428 big block in his car. The GT500 (again, the number was chosen because it "sounded right" and also because no other performance car had a number larger than 440 as part of its name) was offered in 1967 as a separate model from the GT350. Both were available only as fastbacks, but for 1967 a complete fiberglass hood and nose section as well as a spoilered tail with large sequential taillights were incorporated. Interiors carried real roll bars and inertia-reel shoulder harnesses. Automatics were optional and power steering and brakes were mandatory. Initially, 1967 Shelbys were built at the San Jose plant and completed at Shelby's Los Angeles location, but the last cars (approximately 25%) were built in Michigan. Shelby production was moved to Ionia and the A. O. Smith Company was contracted to make the conversions. These cars were supplied from San Jose or Metuchen, New Jersey. No racing versions of the 1967 Shelby were built as the cars were too heavy to compete effectively.

For his 1968 model, Shelby updated the hood and nose sections, added newly-styled taillights (sequential units) and installed rectangular driving lights in the grille. In mid-year, the GT500KR was introduced. This car replaced the GT500 and carried a 428 Cobra Jet engine as well as a few other changes. The "KR" designation stood for "King of the Road" and was used because Shelby heard that GM was about to use that name on a big-block Camaro. Being smaller and more flexible, Shelby beat GM to the punch. The 1968 GT350s carried Ford's new 302 c.i.d. small block and 1968 also saw the introduction of a convertible into the Shelby line. The roll bar was retained, but was encased in a sculptured cover. About one-third of the 1968 production was convertibles.

SHELBY SHOPPING LIST: Here are some of the "Shelby-unique" parts that each model year car had. If a particular cars lacks these parts, it is generally deemed not quite as valuable.

1965	1966	1967	1968	1969	1970
Fiberglass hood	–	Fiberglass hood	Fiberglass hood	Fiberglass hood	Fiberglass hood
–	–	Fiberglass nose	Fiberglass nose	Fiberglass nose	Fiberglass nose
–	–	Fiberglass gravel pan	Fiberglass gravel pan	Fiberglass gravel pan	Fiberglass gravel pan
–	Plexiglass side windows	–	–	–	–
–	Side brake scoops	Side brake scoops	Side brake scoops	Side brake scoops	Side brake scoops
–	–	Upper scoops	Upper scoops	–	–
–	–	Fiberglass rear spoiler	Fiberglass rear spoiler	Fiberglass rear spoiler	Fiberglass rear spoiler
Optional wheels: 5 spoke aluminum center/ steel rim - 15"	5 spoke aluminum center/ steel rim - 15" (early) Chrome or flat gray Magnum "500" - 14" (standard) "Honeycomb" aluminum alloy 14" (optional)	Kelsey Hayes 5 spoke "Mag Star" - 15" (optional)	"10 spoke" aluminum alloy 15" (optional)	5 spoke aluminum center/steel rim 15" (standard)	5 spoke aluminum center/steel rim 15" (standard)
Koni shocks (standard)	Koni shocks (standard early) (Optional late)	Koni shocks (optional)	Koni shocks (optional)	–	–
Over-ride traction bars	Over-ride (early) Under-ride (late)	–	Staggered rear shocks	Staggered rear shocks	Staggered rear shocks
"Monte Carlo" bar Fender stabilizer	"Monte Carlo" bar Fender stabilizer	–	–	–	–
"Export" brace Shock tower/cowl	"Export" brace Shock tower/cowl	–	–	–	–
Wood rimmed 3 spoke steering wheel w/"Cobra" plastic cap	Mustang GT simulated-wood wheel w/"GT350" center	3 spoke wood wheel w/GT350 or GT500 cap	–	–	–
3" comp. seat belts	3" comp. seat belts	–	–	–	–
"CS" tach & oil pressure gauge in bezel on dash	"Cobra" tach on dash	–	–	–	–
–	–	Roll bar	Roll bar	Roll bar	Roll bar
–	–	Oil pressure/amp gauges	Gauges & console	Gauges & console	Gauges & console
"Cobra" alum. valve covers	"Cobra" alum. valve covers	"Cobra" alum. valve covers	"Cobra" alum. valve covers	"Cobra" alum. valve covers	"Cobra" alum. valve covers
"Cobra" alum. intake manifold	"Cobra" alum. intake manifold	"Cobra" alum. intake manifold	"Cobra" alum. intake manifold	"Cobra" alum. intake manifold	"Cobra" alum. intake manifold
"Cobra" alum. oil pan	"Cobra" alum. oil pan	"Cobra" alum. oil pan* *GT350 - early car	–	–	–
Side exiting exhaust	Rear exiting exhaust	Rear exiting exhaust	Rear exiting exhaust	Center exhaust	Center exhaust
Chrome air cleaner	Chrome air cleaner	GT350-Chrome air cleaner GT500-Alum. oval air cleaner	Both alum. oval air cleaner	–	–
–	GT350 gas cap	GT350 or GT500 gas cap	GT350 or GT500 gas cap	–	–
Fiberglass rear panel	Fiberglass rear trim (back seat optional)	–	–	–	–
Detroit locker rearend (standard)	Optional	Optional	Ford unit optional	Ford unit optional	Ford unit optional

Rear end treatment was changed slightly for 1968. Silver rear panel carried 1965 Thunderbird sequential taillights. Side reflectors were the result of 1968 Federal requirements on all cars. Midway through 1968 production, the GT500KR replaced the "standard" GT500. Lucky owner of this car is Carl Mentz of Reamstown, Pennsylvania.

Mustangs changed, again, in 1969, becoming longer and slightly lower. Shelby's addition took advantage of this by lenghtening the nose even more. Fiberglass front fenders were incorporated and the overall effect was that of a total package. More cars sold with air-conditioning and other "luxury options" rather than performance equipment. Again, convertibles were also produced. By the end of the 1969 model year, Carroll Shelby saw the writing on the wall. The government was taking a dim view of "performance advertising", insurance companies were clamping down by charging higher premiums and competition was much more keen. The Shelby GTs had evolved to the point where they were not all that unique, so

SHELBY PRODUCTION FIGURES

1965	Mustang GT350	562*
	*includes 23 "R" competition models (number approximate; based on actual visual confirmations to date)	
	1965 TOTAL	562
1966	Shelby GT350	1438
	Shelby GT350 Hertz model	936
	Shelby 350 convertible	6
	1966 TOTAL	2380
1967	Shelby 350 fastback	1175
	Shelby GT500 fastback	2050
	1967 TOTAL	3225
1968	Shelby Cobra GT350 fastback	1253
	Shelby Cobra GT350 convertible	404
	Shelby Cobra GT500 fastback	1140
	Shelby Cobra GT500 convertible	402
	Shelby Cobra GT500KR fastback	933
	Shelby Cobra GT500KR convertible	318
	1968 TOTAL	4450
1969	Shelby GT350 fastback	1085
	Shelby GT350 convertible	194
	Shelby GT500 fastback	1536
	Shelby GT500 convertible	335
	1969 TOTAL	3150
1970	Shelby GT350 fastback & convertible	315
	Shelby GT500 fastback & convertible	286
	1970 TOTAL	601
	TOTAL SHELBY PRODUCTION (1965-1970)	14,368

RECENT PUBLICATIONS CONTAINING SHELBY INFORMATION AND DATA:

Shelby's Wildlife	A good general overview of the Shelby story, but contains a great many factual inaccuracies and errors.
Mustang Does It!	Over 1,300 detail photographs are combined to show each years' changes and similarities; everthing from trim items to engine parts.
Shelby American Guide	Published by the Shelby American Automobile Club, it is an extremely detailed technical and developmental history of all Shelby cars and is based on factory data.

Shelby asked Ford Executive Vice President Lee Iococca to cancel the program. The spigot was turned off and the cars that remained in the pipeline or on dealer's lots after the 1969 model year were retitled as 1970s and given black-painted hoods and chin spoilers. A total of 601 such cars were produced.

Each year and model Shelby has its own handling and performance characteristics, based on a combination of horsepower, weight and front-to-rear weight distribution. As a rule of thumb (or right foot), the early Shelby's – 1965s and early (the first 250) 1966s – were the most responsive. A price was extracted, because they were also the least comfortable. The 1967 GT350s carried the same engine as the 1966s, but they weighed about 200 lbs. more. The GT500s had more horsepower, but their heavier front weight distribution made them less responsive in the handling department. Early cars were loud, rough-riding and had minimal creature comforts. Later cars were much more civilized, to the point of becoming thought of as "luxury GT" cars. Shelby also had the habit of offering all race options across the counter. Such parts could be ordered on a particular car and could be installed at the factory. Although this happened, for the most part, on the earlier cars, it should be remembered that such parts were legitimate options.

Since the earlier Shelbys were virtually hand-built once they reached Shelby's plant, finding cars with parts "missing" or slightly different from the next car is no trick. Parts were used in relation to their availability. If supplies ran out, the "production line" was not stopped until more arrived. The cars were either produced without them (if it was a non-essential part) or other parts were substituted. Also, many

In 1969, the fastback was officially called a "sportsroof". Since the 1969 Mustangs were lower and longer, so were the Shelbys. Fiberglass front fenders joined the fiberglass nose and hood in an attempt to keep the weight down. Hoods sported three NASA-type intake ducts and two exit ducts. 15" mag-style wheels were standard; chin spoiler borrowed from the Boss 302, was optional.

1968 Shelby GT350 showing the distinctive vinyl-covered rollbar.

parts were updated or changed during the model run. If a change was made (such as switching from over-ride traction bars to under-ride units) the old supplies were depleted before the change was made. Thus there was no *one* transition car. Production on the 1967 and later cars was more consistent, but by no means was each of those cars exactly alike, either.

The original GT350 listed for $4,547 in 1965. Each subsequent GT350 was cheaper, and their prices formed a textbook bell-shaped curve that ended with normal inflation causing the 1970s to be priced at almost the same level as the first generation, five years earlier. As happened with most mid-sixties performance cars, once the original owners began selling, the prices dropped. This was especially true with the early cars because they looked much more like stock Mustangs than the later cars. Shelby Mustangs reached their lowest point in the early 1970s, when it was possible to buy a decent car for under two thousand dollars. The year saw the formation of the now-defunct Shelby Owners Association, and although at its highpoint, three years later, it had only about 1,200 members, it was responsible for the beginnings of interest and enthusiasm in these cars. The value of Shelby Mustangs has climbed steadily since then. Today, asking prices between $4,000 and $8,000 are the norm, and stories of "thousand dollar-Shelbys" are not heard as much, today, as they were in the recent past. At any rate, four thousand dollars leaves a lot of leeway, so it may be helpful to discuss what factors can cause a car to be placed at either end of the spectrum.

GENERALITIES:

1. **EARLY VS. LATE.** Generally speaking, early cars are bringing higher prices than late cars. This is primarily due to the fact that there were less of them produced (the law of supply and demand); they were the only Shelbys to be actually raced (heritage); and because they were more responsive (lighter weight and less compromising suspension). Lately, however, the last generations have shown a marked increase in value, probably due to the fact that they were the end result of the Shelby evolution, and also because they have been around the

shortest length of time and, therefore, have had less time to rack up mileage and abuse.

2. **FASTBACKS VS. CONVERTIBLES.** This area concerns the 1968-1970 models, as these were the only years that Shelby offered convertibles to the public. Since true convertibles are no longer being produced, they are, as you would expect, more desirable. All other things being equal, a convertible is likely to fetch a higher price than a fastback.
3. **BIG BLOCK VS. SMALL BLOCK.** Prices here are roughly the same. Buyers all have their preferences, and both cars have advantages and disadvantages. Small-block Shelbys are easier to work on under the hood, are more economical to operate, and usually handle better. Big-blocks accelerate quicker and have more of a *macho* image connected with them.
4. **ORIGINALITY VS. MODIFIED or "IMPROVED".** All things considered, an original car will bring a higher price than a modified one; the exception being a car that has been modified with Shelby racing parts (i.e. Weber carburetors, "R" model front apron, Shelby wheels, competition specification roll bar, etc.). These cars so easily lend themselves to personalization and modification that probably less than a third are still original. As values rise, many are being brought back; also, many parts are beginning to be re-manufactured. However, current owner and enthusiast organizations do not

STANDARD OPTIONS

	1965	1966	1967	1968	1969	1970
Mag-style wheels	–	–	–	–	–STANDARD–	
Rallye stripes*** (Dealer installed)	–	–	–			
Rear seat (mandatory option*)	None	–	–	–		
Detroit no-spin rear axle	Standard	–	–	–FORD UNIT OPTIONAL–		
Paxton supercharger		–	–	–		
Automatic transmission		–	–	–	–	–
Air conditioning	**	**	–	–	–	–
Tinted glass			–	–	–	–
Tilt-away steering wheel				–	–	–
Intermittent W.S. wipers					–	–
Power ventilation					–	–
H.D. Battery					–	–
Colors	White	White Black Red Blue Green	White Red Lt. Blue Med. blue Dk. blue Lime Dk. green	White Black Red Blue Lime Green Gold	Off-white Black Red Maroon Blue Aqua Silver –Grabber Colors– Blue Green Orange Yellow Competition Orange	Off-white Black Red Maroon Blue Aqua Silver
Interiors	Black	Black	Black Saddle	Black Saddle	Black White	Black White
Convertible tops				Black White	Black White	Black White

*Early 1966 cars had rear shelf instead of seat
**Dealer installed
***Over-the-top stripes; all side stripes standard

stress *concours* originality; they have a "do your own thing" philosophy that accepts anything – something that is refreshing amid many clubs that are originality-oriented to the point of excess.

. **SPECIAL MODELS.** Almost each year saw "special" models built by the factory. In 1965 it had its "R" models – bona fide SCCA B/Production racers, and almost all have been located and are either still being raced (having received all of the "updates" required to remain competitive) or have been (or are in the process of being) restored to 1965 racing specifications. Prices run between $10,000 and $25,000, depending on the condition and originality. Most have found permanent homes. As mentioned, the six GT350 convertibles that were built in 1966 are in a class by themselves. All have been located and prices are likely to start at $10,000. There are also a handful (the actual number built was thirteen) 1965 and 1966 factory-built drag GT350s. These cars were set up for the quarter mile and probably few have survived intact throughout the years. Depending, again, upon originality, prices might start at about $12,000. In 1967 and 1968, a limited number (currently unknown, but thought to be less than 50 for each year) of special 427-equipped GT500s were produced. Authenticating such a car is almost impossible because the engines were changed after production and no special code or identification was thought to have been assigned to these cars. Probably few have survived with their original engines still in them, 427 race engines being what they are.

SPECIFICS:

. **"SHELBY-UNIQUE" PARTS.** A high percentage of these parts are currently available, with more coming on the market every day. Depleted stocks have been re-manufactured, existing supplies have been discovered, and while bargains are rare, at least the stuff is obtainable. Each model, however, has some parts that just cannot be had, for love or money. If a car lacks these parts, but is otherwise complete, this is usually taken into consideration in the price. If a car lacks parts that *can* be obtained, it is not uncommon for the buyer to negotiate the price of obtaining them into the final price. (Examples: cast aluminum "COBRA" engine parts: valve covers – $100 a pair, oil pans – $350, intake manifolds – $150: original equipment mag-style wheels – $75-$100 each.

2. **MILEAGE.** As with any car, low mileage is an asset, but in many cases it is difficult to document.
3. **CONDITION.** Since the cars were built using mostly Mustang parts, replacement of these items isn't too difficult, unless they are no longer available from Ford. Many Shelby-unique parts are being re-manufactured, but overall condition is reflected in the price. Shelbys are rust-prone in the same places Mustangs are – check quarter panels, inner fender panels and the bottom edge of both doors.
4. **ORIGINALITY.** Relatively few Shelbys have made it this far along without some owner drilling a few holes or adding or changing something. In most cases, these modifications may be removed or replaced, but an original car is indeed rare – and thus valuable.

AUTHENTICATION:

All Shelbys have Shelby-unique consecutive serial numbers, although late models have a combination Shelby/Ford number (1968) and special Shelby codes (1969-1970). All early cars also have Ford serial numbers that must correspond to a particular code in order to be Shelbys. Many fakes are spotted in this manner as it is the most obvious giveaway. There are other more subtle clues.

It may appear fairly simple to "produce" or fake a Shelby, but in point of fact, it is not. To do so in such a manner that detection would be impossible would cost more than the car is likely to be worth. However, as prices rise on these cars, the possibility of "counterfeiting" becomes more likely. Many counterfeit Shelbys surface after the unsuspecting owner shows up at an enthusiast meet and only then discovers – painfully – that he has acquired a customized Mustang.

The Shelby American Automobile Club is currently in the midst of compiling a complete registry of all known serial numbers for these cars. Such a listing will help in locating stolen cars as they surface and also make duplication of serial numbers difficult. SAAC is always willing to help members and non-members (read *prospective* members) authenticate any given car prior to its changing hands.

Side stripes, up from the rockerpanels for the first time, were color-keyed to the car's color and were reflective. Taillights were, again, 1965 T-Bird. Exhaust was routed to the center and out cast aluminum ports. This stunning 1969 belongs to noted automotive photographer Ray Hofmann.

1970 cars were actually left-over 1969s. Flat black panels were painted on the hood to "update" them and serial numbers were also changed to reflect this. Convertibles were first offered in 1968 and continued through production, accounting for about one-third of all cars sold. Side scoops were lower than those on the fastbacks because of convertible top mechanism. Roll bars were covered with sculptured plastic covers. This black beauty is owned by John Donovan of Trenton, New Jersery.

David uses his Mustang! He drives it to and from meetings as well as competing in Post Historic Road Sports events

Riding thunder

Mike McCarthy discovers how the Shelby Mustang got its muscle car image . . . a road-going sportscar with drag-strip performance

When David Baraclough was 21 (ish) he used to have great fun racing Minis against the likes of Gerry Marshall and Bill McGovern. Then the scene turned rather commercial for his tastes, so he pulled out, and eventually went to live in America for a few years in the early seventies. What more logical car, then, to return to racing in than a Shelby Mustang? An ad in *Autosport* last summer found him buying a genuine Shelby G.T.350.

David's car is a 1965½ model in that it has some 1965 components (such as the 'overrider' trailing arms) and a 1966 body with the plastic rear three-quarter window. When he bought it, the car had seen quite a few owners from the time it was first registered in this country in 1967, most of whom are lost in the annals of history and the Swansea computers. It was painted black with a vast gold snake on the bonnet – "quite tastefully done, actually!". But it was the real thing, and could be converted back.

Stripped prior to rebuild

Naturally the car was totally stripped and the rebuild began. Geoff Sullivan looked after the bodywork, removing layer after layer of paint, underneath which were still discernible traces of the original white and blue colour scheme. Starting with a bare shell, the wheel arches, under-floor panels, bottoms of the doors, boot floor and assorted other sections were replaced, using as little lead as possible – the G.T.350 is heavy enough as it is! Once all this was done the car was re-finished to its present gleaming condition.

The suspension and mechanical components were the preserve of Mick Sherwin of S, S & L. He sent David shopping in America a few times, and working with what was on the car, some Shelby drawings of the competition set-up, and the bits David brought back, re-assembled it all, lowering the pick-up points on the front upper wishbones *à la* Shelby racing. In fact David is still playing around with the suspension settings as he gains experience with the car on the track: a rear anti-roll bar is a recent fitment, for example. Incidentally, it was while the car was at S, S & L that the original ten-spoke Shelby wheels were stolen: if you're offered a set we suggest you contact your local police as soon as possible. Thus the wheels currently fitted are Fenton Vecmaster units dating from 1967, so are of roughly the right period, and for racing David uses Michelin TB15 tyres.

The brakes were fitted with all-new but genuine Shelby components. However the first race saw dramatic fade, so the Kelsey Hayes master cylinder gave way to a Lockheed/Girling unit, and another servo replaced the original.

The only major change from original is that the engine has been replaced with another, built up by Mathwall Engineering. The Cobra/Shelby unit was pretty tired, anyway, so David had a choice: rebuild it for racing or keep it as a road goer. Since racing can impose great strain on an engine David opted for the latter, but has kept the original and had it restored by Competition Engine Services.

So to the car. When you climb in, over the high-sided driving seat, the first impression that strikes you is just how civilised – no, *garish* is the word – the interior is for a racer. Remember it's still Mustang at heart, and there arrayed in front of you is a typical mid-sixties American facia! Only the small, leather-trimmed steering wheel and additional gauges down by the gear lever tell you this is 1982. Chrome knobs even! Good grief!

The *real* Mustang heart, however, comes out when you switch on. Unlike the GT40 I drove recently (*C&S*, July) which had the crossover exhaust system and sounded like a genuine racing engine, the exhaust note of the G.T.350 is pure American Dragster. There is that typical V8 beat, but magnified thanks to the straight-through exhaust and lack of sound-deadening.

Red line seems superfluous

The engine is really quite remarkable. At low speeds it's a stump-puller, and you can trickle the car along at 1500rpm or so in top with nary a qualm. Floor it and by 2000rpm you're travelling: by 3000rpm you're travelling very quickly indeed! There is so much torque that the 6500rpm red line seems superfluous unless you're trying to get the most from the engine under racing conditions. The gearing in the car at present gives roughly 20mph/1000rpm, so top speed isn't all that high – around the 120mph mark. But you get there quickly . . . it's easy to see where the term 'muscle car' came from, and the attractions that drag racing (either on street – illegal – or strip – legal) has for so many Americans. It's like riding thunder.

The major controls work as they should in a racing car. The clutch is heavy and of the on-off variety, the brakes nicely weighted and superbly progressive. The gear change is notchy at all times and baulky if you try casual changes, but ram it through and it takes on a whole new character. Snatched changes are a case of slam, bam, and you're glancing at the rev counter preparing for the next.

At low speeds, too, the steering is heavy while the limited slip differential clonks and bangs away behind you. What's more, it keeps trying to push you onto the straight and narrow so the initial impression is one of strongish understeer. The faster you take a corner, however, the more precise the steering becomes, while the limited slip diff stops its straight-on tendencies and works properly. On the road, and fitted with the Michelin racers, the G.T.350 is nicely balanced: with all that power, though, it can become an instant throttle steerer, and the angle that the tail adopts to the direction of travel begins to depend on how far down the accelerator pedal had been depressed. And in spite of those Michelins, the ride is surprisingly good!

I came away from David's car wondering whether the Mustang was a dragster that could go around corners or a sports car that has dragster performance. The answer appears to be both which, I suppose, is just what Carroll Shelby would have wanted.

This is a racing car? With all the garish chrome inside the Mustang looks far from being a real performance machine

HOW TO BUY A USED SHELBY

A GOOD ONE IS OUT THERE IF YOU KNOW WHERE AND HOW TO LOOK

If you're looking to buy a Shelby Mustang, you're not likely to hear the piercing voice of a straw-hatted barker in a brightly striped jacket and bow tie, slapping his cane on the table top as he answers questions you never asked. But if you don't know very much about Shelbys, you might just think you dropped in on a carnival side show. You *can* buy a Shelby without being conned or fleeced if you know a little about them. And, as far as those wild claims made about Shelbys—like immediate recognition, sudden good looks, and an increased social life—they're all true!

Carroll Shelby's version of the Mustang is currently one of the most popular of the 1960s performance cars. They have all the right characteristics for a true automotive pedigree: racing history, limited production, and unique parts. They were also extremely fast, agile, and responsive. Shelbys continue to appreciate every year. New, they sold for about $4500; today, you'll have to pay twice that to own a good one. How can you find a "good" one? Better yet, how can you tell that one is "good" after you've found it?

For openers, the best way to acquire Shelby knowledge is to read about the cars, look at as many of them as you can, and talk to people who own them. The best way to do all three is to join a Shelby enthusiast organization. Currently, there is only one national Shelby club. The Shelby American Automobile Club (SAAC), 24-C April Lane, Norwalk, CT 06850, has almost 6000 active members across the country. It sponsors, through dozens of local regions, all types of events from small picnics to open tracks at race circuits to a gargantuan annual national convention that brings together thousands of enthusiasts and hundreds of cars of all types. The SAAC also publishes a professional-quality, bimonthly magazine as well as other in-depth books, like a definitive technical and developmental history of all the cars from Shelby American and an owners registry. (Let's say there's no shortage of reading material.) SAAC membership is open to anyone interested in these cars, and yearly dues are $21.95. A detailed brochure is yours for the asking.

Finding a Shelby for sale is no trick, but finding a good one is. The primary sources of cars for sale are classified ads. Naturally, the longer the lead time (the time it takes to compile, print, and distribute a magazine), the less current the ad will be. Larger magazines like HOT ROD or *Road & Track* require three months lead time; monthlies like *Hemmings Motor News* and *Cars & Parts* are half that, newspapers are almost daily. Also, stay on top of publications that cater to Shelby and Mustang enthusiasts (the SAAC's *Mustang Monthly:* P.O. Box 5917, Lakeland, FL 33803; *Super Ford Magazine.* P.O. Box 2011 Ovid St., Seneca Falls, NY 13148).

A combination of these will give you a good feel of the current market and will enable you to spot ads that look promising. Be prepared to jump on such an ad and follow it up *immediately.* It should go without saying that "good cars priced low go quickly." Another excellent source is word of mouth, in fact, it is often the best source for good buys. Fellow enthusiasts are always passing tidbits along, long before they ever reach

USED SHELBY

1

(1) A Shelby American Automobile Club convention is an excellent place to look over the various conditions and styles that Shelbys are available in. Virtually all possible combinations turn out.
The 1968 Shelbys came in three models: GT 350, GT 500, or GT 500KR (for "King of the Road"). The KRs actually replaced the standard GT 500s at midyear using the 428-cubic-inch engine with 427-style heads.

(2) The original Hertz "rent-a-cars" didn't have Paxton superchargers; however, the Paxton was a legitimate Shelby option between 1966 and 1968, and in this case it adds to the value of the car.

(3) This GT 500 has a 427-cubic-inch engine, which the owner claims came stock, as a special order, from the dealership. Some did come this way, but unless the original paperwork is present, there is no way to tell for sure.

(4) The early 1966s had all of the desirable 1965 features—lowered A-arms, Koni shocks, and 15-inch Shelby Cragar wheels, along with distinctive '66 features like side scoops and rear quarter windows.

print. Also, detective work, based on tips from non-enthusiasts, can yield previously unknown cars at amazingly low prices. The key is to follow up any leads, no matter how weak or unpromising. You may just find a low-mileage Shelby tucked away in a barn under a foot of dust whose owner will be slapping his knee and laughing for hours after you've bought his 15-year-old car for $1000 and driven off.

If you're like everyone else who has thought about stepping into a Shelby Mustang, a few other questions have probably popped into your mind. Like, which years and body styles are rarer or more valuable? Production figures from Shelby American and Ford indicate that each year's Shelby production increased by a thousand cars. This was due to the car's success. So they certainly qualify as "limited production." (Rarity depends on your definition.)

If we use the average advertised price over the past year or two as a yardstick of value, things become a little clearer. All things being equal (condition, mileage, originality, etc.), the first year's GT 350s, sold in 1965, generally carry the highest price tags. This is a function of numbers (only 562 were produced) and of concept (the '65s had the best performance in all areas), as well as racing history (perhaps 30 or so were built as factory competition models, and there was a startling similarity between them and the street cars). It goes without saying that any of the '65 GT 350 "R" models (for "racing") are worth two to three times what street cars are worth.

Historically, the 1966 Shelbys have been the most desirable. There were six convertibles built at the end of 1966 production. These were given away by Carroll Shelby to friends and associates, and are worth probably twice as much as a fastback—if you can find one for sale. Almost a thousand special GT 350 Hertz models were built in 1966. These cars have their own following of owners who enjoy the fact that their cars were actually for rent when they were new (another part of the Shelby legend). Hertz price tags are usually no different than those of other Shelbys.

While 1967-and-later Shelbys always seemed to lag behind the earlier ones in price (by about 25 percent), the last couple of years has seen their value rise to where they are now of equal value with the earlier cars. There is, likewise, little difference between the asking price of a small-block GT 350 and a big-block GT 500. Both have their advantages and drawbacks. The small-blocks, because of their lighter weight and more favorable distribution, handle better and are easier to work on. The big-blocks are faster, and changing spark plugs is easy—if you happen to be an 85-pound, double-jointed contortionist. In general, convertibles fetch more than hardtops because they are extinct today. This is not unique to Shelbys.

Until now, we've been talking about basically stock Shelbys. But plenty of them have been modified since they were new. How do such cars fare value-wise? Not as well as those that are stone-stock. But depending on the level of modification, the ease of returning to original, and the craftsmanship and quality, the value tends to taper off accordingly. Generally, the value of a modified Shelby is equal to the amount a perfect example would be worth, minus what it would take to

2

3

4

get the car to that point.

There is no shortage of buyers' opinions and sellers' rationalizations regarding which Shelbys are "more valuable." As with any series of cars, there are plenty of "firsts," "lasts," "only ones," and "one of a few(s)," due to the fact that all Shelbys carry consecutive serial numbers and because there were running changes made during each model year's production.

And what about fakes? As the value of Shelbys rose over the past few years, it became financially feasible for individuals or companies to remanufacture parts for these cars. Today, almost all Shelby-unique parts are available. Concurrently, rising values give way to the nagging fear that these parts would, sooner or later, start finding their way onto Mustangs which would then be sold as Shelbys—for Shelby prices. Happily, this has not been the case. While a few crude imitations have turned up here and there, they are obvious counterfeits. If such cars do change hands, the victims are unknowledgeable and usually driven to fast action by a seductively low "once-in-a-lifetime" bargain price.

Since all Shelbys have unique serial numbers, attempts to alter or change them are obvious. Such tampering is a felony as well as a federal offense. To take a Mustang and make it into a perfect Shelby forgery that would be undetectable would require more time, effort, and money than a Shelby would be worth.

Obviously, the more you know about Shelbys and the more of them you've seen, the better able you'll be to judge whether the car is genuine. As a further safeguard, SAAC keeps track of as many serial numbers as they can locate and publishes an owners registry containing detailed information. You don't hear about too many stolen Shelbys anymore. Because the cars are so valuable, there is less chance of their being left unattended or unsecured, and professional car thieves shy away from them because of their unique serial numbers. Stolen cars are reported in SAAC's magazine and, with 6000 people across the country aware of them, the owner is notified and action is taken as soon as they turn up.

One final question that almost always comes up after a Shelby has been purchased is "Can I insure it for what I think it's worth?" Recently, SAAC has discovered insurance carriers who have set up special policies for club members. This insurance is purchased as a rider (on top of your standard liability policy—which is no problem to obtain from the company you use now) and is "stated value." That is, the cost is pro-rated, based on the dollar amount you assess your car at. It covers collision, fire, and theft, and there are categories based upon annual mileage. Figure on about $275 for a $9000 car (on top of your normal coverage).

There are a few hidden costs of owning a Shelby that we didn't mention. Figure on an unlisted telephone number (to keep the number of incoming female calls manageable), a fat "black book" ("Don't call me, I'll call you"), and plenty of money in the checking account to pay for those speeding tickets (at anything under 55 mph, a Shelby is just idling). Are you sure you want to live this way? SM

KING OF THE ROAD

SIMPLY ONE OF THE FINEST GT500KR'S IN THE WORLD

BY ALEX GABBARD

Shelby GT500 Mustangs were introduced for the first time in 1967 as the big-block brother to the GT350s. They were plush, top-of-the-line Mustangs with special Shelby features: characteristic trim, scoops, and stripes that continued the image of pure performance.

Then for 1968, a convertible model was offered, and by midyear, a higher-performance version was introduced as the GT500KR. The "KR" stands for "King of the Road."

Ford Motor Co. had decided to offer a new engine for 1968 which they called the Cobra Jet. It was essentially a 428-cubic-inch block like previous Police Interceptor engines, but it was fitted with larger-port 427 heads and an intake manifold carrying a Holley 735cfm 4-barrel. The GT500KR was the replacement for the GT500, but how it got its name (KR) is a corporate *coup* for Ford over GM.

GM was planning to introduce a big-block Camaro at midyear called King of the Road. Ford, however, was able to pull a fast one by introducing a KR first through Shelby's small-company flexibility. It was the GT500KR, but it wasn't just a GT500 Shelby with a new name.

They had all sorts of changes such as staggered rear shocks, wider rear brakes and drums, and other heavy-duty equipment. But for the first time, Shelby cars (including the KR) were not actually built by Shelby American, Inc., in Los Angeles. Shelby had shut down his plant in December of 1967, and the cars were built on contract to A.O. Smith in Ionia, Michigan.

A total of 933 GT500KR fastbacks and 318 convertibles were built as 130mph GT cars. But they were equally at home on dragstrips. The 428 Cobra Jet engine in drag cars (the same powerplant as in the KR) grabbed Super Stock Eliminator honors at the Pomona Winternationals. HOT ROD Magazine called the KR "the fastest running pure stock in the history of man." The 428 was advertised to put out 355 hp, and the '68 optional engine (the 427) was rated at 400 hp—all for the street.

But that was 15 years ago, when a mere $4472.57 could put you in the driver's seat of a KR fastback. That was exclusive territory then, and they're even rarer now. So when you see a GT500KR, look closely; it may be the only time you'll ever see one.

So you'll know what a perfect GT500KR should look like, check out Jeff Murr's. Jeff and his father spent months restoring the car, which has since taken 12 trophies in eight shows and got 298½ out of 300 possible points at the 1982 Shelby American Auto Club National.

Jeff bought his KR from a dealer in Republic, Missouri, and spent 1½ years getting the parts needed for proper restoration. It's tough to restore a car to showroom-perfect condition, but it can be done with enough searching.

Jeff found original-equipment Lucas foglights in Florida, shocks in South Carolina, Autolite battery and oil filter in North Carolina, tires in Wisconsin, wheels in Pennsylvania, steering wheel in Iowa, exhaust system in Alabama, and sequential flashers in Georgia. Helping in the project and painting the car was fellow KR-owner Lamar Allen of Cleveland, Georgia, whose KR Shelbys have also been National winners.

For Jeff Murr and his father, Howard, what resulted is the finest GT500KR we know of. Something to be proud of for sure. *M*

Only 933 GT500KR Shelby fastbacks were made in 1968. Jeff Murr's black beauty has to be one of the finest examples in the world. It's 100 percent stock.

PHOTOGRAPHY: ALEX GABBARD

The KR stood for King Of The Road, and referred to the awesome powerplant; a 428-cubic-inch Cobra Jet, big-block Ford. This one is mated to a C-6 automatic transmission.

An interesting feature of the all-black stock interior is the tilt, swing-away steering wheel. The car is fully instrumented with tach, oil pressure, clock, water temperature, and ammeter gauges.

The '68 Shelby featured a rear deck spoiler, twin-tipped exhaust outlets, and sequential turn signals from '65 Ford Thunderbirds.

Even the trunk of the Murr Mustang is flawless. The spare tire is an E70-15 Goodyear Wide Tread GT, just like the original. The spare wheel matches the rest of the 15×7-inch Shelby ten-spoke wheels.

THE SHELBY RACER THAT TURNED SHOW WINNER

Text and Photography by Don Emmons

Ford's Mustang was a crowd pleaser from Day One and the Shelby GT350 was Ford's "Rambo." The GT350 was a bold move on Ford's part that gave the crowd something to really get stirred up about.

The blue-and-white striped bundle of terror was introduced in 1965. Some 20 years later, the car is more recognizable to the masses than it was at its inception—and the price is reflective of that recognition. Very few of these cars were built, but they made a big name for themselves, thanks to an active group of Ford enthusiasts, both in and outside the factory.

The early history of this particular car is not known, but its first accountable venture was its entry in the Cannonball Race by a couple of Shelby enthusiasts. The Cannonball Race was a natural for the GT350; a race guaranteed to quicken the pulse rate of any Shelby owner. Though the car was a strong finisher in the Cannonball, it suffered a slight setback due to some downtime after an encounter with law enforcement types.

Shortly after the race, the car was sold to Bud Williams, a Mustang enthusiast from Southern California. Bud had been involved in racing of one sort or another most of his adult life, and he'd just finished restoring a 1970 Shelby. After buying the car, he drove it around for a while before deciding to restore it.

Bud was in a partnership with friend (and well-known drag racer) Art Chrisman, for the purpose of restoring cars. Art and his son did the actual work, while Bud supplied the financial backing.

Their goal was to make the car a 100-point restoration and the Shelby was accordingly stripped down to its last nut and bolt. It was about then that

CANNONBA

Bud started bench racing some friends and made a few rash statements about winning a very popular upcoming car show held annually by the Pantera Club in Newport Beach, California.

All he had to do now, was back up his words . . .

The sheetmetal part of the car was first-class, but the fiberglass hood was not show quality (as they were never intended to be from the start), and years later this one was in grim shape. Bud feels that Steve Davis is the best sheetmetal man in the race car business, so he called on him for help with the hood. Steve took a new steel '65 Mustang hood, separated the underside reinforcing structure from the top side, and proceeded to make an aluminum top that duplicated the hood complete with scoop.

Paint was handled right at Chrisman's shop. The blue stripes were painstakingly laid out to Shelby specs, ensuring that they were exact.

The interior was done with a total restoration in mind, so all original items were obtained. Where needed items were not available new, the old ones were rebuilt. Upholstery was turned over to Len Barnett at Thornton Upholstery in Santa Ana, California.

Most of the subtle refinements were held to the underside. They basically adhered to the original, but Bud's quest for the ultimate car reared its head. He decided to change the rearend to a 9-inch Lincoln Versailles, which was a bolt-on change that provided disc brakes. They discovered that Versailles front ends were also interchangeable with the original. So with new ball joints installed, the Versailles disc brakes went on the front, giving the car a late-model braking system. Add a power booster, synthetic fluid, and it stops almost before you get your foot planted. Bud says this braking system is second to none.

Nearly everything but the exterior has been powder-coated, a popular trend for chassis parts these days. Powder coating is an extremely durable finish that can be done in nearly any color desired. Orange County Plating did the work.

The engine compartment got the same treatment. The stock Shelby-built

LL RUNNER

1. Interior was redone using exact materials or copies of original materials. Shelby cars were equipped with three-inch racing lap belts.

2. A new radiator was built for the car with a polished brass top tank and tubular overflow catch tube to the side.

3. Fine detailing is apparent here. Car has both export brace and Monte Carlo bar done in black powder coating.

4. Shelby add-ons include the tach cluster and Cobra wood-rimmed steering wheel. This tach setup was used only on '65 Shelby cars.

5. Engine was gone through by ex-drag racer Art Chrisman. Subtle refinements brought it up to 310 hp. Stock manifold and Holley carburetor were retained.

RUNNER

271hp 289-cubic-inch V8 was left basically stock. Art Chrisman performed his magic internally but didn't do that much at all. This car was to be driven on the street, with perhaps a few turns around the track at Shelby Club meets, but that would be the extent of its extracurricular activities. Externally, the engine was left completely stock with the exception of aluminum pieces that were polished or

There's not a sharper GT350 anywhere than this one owned by Bud Williams of Anaheim, California. Thousands of hours of work by Art Chrisman went into restoring this car.

Front brakes were changed over to '79 Lincoln Versailles units. These are interchangeable from front ball joints outward and provide the same braking system all around.

The 1979 Versailles 9-inch rearend comes complete with disc brakes. Dimensions are exact to Mustang's and bolts right in. Only emergency brake cables need to be altered. This car is plumbed throughout with polished stainless tubing or braided hose.

chromed. The only addition to this area is a specially built radiator which helps cool the unit better than the orignal.

All through the rebuilding of this car, emphasis was put on the super detailing needed to make this a show piece. The topper is that the car was finished the day before the Pantera show where it was entered in the American Sports Car class. When the smoke cleared, the GT350 came out on top. It was a tribute to Chrisman Automotive of Fountain Valley, California, and Bud Williams' desire to build it his way. *M*

In keeping with the flawless exterior, this metal hood was fabricated to replace the stock fiberglass Shelby hood. It's very good looking but the price is high.

In order to meet standards for SCCA's B-Production Racing class, over a hundred cars had to be "series-produced" and the car had to be a two-seater. Hence the flat panel and spare mounted in rear seat area.

No area was left untouched in this restoration. Underside also saw its share of detailing. Most items were powder coated in black. Exhaust system, including headers and mufflers, were made up using stainless steel tubing. Cobra eight-quart oil pan was used. Beefier Lakewood scattershield and driveshaft loop were added.

Another refined detail is the side stripes. No decals here; these are painted on. Stock Cragar wheels were re-rimmed and run Goodrich Competition TAs, 215/60-15s front and 235/60-15s on the rear.

RARE GOLD

THIS SHINING GT350-H OFFERED CIVILIZED SHELBY PERFORMANCE AUTOMATICALLY

Text and Photography by Jerry Heasley

Early Mustangs with K-code 289 solid-lifter engines are hard to find with automatic transmissions, but there *is* one car that has this engine/tranny combination in a fastback body style, and it's a Shelby. It's the '66 GT350-H, of course.

For 1965, whether it was a Shelby or a Mustang, the Hi-Po engine was strictly for a four-speed car. In 1966, after many requests, Ford announced that it would finally bolt

Although the engine is not detailed for concours showing, it is clean, and will make a good Shelby entry in the street-driven class.

a Cruise-O-Matic behind its high-performance small-block in the Mustang.

Carroll Shelby might not have done the same for his fastback Mustang, had it not been for the special deal with the Hertz Rent-A-Car people. Shelby American's Peyton Cramer presented the idea of a special Shelby sports car to Hertz for its popular Sports Car Club. According to Rick Kopec's *Shelby American Guide*, Cramer was hoping for an order of about 25 to 50 cars. However, most likely because he suggested that the cars be painted black and gold (the same as Hertz' own cars back in 1927), Cramer came away from the meeting with a surprising order for 1000 GT350s! Needless to say, Carroll Shelby was astounded.

Thus, the very first Hertz Shelbys were black and gold GT350s, with four-speed transmissions, override traction bar, lowered A-arms, and 15-inch Shelby/Cragar wheels.

Del and Lois Proudfit of Brush Prairie, Washington, have acquired a portion of that esteemed history. Their '66 Hertz car is a rare white with gold stripes model featuring an automatic transmission. Bearing serial No. 6S 490, their GT350-H was

Notice the competition seat belts and dash-mounted tach, along with the floor-mounted C4 automatic transmission shifter. The seats are standard Mustang buckets. The center of the dash displayed a label warning, "This vehicle is equipped with competition brakes. Heavier than normal pedal pressure is required."

RARE GOLD

The Proudfit Mustang collection includes a K-code convertible, fastback, and this GT350-H. The GT350-H is Del and Lois' only K-code pony with automatic transmission.

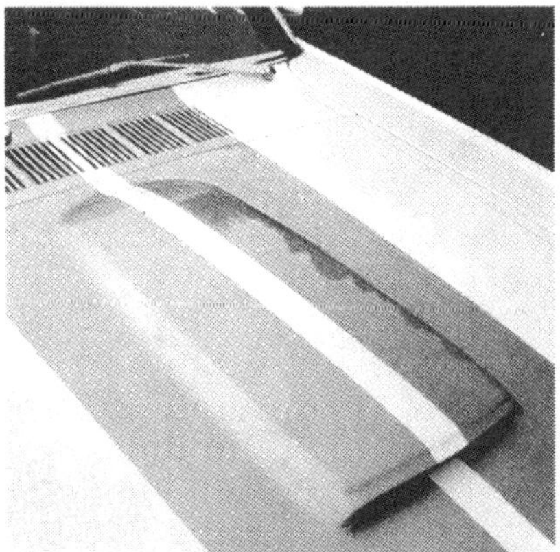

The hoodscoop was functional and standard.

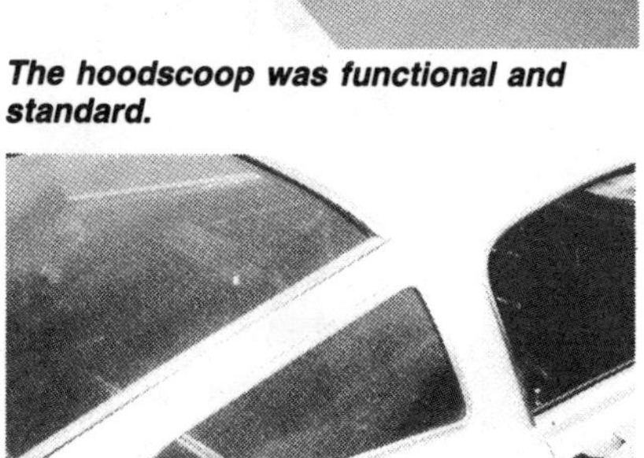

Rear quarter windows were standard for '66 Shelbys.

The ID plate reads, "SFM 6S 490"—"SFM" is for Shelby Ford Mustang, "6" is for 1966, "S" is for "street," and "490" is the consecutive unit number.

built early in the production run. (The very first Hertz car was built around No. 450.) Thus, their Shelby demonstrates that these cars were quickly "civilized" for use by a wide range of Hertz customers.

Del enjoys pointing out the so-called "piggyback" master cylinder, which lowered brake pedal pressure. Then, there's the C4 automatic, with high-performance shifting abilities, mated to a slightly smaller 595cfm Autolite carburetor. To smooth the ride, most Hertz Shelbys came with underride traction bars. Front ends didn't have lowered A-arms. Obviously, Shelby-American was giving Hertz more of what it wanted and saving money in the process.

The result was an auto that was, without question, a full GT350, but with slightly more civilized manners. This suits Del and Lois just fine as they cruise the sparsely traveled roads around their rural home. *M*

The Christmas Flower Project Book

Joanna Sheen

Photography by Jacqui Hurst

MEREHURST

Published in 1995 by Merehurst Limited,
Ferry House, 51–57 Lacy Road, Putney, London SW15 1PR

ISBN 1 85391 352 9

A catalogue record of this book is available from the British Library.

Edited by **Heather Dewhurst**
Designed by **Lisa Tai**
Styling by **Jacqui Hurst**

Typeset by Servis Filmsetting Ltd., Manchester
Colour separation by Global Colour, Malaysia
Printed in Singapore by Craftprint

Contents

Introduction

Christmas is not only a time for families and celebrations, but also for flowers and decorations. Whether you are spending Christmas in a small apartment or a big family house, there is always a place to hang a decoration or two.

Homemade Christmas gifts are very precious and special wrapping adds to the overall effect. In this chapter there are ideas for packaging parcels and decorating crackers, and making all sorts of arrangements from holly rings to topiary trees, all of which will add extra sparkle.

As Christmas is a time for sharing and joining in, many of these projects could by made by younger flower arrangers, but with plenty of adult supervision, of course!

So whether you are planning your floral decorations for Christmas well in advance or looking through for last-minute ideas, I hope you enjoy the projects and are inspired to move on to other creative ideas and designs.

Preparing for Christmas

Many of the items used in these projects can be bought all year round, but some may need a little advance preparation. Think ahead during the summer days and harvest flowers and seed heads for drying at the right time, and store them away in a warm, dark place.

Planning ahead

The projects using dried or preserved items can be made in advance and stored. Make sure, however, that you pack them away carefully – preferably in the dark. It is all too easy for another member of the family to damage something that has been carelessly left in the spare room, unless they are all warned in advance!

Fresh ingredients such as blue spruce and holly can be cut a little while before Christmas, but make sure they are kept cool and damp. Spruce lasts very well if it is kept outside with the occasional rain shower keeping it watered!

Spraying with gold and silver paint

Spraying plant material, such as dried seed heads, with gold or silver paint can add a truly festive touch to arrangements. There is no need to stop at seed heads or other dried flower bits and pieces when you are thinking of things to spray. Artificial fruit looks wonderful with a coat of gold or silver – and that includes the cheapest, most unrealistic plastic specimens that you never thought would be useful at all. Bargain baskets and containers that were a hideous colour look quite transformed when they are given a liberal coating of gold or copper metallic spray.

Spraying with gold paint can be a very messy affair. The easiest way is to lay the items you want to spray in a large shallow fruit box – the sort used by supermarkets. Spray one side and then once they have dried off a little, turn them over and spray the other. If you decide to take items out in the garden to spray, do take care to protect your hands and shoes. The spray drifts quite a lot and I have ruined a couple of pairs of shoes with the addition of gold paint speckles!

Making a wreath base

As a change from using shop-bought wreath bases, you might consider making your own wreaths or rings. You could try using lengths of vine or clematis stems, but for a really sturdy wreath I have found the trimmings from a willow tree the most successful material.

Make up a board with nails placed in the pattern as shown in the diagram opposite. The size of the board will depend on the finished size you want your wreath to be. Wind the stems in and out of the nails on the board making a good circular shape, easing each piece around the bends. Continue with plenty of stems until you have a sturdy wreath. As the depth of the ring increases, try to wind some of the stems through the others to make the wreath even more secure. Once you feel that you have a ring of the shape and size that you are happy with, slide some lengths of wire under the ring in various places around the circle and knot or twist the wire so that it holds the depth of the ring in shape.

Leave your wreath on the board to dry in a really warm place, such as a boiler cupboard, until the stems have dried completely. This will take varying amounts of time depending on the variety of plant material used and the dampness.

Making a bow

Bows always feature largely in Christmas arrangements and packaging and making a wired bow is much easier than it looks and is a technique worth trying. The diagram on the right illustrates where the wire should be tied. Take a length of ribbon and wrap it around like wrapping a shawl around your shoulders. Gather up the middle, making a bow and tails and bind the middle with some wire. The wire can be hidden by tying a small piece of ribbon around it or by covering it with flowers or other decorations.

The overall effect at Christmas should be one of abundance and cheerfulness. So don't worry too much about the technical side of your ideas or arrangements, just have fun, rope as many of the family in as possible and make the house look happy and friendly.

Collect an assortment of useful Christmas decorations for your arrangements (above). Make a wired bow in two ways, and a board for a wreath base (below).

Holly & ivy ring

Christmas wreaths and rings are a welcome sight on doors during the festive season. This idea is a little different with the golden eggs adding an extra sparkle. The basic ring can be made at home or you can use a commercial wreath as the base.

Holly & ivy ring

A traditional holly ring always looks pretty at Christmas and will last well against the elements. This combination of holly and ivy will last for quite some time and the feature of the twigs and gilded eggs adds an unusual twist to a very popular Christmas decoration. You will need plenty of sprays of holly and some very long trails of ivy. The eggs are emptied of their contents by piercing them at each end with a large needle, waggling the needle inside the egg to break down the white and yolk, and then blowing gently through one end. Once the eggs are empty, rinse the shells well with clean water to make sure all the contents have been removed and leave them to drain overnight.

INGREDIENTS

Glue gun and glue

Holly and ivy, see above

A twiggy ring, 30cm (12in) in diameter

A bundle of lichen-covered twigs

5 eggs

Gold spray paint

1 Glue the sprays of holly all around the ring, covering it evenly but not so densely that you cannot see the twiggy ring underneath. Both plain holly and variegated holly look very attractive, or you could use a mixture of them both. If you cannot find any holly, any other evergreen leaf could be substituted.

2 Wind the trails of ivy around the ring, bending them through the holly leaves and in and out of the centre of the ring. This will hold them securely enough, but you could glue each end to keep them in a particular position. Make sure the ivy has been given a long drink in water before you use it to help the ring stay fresh for as long as possible.

3 Glue the lichen-covered twigs in position. You could either add the twigs to a central point or offset to one side. Spray the eggs with gold paint and, when dry, glue them to the ring cluster. Place a few ivy leaves between each of the eggs to complete the festive decoration.

*F*ireside basket

This unusual display of bayleaf balls and fresh orange pomanders adds a lovely spicy aroma to the room. The arrangement also looks attractive when dried. Extra essential oil can be rubbed into the balls or pomanders to strengthen the scent.

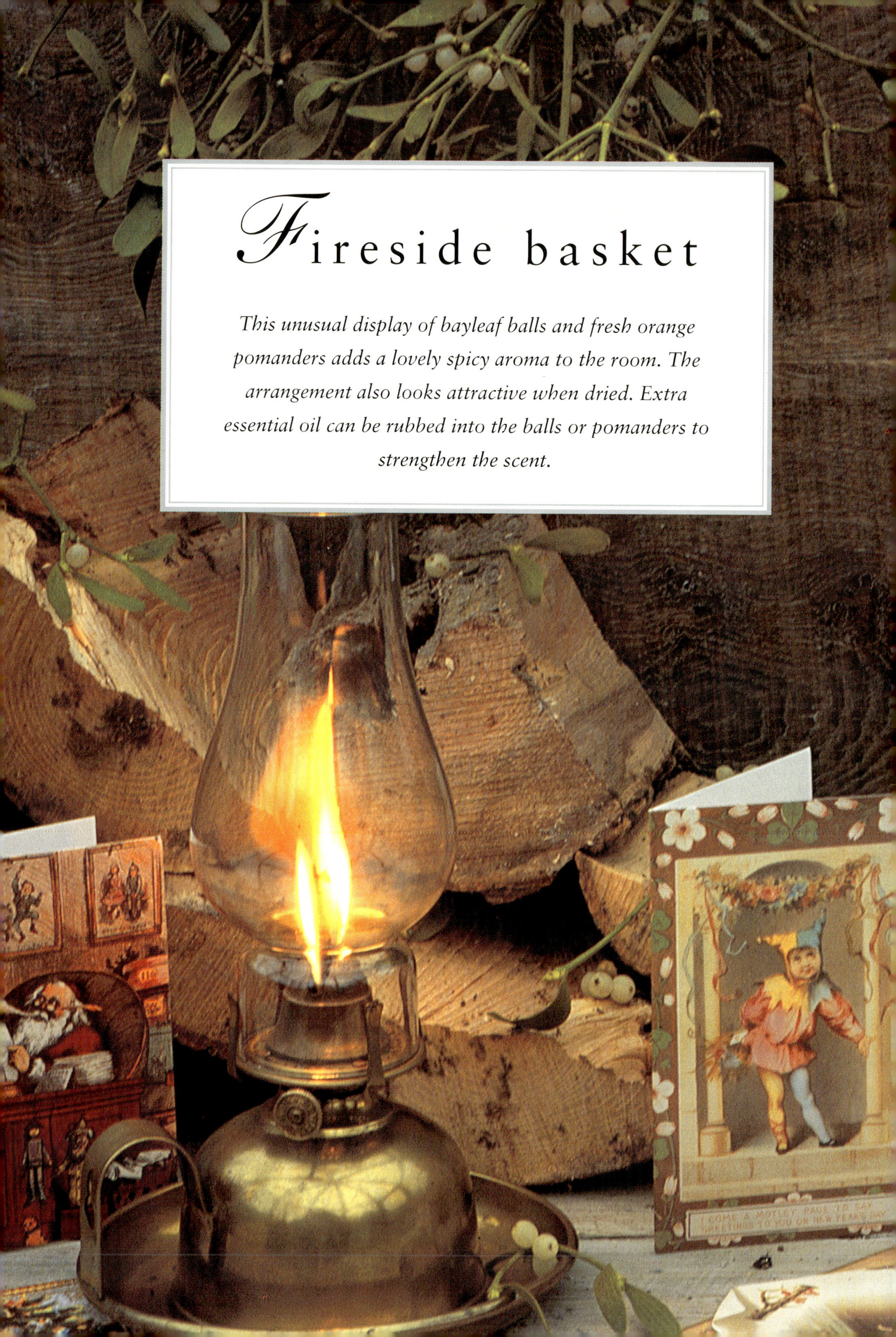

A
JOLLY

Fireside basket

Whether you have a real log fire or just a radiator, the warmth will help release the lovely natural perfume of these pomanders. The bay leaf balls are a little fiddly to make but last well and can be enjoyed for ages after they have been made. The orange pomanders can be enjoyed while they are drying and then arranged with fruit or as a decoration on their own. You will need a carrier bag full of fresh bay leaves – more than you might think you need, as many may well be damaged. The oranges also take quite a few cloves so it is best to buy a large pack or two from a herb supplier.

INGREDIENTS

Bay leaves

0·71mm (22 gauge) stub wires

Assorted foam balls, 7cm (2¾in) and 9cm (3¼in) in diameter

Glue

Cloves

Oranges

1 Sort out the bay leaves, discarding any that are damaged. Cut the wire into approximately 5cm (2in) lengths and bend each in half to make a 'hairpin'. Use these hairpins to pin the bay leaves onto each foam ball, so the entire surface is covered. The patterns you create are up to you – you can start from the top or middle, so long as the leaves overlap each time.

2 Continue to cover the ball until the foam is completely covered. Cover the joins between the rows of leaves by changing direction and pinning leaves across each other. The last few leaves can be glued on with a small amount of glue to prevent the wires being seen.

3 Tip the cloves onto a flat surface and press them one by one into the skin of each orange. You can make many different patterns with the cloves – swirls and stripes or patches, circles and sections. Place the decorated oranges into the basket together with the bay leaf balls.

Christmas pot pourri

Memories of Christmas are as much about scents and aromas as they are of tastes and sights. Spiced wine mulling in a large pan, aromatic cinnamon and tangy oranges – all of these bring Christmas memories to mind. A festive pot pourri can encapsulate many of these fragrances. For this recipe, you will need 10 slices of dried orange, 10 cinnamon sticks, 10 nutmegs, a cup each of dried marigold flowers, dried green leaves, rose hips and costus flowers or small pine cones. For the essential oils, I used a mixture of cinnamon, ginger and orange.

Ingredients

Dried ingredients, see above

Bowl and metal spoon

15g (½oz) orris root

10ml (2 tsp) essential oils of your choice

Large plastic bag and elastic band

Container for pot pourri

1 Assemble all the ingredients; put aside some orange slices and cinnamon sticks to decorate the top, but cut or break the rest. Place the ingredients in a bowl, add the orris root and essential oils, and mix well with a metal spoon.

2 Tip the mixture into a large plastic bag and shake well. Secure the top with an elastic band and store for at least a week for the perfumes to mature and blend. Then tip the pot pourri into the container and decorate the top with the reserved orange slices and cinnamon sticks.

For Auld Lang Syne
BE THINE
A very hearty
wishes of friendship

Floral parcels & Decorated crackers

Beautifully gift-wrapped parcels need not take hours to do or cost a lot. These floral giftwrap ideas are inexpensive and make a parcel really special. Christmas crackers, too, can look really sumptuous, simply by adding extra decorations.

Floral parcels

Half the fun of receiving presents at Christmas is looking at them under the tree or appreciating the wrapping before you open the gift. Anticipation is certainly one of the best parts of any celebration! You can add small floral decorations very simply to Christmas and other gifts and it makes the present mean so much more. There are three different ideas featured here, but you will be able to think of many more once you have the flowers to hand together with a box of 'useful bits' that you have saved throughout the year. To make these decorations you will need 2 yellow roses, 2 sprays of carnation, a spray of solidago and spruce for the first design, 2 or 3 long strands of ivy for the second and a small piece of spruce and a poinsettia flower for the third.

INGREDIENTS

Plain wrapping paper

Flowers and foliage, see above

Assorted wires

1m (40in) orange chiffon ribbon (1st design)

Glue

3m (120in) gold curling ribbon (2nd design)

50cm (20in) thick red and gold cord (3rd design)

1 After wrapping the parcel, make a small bunch with the roses and spray carnations and wrap some wire firmly around them. Add the spruce at the back of the bunch and some pieces of solidago at the sides. Wire the whole bunch firmly. Tie the ribbon around the bunch in a pretty bow. Fix the bunch to the parcel by either gluing it down or tying it in position with ribbons.

2 Wrap the parcel firmly, making the corners as neat as possible. Take a long strand of ivy and tie it around one side of the parcel. Using small strands, tie those in to the first strand so that they come out at various angles. Cut the curling ribbon in half, then tie it in the middle of the ivy and curl by winding it around a pencil.

3 Take a piece of spruce as a backing and wire it together with a poinsettia flower. Once the wire is bound tightly around the flower, trim away any excess and tie the piece of cord into a bow around the posy. Attach to the parcel either with a little glue or by tying it on.

Decorated crackers

Most crackers are bought for their decorative value rather than for their contents. Sometimes the jokes can liven up a slightly quiet gathering but mainly you can rely on getting a plastic monkey, while the youngest child gradually collects all the contents of the crackers. These crackers may well be no improvement inside but the *outside* can be helped along with a little flower know-how! You will need some commercial crackers, ones that have a basically nice paper base but they do not need to have very special decorations. You will also need some gold-sprayed pteris leaves or other dried foliage, cinnamon sticks, small pine cones and orange slices. Obviously the exact quantities will depend on how many crackers you are decorating.

INGREDIENTS

Crackers

Glue gun and glue

Flowers, foliage and spices, see above

Gold-wired cord

1 Remove all the commercial trimmings from the crackers, unless they blend with the decorations you are adding, like a twist of gold tinsel or something similar. Cut two or three small sprays of gold leaves and carefully glue them onto the centre of the cracker – use the minimum amount of glue so that it does not show.

2 Add two or three small pieces of cinnamon stick, gluing them on individually. Then place a few small pieces of orange between the cinnamon and the cracker, leaving room for the small pine cones.

3 Glue into position on the cracker two or three very small pine cones, or you could use costus flowers which look very similar. Loops of gold-wired cord can be added to give a stylish finishing touch. You could substitute many other ingredients for the decorations, depending upon the oddments you have to hand.

Yuletide table centre

All that does not glitter can easily be sprayed gold! Gold materials look very effective in Christmas displays and when combined together, as in this arrangement, they make an excellent focal point.

Yuletide table centre

The main thing to aim for with a Christmas table centre is a bright, cheerful centrepiece that will add some festive sparkle. The table is always crowded with glasses, plates and other dishes so it is best to keep any floral decorations fairly compact. Most of the items in this arrangement have been sprayed with gold paint which adds a special glow to the display. You will need some preserved oak foliage, 5 cardoon heads, 15 gilded pine cones, 15 natural walnuts and 9 golden mushrooms.

INGREDIENTS

Glue

A plastic 'frog' fixing

Oval cork base, 25cm × 20cm (10in × 8in)

Half-block of dried flower foam

Secateurs

0·71mm (22 gauge) stub wires

1m (40in) bronze chiffon ribbon

1 Glue the green frog fixing to the cork base. Once it is firmly fixed in position, impale the dried flower foam over the prongs of the frog. Cover the surface of the foam with sprays of gold oak leaves, keeping the oval shape of the arrangement. If the foam seems too high, cut some off the top with a knife to make the arrangement slightly shallower.

2 Push the gilded cardoon heads into the foam, making sure they are well spaced and not too close together. You will need very short stem lengths on these or they will stick out too far. Wire up the pine cones by winding some wire around the base petals of the cone, then bend down the wire and use as a stem. Add them to the arrangement.

3 Fill in the arrangement with wired walnuts (if you prefer you could glue them into position) and wired golden mushrooms. Take some chiffon ribbon and make a few loops, holding the base of the loops between your thumb and forefinger. Then wrap wire around the base to secure (see diagram on page 7), bend down the wire and press into the arrangement. Place loops on both sides of the table centre.

Traditional arrangement

This very traditional combination of poinsettias, holly and conifer sprigs always looks attractive at Christmas time. If you prefer a larger version, the poinsettias could be used in their pots rather than in foam.

Traditional arrangement

Although it is fun to make modern or unusual flower arrangements and come up with new ideas as often as possible, it is also very satisfying to make a really traditional Christmas arrangement. This particular idea reminds me of many Christmas cards I have received over the years and I like the comforting feel of an arrangement one almost seems to recognize! For the foliage, I used a mixture of blue spruce and variegated holly, about 9 stems of each, and 3 large red poinsettia flowers.

Ingredients

Half-block of fresh flower foam

Long shallow container, approximately 20cm x 10cm (8in x 4in)

3 candle holders

3 ivory candles

Foliage and flowers, see above

2m (80in) red ribbon, 2·5cm (1in) wide

2 x 0·71mm (22 gauge) stub wires

❧

1 Soak the foam in water until it is saturated. Wedge the foam into the container. Press the three candle holders into the foam wherever you would like them. Trim two of the candles so that they are all differing heights. Fix the candles into the candle holders. Place 15cm (6in) long sprays of both foliages into the arrangement.

2 Cut the poinsettia stems to a suitable length and add them to the arrangement. Keep them spaced well apart so that they do not look too crowded.

3 Make two wired bows from the ribbon (see page 7) and place them at either side of the arrangement so that the display can be viewed from both sides. Take care never to leave the arrangement unattended when the candles are alight as indoor bonfires are most unwelcome!

Topiary tree & Indoor wreath

The little apple and cranberry topiary tree is a pretty and unusual decoration which is simple enough for children to make. The wreath is an idea for indoors as the ingredients are definitely not weather-proof.

IX
KIND

Topiary tree

There are many differing designs for topiary trees but this one uses the unusual combination of fruit and foliage instead of flowers. As most of the ingredients are dried it should last well; the cranberries will shrivel or go mouldy after a while but they can be replaced by fresh or artificial berries. The clear glass container makes a change from a terracotta base but a basket or pot could be used if you prefer. You will need about 60 apple slices to include the base, a pack of fresh cranberries, a bunch of preserved beech leaves and some hazelnuts.

INGREDIENTS

Half-block of dried flower foam

Knife

Glass base, approximately 12·5cm (5in) square

Fruit and foliage, see above

Chunky cinnamon stick for trunk

7cm (2¾in) foam ball

Assorted wires

1 Cut the foam to size using a sharp knife, leaving room for the apple slices, and insert in the glass base. Camouflage the foam on all sides with apple slices; arrange them so that they overlap or touch. Choose a fairly chunky cinnamon stick, about 20cm–22·5cm (8–9in) high, and push this into the centre of the foam.

2 Cover the surface of the base with hazelnuts. Impale the foam ball onto the cinnamon stick and cover it with sprays of beech leaves. If necessary, wire together bunches of individual leaves if the ends of the sprays are not bushy enough.

3 Wire the apple slices individually and then push the ends of the wires into the foam tree, placing them evenly among the beech foliage throughout the arrangement. Cut shorter pieces of wire for the cranberries. Insert one end of the wire into a cranberry and the other into the tree. As before, dot the cranberries among the foliage. Finally, sprinkle some cranberries among the hazelnuts at the base of the tree for extra colour.

Indoor wreath

Many people have a welcome ring or wreath on their front door at Christmas. This design is intended for indoor use as the dried flowers simply would not survive the elements. It could be hung in a hallway or on an inside door, depending where you have a space. Again, some of the ingredients have been sprayed gold and it is interesting to see how well they blend with the naturally coloured material. To make this indoor ring, you will need a bunch of eucalyptus and some blue spruce, a bunch of gilded poppy heads, a bunch of preserved and gilded foliage, such as box, and 20 pink roses.

Ingredients

Twiggy ring, approximately 30cm (12in) in diameter

Flowers and foliage, see above

Glue

1·5m (62in) ribbon, 7·5cm (3in) wide

Assorted wires

1 Cover the ring roughly with blue spruce and glue this firmly into position. Add pieces of eucalyptus to thicken up the background. Make a wired bow with the ribbon (see page 7).

2 Glue the bow onto the ring. Add in the gilded poppy heads and box, curving the ingredients around with the circle, covering about three-quarters of the ring.

3 Finally, glue the pink roses onto the ring. These are best placed in groups of three or five in between the gilded poppy heads and box foliage. This design would look equally pretty if the poppies and box were sprayed silver instead of gold.

Festive garland

This luxurious spruce garland can be made at home for a fairly reasonable price. Although it takes a little while to make, it does look wonderful!

Festive garland

Spruce is a very attractive foliage to use at Christmas and it has the added benefit of lasting very well. It will not drop like some conifers or traditional Christmas trees – it just dries and turns a greyish-blue instead of bluish-green. It will take a reasonable amount of spruce to make this garland – about 50 sprigs to make a 1·5m (62in) garland. It is hard to be accurate about the exact number required as it depends on the quality and density of the spruce.

Ingredients

1·8m (6ft) twisted paper ribbon

A reel of binding wire

Spruce, see above

1m (40in) wired ribbon, 6·5–7·5cm (2½–3in) wide

Glue gun and glue

Assortment of gilded poppy heads, cinnamon sticks, walnuts, dried orange slices, pine cones and gilded lotus heads

Spray gloss varnish

1 Bend some twisted paper ribbon over at both ends and wire firmly to make a hanging loop at each end of the garland. Twist a small piece of wire in the middle of the garland to remind you where the centre is. Starting from one end, place a 10cm (4in) sprig of spruce over the loop and wire it onto the twisted paper ribbon. Keep adding pieces of spruce, fanning them out so the garland is fairly flat rather than chunky. Continue until you reach the marked middle point, then turn the garland around and start from the other end.

2 Once you reach the middle point, cut some smaller pieces of spruce and place them at right angles to the ribbon and wire on to fill the gap in the middle. Use plenty of wire as it is more important to have a secure garland than a neat one, and the glued-on bits and pieces will cover most wiring problems. Make a large bow with the ribbon (see page 7) and attach it to the centre with glue.

3 Glue on all the bits and pieces you have chosen. Place larger pieces like the lotus heads on first, and the cinnamon sticks through the centre of the bow and at intervals along the garland. Once you have all your nuts and cones glued on, give them a coat of spray gloss varnish as this will highlight their colours and give the nuts in particular an attractive shine.

Christmas tree decorations

We all have many beloved family tree decorations that have been around for years – some may be very special as they were made by the children. Although there are always new, different and exciting commercially-made tree decorations on the market, these homemade decorations are a little bit different, and something perhaps the whole family could make together a couple of days before Christmas. You will need some sea lavender, solidago, spray carnation, 1 or 2 roses, and any fresh foliage from the garden.

Ingredients

Gold wire or cord

Terracotta flowerpots, 2·5cm (1in) in diameter

Fresh flower foam

Flowers and foliage, see above

Silver rose wire

Cellophane

Curling ribbon

1 Thread two loops of gold wire or cord through the hole in the base of each flowerpot to act as hanging loops for the decoration. Fill the pots with foam that has first been soaked in water and make a miniature arrangement with foliage, sea lavender and solidago. The ingredients can be altered according to what is available.

2 Make some small bunches of carnations or roses with sea lavender or solidago, and wire into a small posy. Add a wrapping of cellophane and secure with curling ribbon, adding a loop to hang them from the tree.

Dramatic table centre

In complete contrast to the traditional red, green and gold Christmas colour schemes, these orchids look absolutely stunning in this unusual table centre arrangement. The twigs can be found lying on the ground in woods and fields.

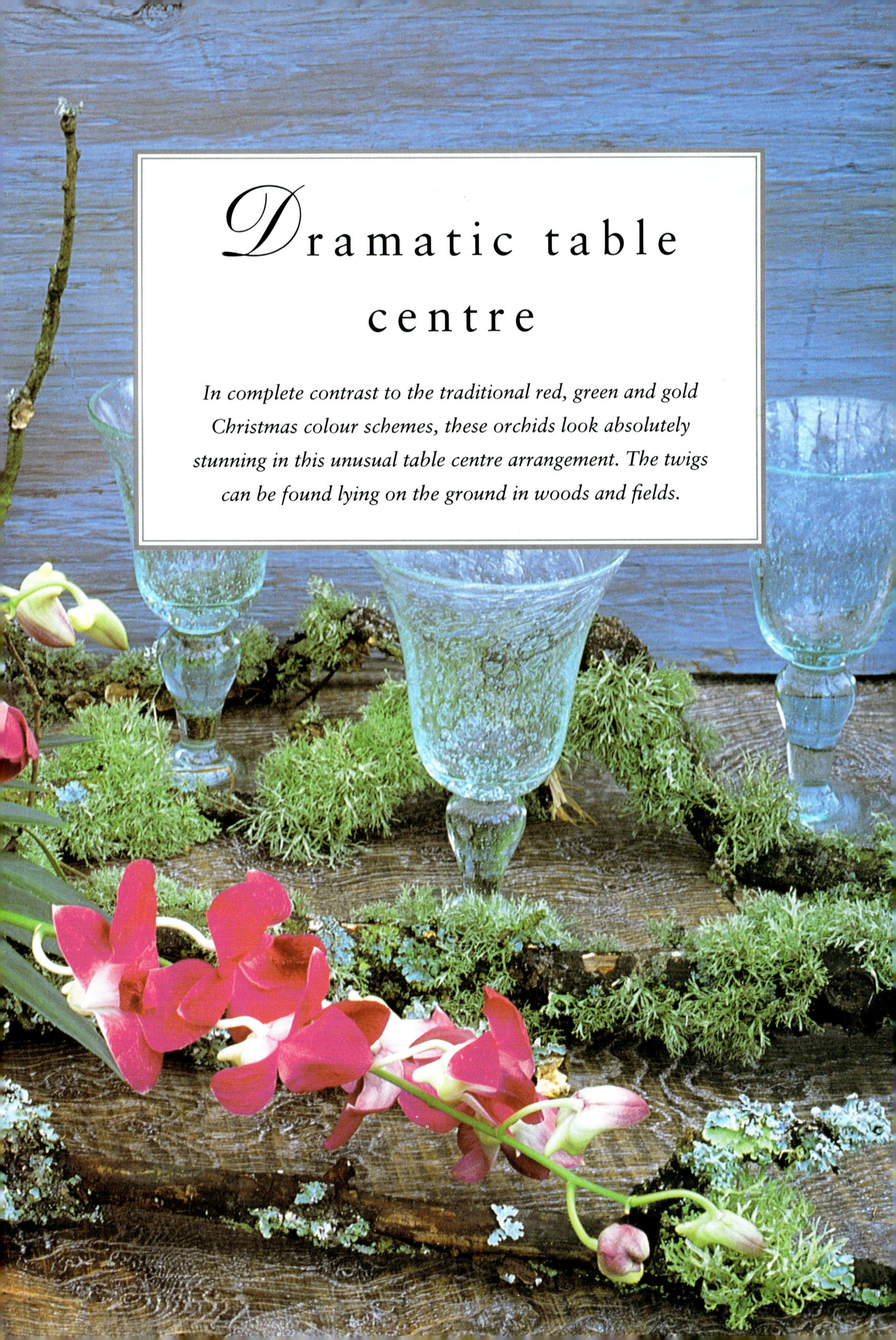

Dramatic table centre

The colouring of this arrangement is by no means traditional, but it is certainly dramatic and would be ideal for a New Year's Eve celebration. Although it looks very impressive, the actual arrangement is very easy and will only take a short while to make. The list of ingredients is quite short – you will need 2 small bunches of mauve Singapore orchids, a selection of lichen-covered twigs and some hellebore leaves.

INGREDIENTS

Fresh flower foam

Small baking tin or other container, approximately 15cm (6in) wide

Flowers and twigs, see above

1 Soak the foam in water until it is saturated and fill the tin with it. Lay the twigs in position if you are arranging the flowers in situ; if not, then lay them on the table and add the container with flowers after Step 3.

2 Insert the hellebore leaves into the foam, making the arrangement long or round, depending on the shape of your table. If you cannot obtain hellebore leaves, any other large, dark leaves would be suitable.

3 Next add the sprays of Singapore orchids. If they are very long, break them in half and add both halves towards the centre of the display. These orchids are also readily available in other colours so the arrangement could be varied simply by the substitution of cream or peach orchids, for instance.

Suppliers

For details of dried flowers, craft ingredients, and pressed flower ingredients and oils by mail order; also two-day craft courses on flowers and other crafts:

Joanna Sheen Limited
PO Box 52
Newton Abbot
Devon TQ12 4QH

For pot pourri ingredients and oils, and details of shops throughout the country:

Culpeper Limited
21 Bruton Street
London W1X 7DA

For details of dried flowers by mail order and farm shop:

Caroline Alexander
The Hop Shop
Castle Farm
Shoreham
Sevenoaks
Kent TN14 7UB

ACKNOWLEDGEMENTS

Jacqui Hurst and Merehurst would like to thank the following for lending props for the photographs in this book: David Robertson and Peta Weston for creating the special backgrounds; and Margaret Check, Hazel Hurst and Nesta MacDonald who kindly lent their lace, table linen and crockery.

Conception et réalisation :
Françoise Detay-Lanzmann / Nicole Hibert
Direction artistique : Jean Retailleau
Couverture : Agence Média

Dépôt légal : Avril 1993
ISSN : 1242-532
Loi n° 49-956 du 16 juillet 1949
sur les publications destinées à la jeunesse

ISBN : 2-7404-0265-1

Françoise Detay-Lanzmann / Nicole Hibert

LES ANIMAUX DE L'EAU

Illustrations de Héliadore

MANGO

DE LA SURFACE AUX PROFONDEURS

Belouga

Méduse

Sèche

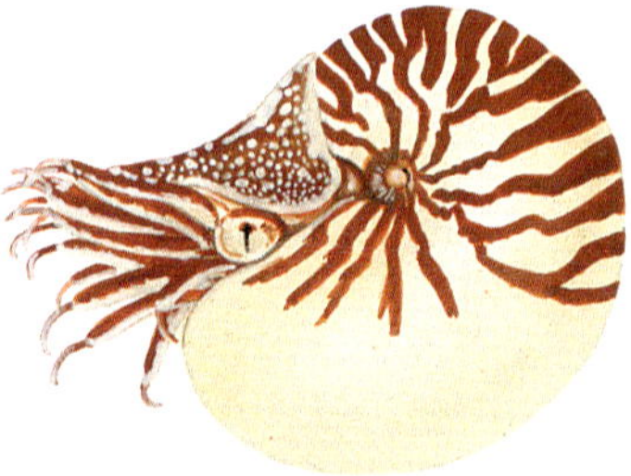

Nautile

Etoile de mer

Tortue marine

Poisson-crapaud

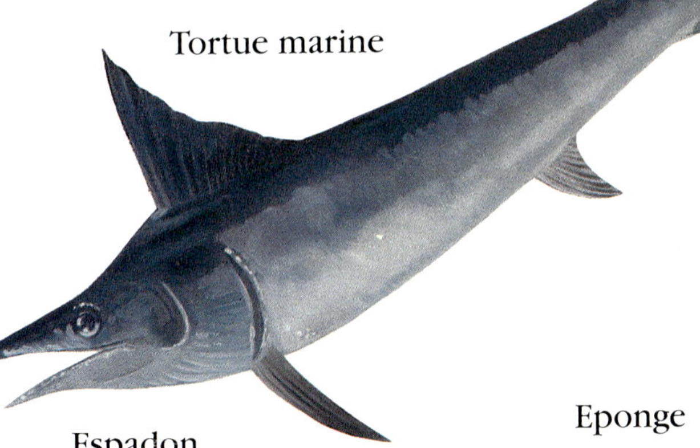

Espadon

Eponge

LA BALEINE

La baleine est le plus grand *mammifère* marin. Elle vit dans les eaux glacées des pôles, qu'elle quitte à la fin de l'été pour rejoindre les mers chaudes, où elle donne naissance à son petit.

Certaines baleines n'ont pas de dents, mais des fanons qui servent à filtrer l'eau de mer pour retenir le *plancton* et de minuscules crevettes : le krill.

Les autres baleines, comme le cachalot et la baleine blanche, ont des dents redoutables. Elles mangent des poissons et des calmars.

Le baleineau, qui pèse 1 000 kilos à la naissance, boit 500 litres de lait par jour ! Sa mère risque parfois sa vie pour le protéger des requins ou des chasseurs.

Comme elle a besoin de respirer, la baleine possède un évent, une ouverture qui lui permet d'inspirer de l'air quand elle monte à la surface. L'évent se ferme quand la baleine plonge.

LE DAUPHIN

Le dauphin est aussi un *mammifère.*
Très intelligent, il est l'ami des marins
et il lui arrive même de sauver des nageurs imprudents.
Le dauphin possède un vrai langage que les hommes
espèrent comprendre un jour.

Le dauphin a une *ouïe* très fine
qui lui permet de *localiser* des objets
éloignés et de se diriger dans la nuit.

Les dauphins ont la peau très sensible.
Ils aiment se caresser et se frotter
les uns contre les autres.

Les dauphins voyagent en groupe.
Les femelles et les petits sont au centre,
protégés par les mâles et les adolescents.

L'orque est un grand dauphin qui
attaque parfois d'autres *mammifères*
marins comme la baleine.

LE PHOQUE

Le phoque est un *mammifère* marin qui vit surtout dans les eaux glaciales du pôle Nord. Son corps est parfaitement adapté à la vie marine, car ses quatre pattes se transforment en nageoires. Il se nourrit de poissons, de mollusques et de crustacés.

Le bébé phoque a une épaisse fourrure blanche qui le protège du froid. C'est pourquoi on l'appelle " blanchon ".

Avant de plonger, le phoque vide ses poumons. Il peut rester cinq minutes sous l'eau sans respirer.

Le morse, l'énorme cousin du phoque, a des défenses de 60 centimètres recourbées vers le bas.

L'éléphant de mer pèse plus de 3 000 kilos et mesure 6 mètres, la taille d'une grande voiture.

LE REQUIN

Le requin a toujours fasciné et effrayé les hommes. Pourtant, la plupart de ces grands poissons ne sont pas dangereux. Certains attaquent parfois, quand ils se sentent menacés ou quand on envahit leur territoire.

Le requin a de puissantes mâchoires et plusieurs rangées de dents tranchantes comme des rasoirs.

Le requin a une *ouïe* et un odorat très développés. Son regard perçant lui permet de voir la nuit.

Le requin-marteau est appelé ainsi à cause de la forme de sa tête. Ses yeux et ses narines sont situés aux extrémités du marteau.

Le requin-baleine est le plus grand des poissons. Malgré ses 310 rangées de dents, il n'est pas méchant.

LA RAIE

La raie nage très vite en faisant onduler ses grandes nageoires que l'on appelle des ailes. Grâce à sa forme aplatie, elle peut se cacher dans le sable pour attendre sa proie. Elle la capture en se couchant dessus, car sa bouche munie de dents se trouve sur son ventre.

La raie cornue est un " poisson volant " qui vit près de la surface et fait des bonds hors de l'eau. Elle peut peser près de 2000 kilos.

La torpille paralyse sa victime en lui envoyant une décharge électrique. Comme une pile, elle a besoin de temps pour se recharger. Ensuite, elle attaque de nouveau.

LA PIEUVRE

Timide et prudente, la pieuvre vit dans de petites grottes naturelles. Elle n'en sort que pour chasser en se dissimulant parmi les roches et les coraux. La pieuvre possède une sorte de bec en corne, assez fort pour broyer la carapace d'un crabe

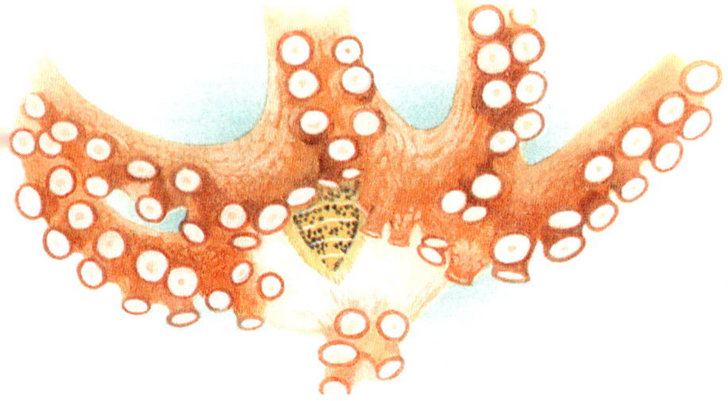

La pieuvre rampe en s'aidant de ses huit tentacules couverts de ventouses.

Suprise par un ennemi, elle lâche aussitôt un nuage d'encre noire, pour se donner le temps de disparaître

LA TORTUE

La tortue-luth est la plus grosse tortue marine. Elle peut nager très vite, grâce à ses pattes de devant qui forment d'immenses nageoires. Armée d'un bec redoutable, elle se nourrit de poissons, de méduses et d'autres animaux au corps mou.

Bien qu'elle passe sa vie dans l'eau, la tortue-luth vient pondre ses œufs à terre. Elle les enfouit dans un énorme trou, puis elle regagne la mer.

Les petites tortues naissent au bout de 7 ou 8 semaines. Elles se précipitent vers la mer pour échapper aux oiseaux qui veulent les manger.

L'HIPPOCAMPE

L'hippocampe, ou cheval de mer, vit parmi les plantes aquatiques auxquelles il s'accroche avec sa queue. Son corps est recouvert d'une véritable cuirasse osseuse. Pour se nourrir, il fonctionne comme un aspirateur et avale les petits crustacés et les petits poissons.

L'hippocampe est le seul poisson qui nage en position verticale, grâce à la nageoire en forme d'éventail qu'il porte sur son dos.

Le mâle a une poche sur le ventre, comme le kangourou. La femelle y pond ses œufs et les jeunes y vivent jusqu'à la fin de leur croissance.

LE SAUMON

Né dans le lit d'un torrent, le petit saumon vit en *eau douce* pendant quelques années, avant de rejoindre l'océan. Il retournera ensuite vers sa rivière natale, après un long et dangereux voyage, pour pondre ses œufs et mourir.

Les œufs éclosent au bout d'une cinquantaine de jours. Beaucoup de petits saumons sont mangés par les poissons ou les oiseaux.

Les barrages sur les rivières empêchent parfois les saumons de remonter le courant. On installe des sortes d'escaliers ou d'ascenseurs pour les aider.

L'ANGUILLE

Elle ressemble à un serpent, pourtant l'anguille est un poisson. Grande voyageuse, elle passe ne partie de sa vie dans les eaux douces des fleuves et des marais, puis traverse l'océan Atlantique pour aller pondre ses œufs.

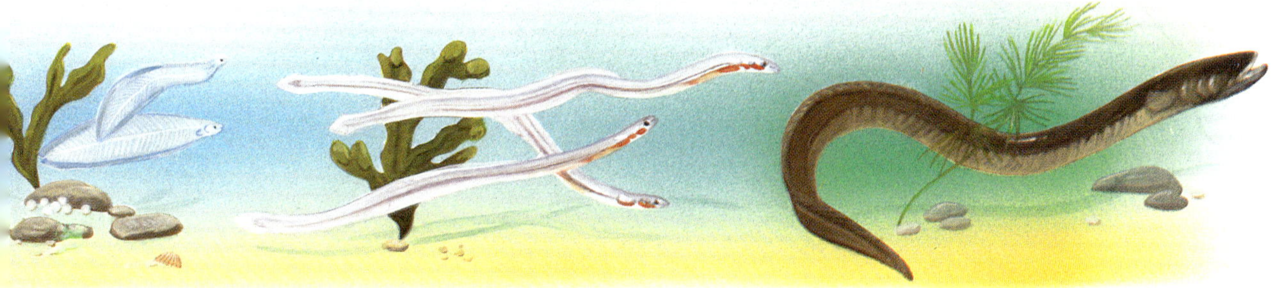

L'œuf donne naissance à une larve qui traverse l'Atlantique en sens inverse. Près des côtes, elle se transforme en civelle transparente et remonte les fleuves.

Elle prend alors sa forme d'anguille. Adulte, elle se modifie à nouveau pour s'adapter aux profondeurs de l'océan : sa peau s'épaissit et ses yeux grossissent.

LES CRUSTACÉS

La plupart des crustacés vivent dans les mers, où ils habitent les zones rocheuses. Le corps des crustacés est recouvert d'une *carapace* articulée. Ils ont des antennes qui sont sensibles aux vibrations, aux goûts et aux odeurs. Certains crustacés possèdent des pinces.

Le crabe marche sur le côté.
Ses pinces aux muscles puissants sont des armes redoutables.

La crevette, comme le caméléon, est capable de changer de couleur pour qu'on ne la remarque plus dans le paysage.

Le bernard-l'ermite a une *carapace* très mince. Il se réfugie donc dans une coquille sur laquelle s'accroche souvent une anémone de mer.

Pour grandir, le homard doit muer. Il sort de sa vieille carapace à reculons puis il la mange, en attendant que la nouvelle durcisse.

LES CORAUX

Les coraux multicolores, qui ressemblent à des plantes, sont en réalité de petits animaux vivant en *colonies.* La plupart d'entre eux habitent les eaux chaudes où ils construisent des barrières rocheuses : les récifs.

Les coraux ont besoin pour se développer de lumière et d'une eau pure qui reste à la même température.

Certains coraux mangent des vers, des larves et des œufs. Ils les attrapent avec leurs tentacules *venimeux* qui paralysent leur proie.

Les anémones de mer, cousines des coraux, vivent sur les rochers ou dans le sable au fond des mers. On les appelle les “ animaux-fleurs ”.

La bouche de l’anémone se trouve au milieu de ses tentacules, comme le cœur d’une fleur. Elle se nourrit de poissons et de crevettes.

LES COQUILLAGES

Les coquillages vivent dans toutes les mers
et se reproduisent en pondant des œufs.
Certains ont une seule coquille, souvent en forme
de spirale, comme le bigorneau ou le nautile.
D'autres ont une coquille formée de deux parties,
comme l'huître ou la moule.

La coquille sert à protéger le corps mou de l'animal qui vit à l'intérieur et qu'on appelle “ mollusque ”.

La porcelaine tigrée repousse ses ennemis en leur lançant un nuage d'acide qui leur brûle les yeux.

La coquille Saint-Jacques nage très bien, en avant comme en arrière, ce qui lui permet de fuir ses pires ennemis : l'étoile de mer et le poulpe.

La coquille Saint-Jacques a de nombreux yeux bleus qui détectent les mouvements. Quand l'un est détruit, un autre le remplace.

LES POISSONS BIZARRES

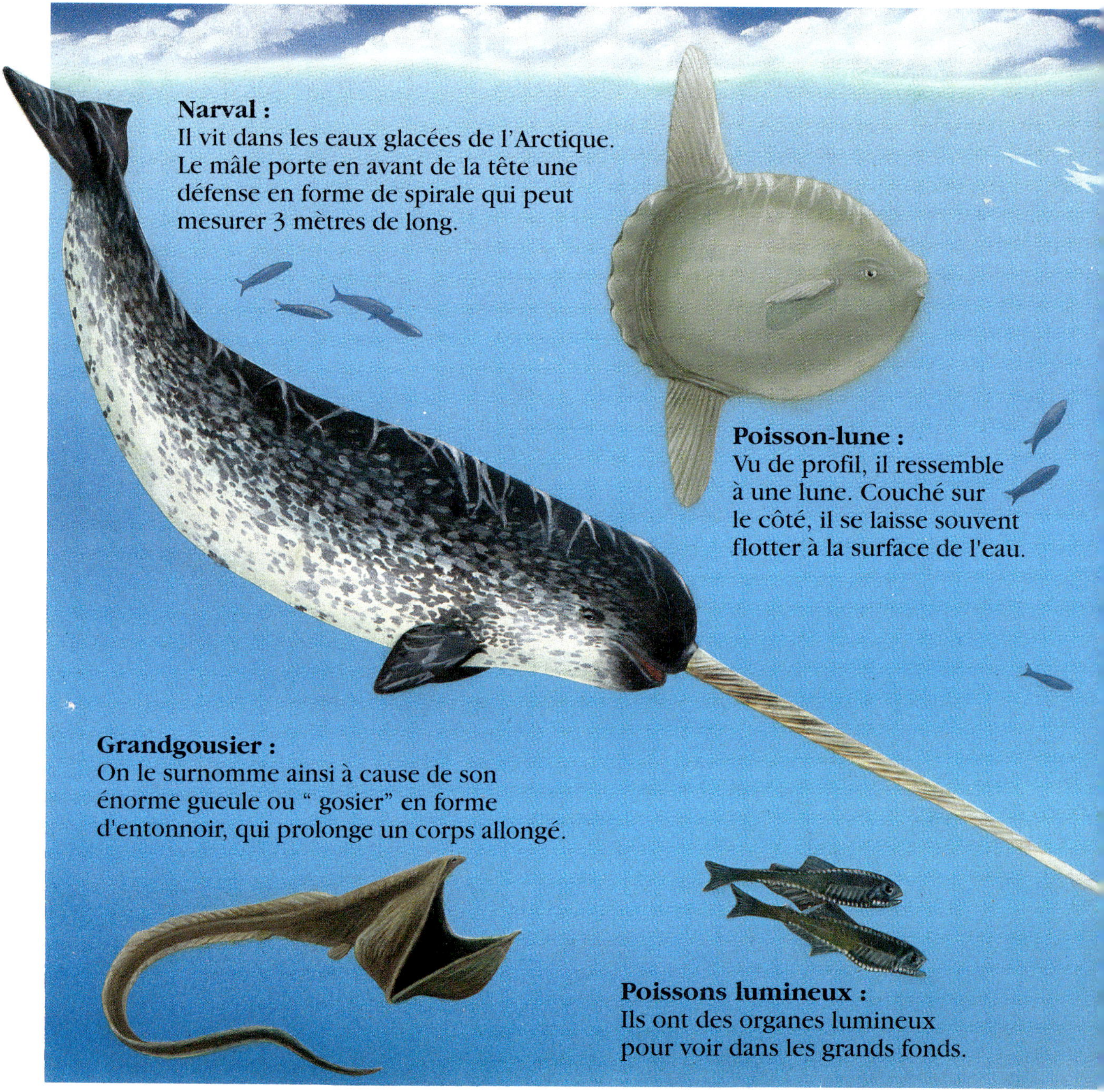

Narval :
Il vit dans les eaux glacées de l'Arctique. Le mâle porte en avant de la tête une défense en forme de spirale qui peut mesurer 3 mètres de long.

Poisson-lune :
Vu de profil, il ressemble à une lune. Couché sur le côté, il se laisse souvent flotter à la surface de l'eau.

Grandgousier :
On le surnomme ainsi à cause de son énorme gueule ou “ gosier” en forme d'entonnoir, qui prolonge un corps allongé.

Poissons lumineux :
Ils ont des organes lumineux pour voir dans les grands fonds.

Poissons volants :
Pour échapper à leurs ennemis, ils jaillissent hors de l'eau et déploient leurs grandes nageoires pour glisser à la surface.

Poisson-scie :
Il se sert de son museau en forme de scie pour racler le fond de la mer et y trouver de petits animaux.

Poisson-pierre :
Il est le plus venimeux des poissons.
La piqûre de ses épines peut être mortelles.

Poisson-porc-épic :
Surnommé le hérisson de mer, il se gonfle d'air pour devenir une boule hérissée de piquants.

The Power of Metaphor: Examining Its Influence on Social Life

Just The facts101 Textbook Key Facts

by Cram101
Texbook NOT Included

Table of Contents

Just The Facts101

Exam Prep for

The Power of Metaphor: Examining Its Influence on Social Life

Just The Facts101 Exam Prep

eAIN 438961

Foundations of Sociology

Sociology is the study of social behavior or society, including its origins, development, organization, networks, and institutions. It is a social science that uses various methods of empirical investigation and critical analysis to develop a body of knowledge about social order, disorder, and change.

:: Political science ::

_________ is the principled guide to action taken by the administrative executive branches of the state with regard to a class of issues, in a manner consistent with law and institutional customs.

Exam Probability: **High**

1. *Answer choices:*

(see index for correct answer)

- a. NOMINATE
- b. Politics of memory
- c. Ethnosymbolism
- d. Political ReviewNet

Guidance: level 1

:: Social ethics ::

_________ is the right to exercise power, which can be formalized by a state and exercised by way of judges, appointed executives of government, or the ecclesiastical or priestly appointed representatives of a God or other deities.

Exam Probability: **Low**

2. *Answer choices:*

(see index for correct answer)

- a. Evolution of corporate social responsibility in India
- b. Dominator culture
- c. Ethicurean
- d. Enlightened self-interest

Guidance: level 1

:: Bullying ::

_________ is an individual's subjective evaluation of their own worth. _________ encompasses beliefs about oneself as well as emotional states, such as triumph, despair, pride, and shame. Smith and Mackie defined it by saying "The self-concept is what we think about the self; _________ , is the positive or negative evaluations of the self, as in how we feel about it."

Exam Probability: **High**

3. *Answer choices:*

(see index for correct answer)

- a. Slut-shaming
- b. Destabilisation
- c. Dehumanization
- d. Ragging in India

Guidance: level 1

:: Liberal democracies ::

_________ , also known as the Republic of _________ , is a country in South Asia. It is the seventh largest country by area and with more than 1.3 billion people, it is the second most populous country as well as the most populous democracy in the world. Bounded by the _________ n Ocean on the south, the Arabian Sea on the southwest, and the Bay of Bengal on the southeast, it shares land borders with Pakistan to the west; China, Nepal, and Bhutan to the northeast; and Bangladesh and Myanmar to the east. In the _________ n Ocean, _________ is in the vicinity of Sri Lanka and the Maldives, while its Andaman and Nicobar Islands share a maritime border with Thailand and Indonesia.

Exam Probability: **High**

4. *Answer choices:*

(see index for correct answer)

- a. India
- b. Marshall Islands
- c. Grenada
- d. Luxembourg

Guidance: level 1

:: Types of communities ::

A _________ is a small or large social unit that has something in common, such as norms, religion, values, or identity. Communities often share a sense of place that is situated in a given geographical area or in virtual space through communication platforms. Durable relations that extend beyond immediate genealogical ties also define a sense of _________ . People tend to define those social ties as important to their identity, practice, and roles in social institutions . Although communities are usually small relative to personal social ties , " _________ " may also refer to large group affiliations , such as national communities, international communities, and virtual communities.

Exam Probability: **Low**

5. *Answer choices:*

(see index for correct answer)

- a. rural communities
- b. Community

Guidance: level 1

:: Shunning ::

_________ is a state of complete or near-complete lack of contact between an individual and society. It differs from loneliness, which reflects temporary and involuntary lack of contact with other humans in the world. _________ can be an issue for individuals of any age, though symptoms may differ by age group.

Exam Probability: **Medium**

6. *Answer choices:*

(see index for correct answer)

- a. Disownment
- b. Outcast
- c. Cold shoulder
- d. Petalism

Guidance: level 1

:: Motherhood ::

A __________ is the female parent of a child. __________ s are women who inhabit or perform the role of bearing some relation to their children, who may or may not be their biological offspring. Thus, dependent on the context, women can be considered __________ s by virtue of having given birth, by raising their child, supplying their ovum for fertilisation, or some combination thereof. Such conditions provide a way of delineating the concept of __________ hood, or the state of being a __________ . Women who meet the third and first categories usually fall under the terms `birth __________ ` or `biological __________ `, regardless of whether the individual in question goes on to parent their child. Accordingly, a woman who meets only the second condition may be considered an adoptive __________ , and those who meet only the first or only the third a surrogacy __________ .

Exam Probability: **Medium**

7. *Answer choices:*

(see index for correct answer)

- a. Mompreneur
- b. Museum of Motherhood
- c. Amah
- d. Mother

Guidance: level 1

:: Demography ::

_________ refers to the population shift from rural areas to urban areas, the gradual increase in the proportion of people living in urban areas, and the ways in which each society adapts to this change. It is predominantly the process by which towns and cities are formed and become larger as more people begin living and working in central areas. Although the two concepts are sometimes used interchangeably, _________ should be distinguished from urban growth: _________ is "the proportion of the total national population living in areas classed as urban", while urban growth refers to "the absolute number of people living in areas classed as urban". The United Nations projected that half of the world`s population would live in urban areas at the end of 2008. It is predicted that by 2050 about 64% of the developing world and 86% of the developed world will be urbanized. That is equivalent to approximately 3 billion urbanites by 2050, much of which will occur in Africa and Asia. Notably, the United Nations has also recently projected that nearly all global population growth from 2017 to 2030 will be by cities, about 1.1 billion new urbanites over the next 13 years.

Exam Probability: **Medium**

8. *Answer choices:*

(see index for correct answer)

- a. Sexually active life expectancy
- b. Child Development Index
- c. Urbanization
- d. Demographic trap

Guidance: level 1

:: Role status ::

A __________ is a set of connected behaviors, rights, obligations, beliefs, and norms as conceptualized by people in a social situation. It is an expected or free or continuously changing behaviour and may have a given individual social status or social position. It is vital to both functionalist and interactionist understandings of society.Social __________ posits the following about social behaviour.

Exam Probability: **Low**

9. *Answer choices:*

(see index for correct answer)

- a. Role suction
- b. Role
- c. gender roles
- d. Role set

Guidance: level 1

:: Mental processes ::

_________ is the means to see, hear, or become aware of something or someone through our fundamental senses. The term _________ derives from the Latin word perceptio, and is the organization, identification, and interpretation of sensory information in order to represent and understand the presented information, or the environment.

Exam Probability: **Low**

10. *Answer choices:*

(see index for correct answer)

- a. Social perception
- b. Divided visual field paradigm
- c. Perception
- d. Homeostatic emotion

Guidance: level 1

:: Communism-based civil wars ::

The __________ , also known as the Second Indochina War, and in Vietnam as the Resistance War Against America or simply the American War, was an undeclared war in Vietnam, Laos, and Cambodia from 1 November 1955 to the fall of Saigon on 30 April 1975. It was the second of the Indochina Wars and was officially fought between North Vietnam and South Vietnam. North Vietnam was supported by the Soviet Union, China, and other communist allies; South Vietnam was supported by the United States, South Korea, the Philippines, Australia, Thailand and other anti-communist allies. The war is considered a Cold War-era proxy war from some US perspectives. It lasted some 19 years with direct U.S. involvement ending in 1973 following the Paris Peace Accords, and included the Laotian Civil War and the Cambodian Civil War, resulting in all three countries becoming communist states in 1975.

Exam Probability: **High**

11. *Answer choices:*

(see index for correct answer)

- a. Vietnam War
- b. Cambodian Civil War
- c. Insurgency of the Communist Party of the Philippines

Guidance: level 1

:: Marriage ::

A __________ is a caregiver of the offspring in their own species. In humans, a __________ is the caretaker of a child . A biological __________ is a person whose gamete resulted in a child, a male through the sperm, and a female through the ovum. Biological __________ s are first-degree relatives and have 50% genetic meet. A female can also become a __________ through surrogacy. Some __________ s may be adoptive __________ s, who nurture and raise an offspring, but are not biologically related to the child. Orphans without adoptive __________ s can be raised by their grand __________ s or other family members.

Exam Probability: **Low**

12. *Answer choices:*

(see index for correct answer)

- a. The History of Human Marriage
- b. Spouse
- c. Aliment
- d. Parent

Guidance: level 1

:: Law enforcement ::

__________ is any system by which some members of society act in an organized manner to enforce the law by discovering, deterring, rehabilitating, or punishing people who violate the rules and norms governing that society. Although the term may encompass entities such as courts and prisons, it is most frequently applied to those who directly engage in patrols or surveillance to dissuade and discover criminal activity, and those who investigate crimes and apprehend offenders, a task typically carried out by the police or another __________ organization. Furthermore, although __________ may be most concerned with the prevention and punishment of crimes, organizations exist to discourage a wide variety of non-criminal violations of rules and norms, effected through the imposition of less severe consequences.

Exam Probability: **Low**

13. *Answer choices:*

(see index for correct answer)

- a. Witness protection
- b. Law enforcement
- c. Body worn video
- d. Police code

Guidance: level 1

:: Organizations ::

An __________ or organisation is an entity comprising multiple people, such as an institution or an association, that has a particular purpose.

Exam Probability: **High**

14. *Answer choices:*

(see index for correct answer)

- a. Organization
- b. International Surf Lifesaving Association
- c. World Association for Disaster and Emergency Medicine
- d. World Cocoa Foundation

Guidance: level 1

:: Penology ::

A _________ , also known as a correctional facility, jail, gaol , penitentiary , detention center , remand center, or internment facility , is a facility in which inmates are forcibly confined and denied a variety of freedoms under the authority of the state. _________ s are most commonly used within a criminal justice system: people charged with crimes may be im _________ ed until their trial; those pleading or being found guilty of crimes at trial may be sentenced to a specified period of im _________ ment. In simplest terms, a _________ can also be described as a building in which people are legally held as a punishment for a crime they have committed.

Exam Probability: **Low**

15. *Answer choices:*

(see index for correct answer)

- a. Changing Lives Through Literature
- b. Holt v. Sarver
- c. Prison Fellowship International
- d. Incapacitation

Guidance: level 1

:: Sociological terminology ::

__________ is an awareness of shared interests, objectives, standards, and sympathies cause creating a psychological sense of unity of groups or classes. It refers to the ties in a society that bind people together as one. The term is generally employed in sociology, and the other social sciences as well as in philosophy, bioethics, and Catholic social teaching. It is also a significant concept in Catholic social teaching therefore it is a core concept in Christian democratic political ideology.

Exam Probability: **Medium**

16. *Answer choices:*

(see index for correct answer)

- a. Reality distortion field
- b. Theoretical sampling
- c. Solidarity
- d. Trust metric

Guidance: level 1

:: Social inequality ::

A _________ is the application of different sets of principles for situations that are, in principle, the same, and is often used to describe freedom that is given to one party over another. A double-standard arises when two or more people, circumstances, or events are treated differently even though they should be treated the same way. Margaret Eichler, author of The _________ : A Feminist Critique of Feminist Social Science, explains that a _________ "implies that two things which are the same are measured by different standards".

Exam Probability: **Low**

17. *Answer choices:*

(see index for correct answer)

- a. Double standard
- b. Structural vulnerability
- c. Social inequality
- d. Distinction

Guidance: level 1

:: Single-party states ::

The __________ , officially the Union of Soviet Socialist Republics , was a socialist state in Eurasia that existed from 1922 to 1991. Nominally a union of multiple national Soviet republics, its government and economy were highly centralized. The country was a one-party state, governed by the Communist Party with Moscow as its capital in its largest republic, the Russian Soviet Federative Socialist Republic . Other major urban centres were Leningrad, Kiev, Minsk, Alma-Ata, and Novosibirsk. It spanned over 10,000 kilometres east to west across 11 time zones, and over 7,200 kilometres north to south. It had five climate zones: tundra, taiga, steppes, desert and mountains.

Exam Probability: **High**

18. *Answer choices:*

(see index for correct answer)

- a. Cuba
- b. Eritrea
- c. Soviet Union
- d. Kingdom of Italy

Guidance: level 1

:: Criminology ::

In criminology, __________ is a theory developed by Edwin Sutherland proposing that through interaction with others, individuals learn the values, attitudes, techniques, and motives for criminal behavior.

Exam Probability: **Low**

19. *Answer choices:*

(see index for correct answer)

- a. Sexual violence by intimate partners
- b. Cui bono
- c. Differential association
- d. Polycrates complex

Guidance: level 1

:: Social philosophy ::

__________ or moral philosophy is a branch of philosophy that involves systematizing, defending, and recommending concepts of right and wrong conduct. The field of __________ , along with aesthetics, concerns matters of value, and thus comprises the branch of philosophy called axiology.

Exam Probability: **High**

20. *Answer choices:*

(see index for correct answer)

- a. Axiom of Equity
- b. Julian Gumperz
- c. Ethics

- d. Historicism

Guidance: level 1

:: MILF pornography ::

In any given society, a _________ is an implicit prohibition on something based on a cultural sense that it is excessively repulsive or, perhaps, too sacred for ordinary people. Such prohibitions are present in virtually all societies. On a comparative basis _________ s, for example related to food items, seem to make no sense at all as what may be declared unfit for one group by custom or religion may be perfectly acceptable to another.

Exam Probability: **Medium**

21. *Answer choices:*

(see index for correct answer)

- a. Maki Tomoda
- b. Savita Bhabhi
- c. Reality Kings
- d. Ayano Murasaki

Guidance: level 1

:: Caste ::

__________ is a form of social stratification characterized by endogamy, hereditary transmission of a lifestyle which often includes an occupational identity, ritual status in a hierarchy, customary social interaction, and exclusion based on cultural notions of purity and pollution. Although __________ systems exist in various regions, its paradigmatic ethnographic example is the division of Indian society into rigid social groups, with roots in India`s ancient history and persisting until today. It is sometimes used as an analogical basis for the study of _________ -like social divisions existing outside India. The term is also applied to non-human populations like ants and bees.

Exam Probability: **Low**

22. *Answer choices:*

(see index for correct answer)

- a. National Campaign on Dalit Human Rights
- b. Al-Akhdam
- c. Half-caste
- d. Caste

Guidance: level 1

:: Sexualization ::

__________ is a person`s perception of the aesthetics or sexual attractiveness of their own body. It involves how a person sees themselves, compared to the standards that have been set by society. The Austrian neurologist and psychoanalyst Paul Schilder coined the phrase body-image in his book The Image and Appearance of the Human Body .Human society has at all times placed great value on beauty of the human body, but a person`s perception of their own body may not correspond to society`s standards.

Exam Probability: **High**

23. *Answer choices:*

(see index for correct answer)

- a. Sexuality in music videos
- b. Pornified
- c. Body image
- d. Exploitation of women in mass media

Guidance: level 1

:: Theories of history ::

Marxism is a theory and method of working-class self-emancipation. As a theory, it relies on a method of socioeconomic analysis that views class relations and social conflict using a materialist interpretation of historical development and takes a dialectical view of social transformation. It originates from the works of 19th-century German philosophers Karl Marx and Friedrich Engels.

Exam Probability: **Low**

24. *Answer choices:*

(see index for correct answer)

- a. system theory
- b. Marxist
- c. Marxism

Guidance: level 1

:: Critical thinking ::

_________ is the process of breaking a complex topic or substance into smaller parts in order to gain a better understanding of it. The technique has been applied in the study of mathematics and logic since before Aristotle , though _________ as a formal concept is a relatively recent development.

Exam Probability: **Low**

25. *Answer choices:*

(see index for correct answer)

- a. Inquiry: Critical Thinking Across the Disciplines
- b. Analysis
- c. Explanatory power
- d. Topical logic

Guidance: level 1

:: Scientific method ::

__________ is the active acquisition of information from a primary source. In living beings, __________ employs the senses. In science, __________ can also involve the recording of data via the use of scientific instruments. The term may also refer to any data collected during the scientific activity. __________ s can be qualitative, that is, only the absence or presence of a property is noted, or quantitative if a numerical value is attached to the observed phenomenon by counting or measuring.

Exam Probability: **Medium**

26. *Answer choices:*

(see index for correct answer)

- a. Construct
- b. Clinical study design
- c. Scientific formalism
- d. Principle of similitude

Guidance: level 1

:: Sociological terminology ::

_________ was introduced from English into American English by William Graham Sumner , an early U.S. sociologist, to refer to social norms that are widely observed and are considered to have greater moral significance than others. _________ include an aversion for societal taboos, such as incest. The _________ of a society usually predicate legislation prohibiting their taboos. Often, countries will employ specialized vice squads or vice police engaged in suppressing specific crimes offending the societal _________ .

Exam Probability: **Medium**

27. *Answer choices:*

(see index for correct answer)

- a. Translation
- b. Mores
- c. Superwoman
- d. Structural inequality

Guidance: level 1

:: Demography ::

_________ , or death rate, is a measure of the number of deaths in a particular population, scaled to the size of that population, per unit of time. _________ is typically expressed in units of deaths per 1,000 individuals per year; thus, a _________ of 9.5 in a population of 1,000 would mean 9.5 deaths per year in that entire population, or 0.95% out of the total. It is distinct from "morbidity", which is either the prevalence or incidence of a disease, and also from the incidence rate .

Exam Probability: **Low**

28. *Answer choices:*

(see index for correct answer)

- a. Mortality rate
- b. Sexual selection
- c. Household
- d. Easterlin hypothesis

Guidance: level 1

:: Concepts in ethics ::

A __________ is a proposition or value that is a guide for behavior or evaluation. In law, it is a rule that has to be or usually is to be followed, or can be desirably followed, or is an inevitable consequence of something, such as the laws observed in nature or the way that a system is constructed. The __________ s of such a system are understood by its users as the essential characteristics of the system, or reflecting system`s designed purpose, and the effective operation or use of which would be impossible if any one of the __________ s was to be ignored. A system may be explicitly based on and implemented from a document of __________ s as was done in IBM`s 360/370 __________ s of Operation.

Exam Probability: **Medium**

29. *Answer choices:*

(see index for correct answer)

- a. Yi
- b. Endowment
- c. Principle
- d. Possession is nine-tenths of the law

Guidance: level 1

:: Motivation ::

A _________ is an idea of the future or desired result that a person or a group of people envisions, plans and commits to achieve. People endeavor to reach _________ s within a finite time by setting deadlines.

Exam Probability: **Low**

30. *Answer choices:*

(see index for correct answer)

- a. Drive reduction theory
- b. Cooling Out
- c. Locus of control
- d. Reinforcing success

Guidance: level 1

:: Medical sociology ::

__________ or medicalisation is the process by which human conditions and problems come to be defined and treated as medical conditions, and thus become the subject of medical study, diagnosis, prevention, or treatment. __________ can be driven by new evidence or hypotheses about conditions; by changing social attitudes or economic considerations; or by the development of new medications or treatments.

Exam Probability: **High**

31. *Answer choices:*

(see index for correct answer)

- a. Sick role
- b. Normalization process model
- c. Paramedicalization
- d. Medicalization

Guidance: level 1

:: Childbirth ::

__________ , also known as labour and delivery, is the ending of a pregnancy by one or more babies leaving a woman`s uterus by vaginal passage or Caesarean section. In 2015, there were about 135 million births globally. About 15 million were born before 37 weeks of gestation, while between 3 and 12% were born after 42 weeks. In the developed world most deliveries occur in hospitals, while in the developing world most births take place at home with the support of a traditional birth attendant.

Exam Probability: **Medium**

32. *Answer choices:*

(see index for correct answer)

- a. Sex after pregnancy
- b. Childbirth
- c. Natural childbirth
- d. Amniotic stem cell bank

Guidance: level 1

:: Liberal democracies ::

_________ , officially the French Republic , is a country whose territory consists of metropolitan _________ in Western Europe and several overseas regions and territories. The metropolitan area of _________ extends from the Mediterranean Sea to the English Channel and the North Sea, and from the Rhine to the Atlantic Ocean. It is bordered by Belgium, Luxembourg and Germany to the northeast, Switzerland and Italy to the east, and Andorra and Spain to the south. The overseas territories include French Guiana in South America and several islands in the Atlantic, Pacific and Indian oceans. The country's 18 integral regions span a combined area of 643,801 square kilometres and a total population of 67.3 million . _________ , a sovereign state, is a unitary semi-presidential republic with its capital in Paris, the country's largest city and main cultural and commercial centre. Other major urban areas include Lyon, Marseille, Toulouse, Bordeaux, Lille and Nice.

Exam Probability: **Medium**

33. *Answer choices:*

(see index for correct answer)

- a. Namibia
- b. France
- c. Vanuatu
- d. Poland

Guidance: level 1

:: Sociological terminology ::

Living organisms including humans are _________ when they live collectively in interacting populations, whether they are aware of it, and whether the interaction is voluntary or involuntary.

Exam Probability: **Low**

34. *Answer choices:*

(see index for correct answer)

- a. Significant symbols
- b. Social
- c. Resocialization
- d. Homosociality

Guidance: level 1

:: Economic inequality ::

_________ covers a wide variety of topics. It can refer to either income distribution or the distribution of wealth . Besides _________ between countries or states, there are important types of _________ between different groups of people.

Exam Probability: **Medium**

35. *Answer choices:*

(see index for correct answer)

- a. World on Fire
- b. Gender pay gap in Australia
- c. Economic inequality
- d. Theil index

Guidance: level 1

:: Economic problems ::

The poverty threshold, poverty limit or _________ is the minimum level of income deemed adequate in a particular country. In practice, like the definition of poverty, the official or common understanding of the _________ is significantly higher in developed countries than in developing countries. In 2008, the World Bank came out with a figure of $1.25 a day at 2005 purchasing-power parity . In October 2015, the World Bank updated the international _________ to $1.90 a day. The new figure of $1.90 is based on ICP purchasing power parity calculations and represents the international equivalent of what $1.90 could buy in the US in 2011. The new IPL replaces the $1.25 per day figure, which used 2005 data. Most scholars agree that it better reflects today`s reality, particularly new price levels in developing countries. The common international _________ has in the past been roughly $1 a day. At present the percentage of the global population living under extreme poverty is likely to fall below 10% according to the World Bank projections released in 2015, although this figure is claimed by scholars to be artificially low due to the effective reduction of the IPL in 2015.

Exam Probability: **Medium**

36. *Answer choices:*

(see index for correct answer)

- a. Poverty line
- b. Benefit shortfall
- c. Overaccumulation
- d. Economic stagnation

Guidance: level 1

:: Sociobiology ::

__________ is a social system in which men hold primary power and predominate in roles of political leadership, moral authority, social privilege and control of property. Some patriarchal societies are also patrilineal, meaning that property and title are inherited by the male lineage.

Exam Probability: **Low**

37. *Answer choices:*

(see index for correct answer)

- a. Patriarchy
- b. Evolutionary ethics
- c. On Human Nature
- d. Queen ant

Guidance: level 1

:: Criminology ::

__________ is any sexual act or attempt to obtain a sexual act by violence or coercion, acts to traffic a person or acts directed against a person's sexuality, regardless of the relationship to the victim. It occurs in times of peace and armed conflict situations, is widespread and is considered to be one of the most traumatic, pervasive, and most common human rights violations.

Exam Probability: **Low**

38. *Answer choices:*

(see index for correct answer)

- a. Penal populism
- b. Deterrence
- c. Crime science
- d. Sexual violence

Guidance: level 1

:: Islamic republics ::

_________ , officially the Islamic Republic of _________ , is a landlocked country located in South-Central Asia. _________ is bordered by Pakistan in the south and east; Iran in the west; Turkmenistan, Uzbekistan, and Tajikistan in the north; and in the far northeast, China. Its territory covers 652,000 square kilometers and much of it is covered by the Hindu Kush mountain range, which experiences very cold winters. The north consists of fertile plains, while the south-west consists of deserts where temperatures can get very hot in summers. Kabul serves as the capital and its largest city.

Exam Probability: **Medium**

39. *Answer choices:*

(see index for correct answer)

- a. Pakistan
- b. Afghanistan

- c. Iran

Guidance: level 1

:: Social work ::

__________ is an academic discipline and profession that concerns itself with individuals, families, groups and communities in an effort to enhance social functioning and overall well-being. Social functioning is the way in which people perform their social roles, and the structural institutions that are provided to sustain them. __________ applies social sciences, such as sociology, psychology, political science, public health, community development, law, and economics, to engage with client systems, conduct assessments, and develop interventions to solve social and personal problems; and to bring about social change. __________ practice is often divided into micro-work, which involves working directly with individuals or small groups; and macro-work, which involves working with communities, and - within social policy - fostering change on a larger scale.

Exam Probability: **Medium**

40. *Answer choices:*

(see index for correct answer)

- a. Social work
- b. Jamnalal Bajaj Award
- c. Raquel busca su sitio
- d. Supportive housing

Guidance: level 1

:: Sociological terminology ::

_________ is a social science theory of the opportunities each individual has to improve their quality of life. The concept was introduced by German sociologist Max Weber. It is a probabilistic concept, describing how likely it is, given certain factors, that an individual`s life will turn out a certain way. According to this theory, _________ are positively correlated with one`s socioeconomic status.

Exam Probability: **High**

41. *Answer choices:*

(see index for correct answer)

- a. Life chances
- b. Policy sociology
- c. Social condenser
- d. Mudsill theory

Guidance: level 1

:: Types of marriage ::

_________ is the practice of marrying within a specific social group, caste or ethnic group, rejecting those from others as unsuitable for marriage or other close personal relationships.

Exam Probability: **Low**

42. *Answer choices:*

(see index for correct answer)

- a. Group marriage
- b. Transnational marriage
- c. Endogamy
- d. Heqin

Guidance: level 1

:: Sociological terminology ::

_________ is the process by which one`s sense of social values, beliefs, and norms are re-engineered. This process is deliberately carried out in a variety of settings such as in many single parent households and military boot-camps, through an intense social process that may take place in a total institution. An important thing to note about socialization is that what can be learned, can be unlearned. This forms the basis of _________ : to unlearn and relearn.

Exam Probability: **Medium**

43. *Answer choices:*

(see index for correct answer)

- a. Middleman minority
- b. Resocialization
- c. Vertical mobility
- d. Social system

Guidance: level 1

:: Sociocultural evolution ::

__________ is a range of economic and social systems characterised by social ownership of the means of production and workers` self-management, as well as the political theories and movements associated with them. Social ownership can be public, collective or cooperative ownership, or citizen ownership of equity. There are many varieties of __________ and there is no single definition encapsulating all of them, with social ownership being the common element shared by its various forms.

Exam Probability: **Medium**

44. *Answer choices:*

(see index for correct answer)

- a. Socialism
- b. Agrarian society
- c. Karl August Wittfogel

- d. Complexity theory and organizations

Guidance: level 1

:: Behaviorism ::

In behavioral psychology, __________ is a consequence applied that will strengthen an organism`s future behavior whenever that behavior is preceded by a specific antecedent stimulus. This strengthening effect may be measured as a higher frequency of behavior , longer duration , greater magnitude , or shorter latency . There are two types of __________ , known as positive __________ and negative __________ ; positive is where by a reward is offered on expression of the wanted behaviour and negative is taking away an undesirable element in the persons environment whenever the desired behaviour is achieved.

Exam Probability: **High**

45. *Answer choices:*

(see index for correct answer)

- a. Chaining
- b. Reinforcement
- c. token economy
- d. reinforcing

Guidance: level 1

:: Child poverty ::

__________ in the United States is a federal and state program that helps with medical costs for some people with limited income and resources. __________ also offers benefits not normally covered by Medicare, including nursing home care and personal care services. The Health Insurance Association of America describes __________ as "a government insurance program for persons of all ages whose income and resources are insufficient to pay for health care." __________ is the largest source of funding for medical and health-related services for people with low income in the United States, providing free health insurance to 74 million low-income and disabled people as of 2017. It is a means-tested program that is jointly funded by the state and federal governments and managed by the states, with each state currently having broad leeway to determine who is eligible for its implementation of the program. States are not required to participate in the program, although all have since 1982. __________ recipients must be U.S. citizens or qualified non-citizens, and may include low-income adults, their children, and people with certain disabilities. Poverty alone does not necessarily qualify someone for __________ .

Exam Probability: **Low**

46. *Answer choices:*

(see index for correct answer)

- a. End Child Poverty coalition
- b. Medicaid
- c. Child Poverty Action Group
- d. Compassion Suisse

Guidance: level 1

:: Crimes against humanity ::

__________ is the act of deliberately inflicting severe physical or psychological suffering on someone by another as a punishment or in order to fulfill some desire of the __________r or force some action from the victim. __________, by definition, is a knowing and intentional act; deeds which unknowingly or negligently inflict suffering or pain, without a specific intent to do so, are not typically considered __________.

Exam Probability: **Low**

47. *Answer choices:*

(see index for correct answer)

- a. Loyalist
- b. The Men of Atalissa
- c. Tokyo Charter
- d. Torture

Guidance: level 1

:: Social philosophy ::

_________ is an economic system based on the private ownership of the means of production and their operation for profit. Characteristics central to _________ include private property, capital accumulation, wage labor, voluntary exchange, a price system, and competitive markets. In a capitalist market economy, decision-making and investment are determined by every owner of wealth, property or production ability in financial and capital markets, whereas prices and the distribution of goods and services are mainly determined by competition in goods and services markets.

Exam Probability: **High**

48. *Answer choices:*

(see index for correct answer)

- a. International Association for Philosophy of Law and Social Philosophy
- b. Grotesque body
- c. Role taking theory
- d. Instrumental rationality

Guidance: level 1

:: Polygamy ::

__________ is the practice of marrying multiple spouses. When a man is married to more than one wife at a time, sociologists call this polygyny. When a woman is married to more than one husband at a time, it is called polyandry. If a marriage includes multiple husbands and wives, it can be called a group marriage or mixed-orientation marriage.

Exam Probability: **Medium**

49. *Answer choices:*

(see index for correct answer)

- a. Polygamy
- b. John of Leiden
- c. Antarjali Jatra
- d. Elisabeth Wandscherer

Guidance: level 1

:: Punishments ::

__________ in criminal law is a period of supervision over an offender, ordered by the court instead of serving time in prison.

Exam Probability: **Low**

50. *Answer choices:*

(see index for correct answer)

- a. Probation
- b. Tarring and feathering
- c. Electronic monitoring
- d. Amende honorable

Guidance: level 1

:: Monotheistic religions ::

_________ is an Abrahamic Universal religion based on the life and teachings of Jesus of Nazareth, as described in the New Testament. Its adherents, known as Christians, believe that Jesus Christ is the Son of God and savior of all people, whose coming as the Messiah was prophesied in the Old Testament. It is the worlds largest religion with over 2.4 billion followers or 31.5% of the worlds populations.

Exam Probability: **Medium**

51. *Answer choices:*

(see index for correct answer)

- a. Vaishnavism
- b. Kathenotheism
- c. Vyasakuta
- d. Satya Mahima Dharma

Guidance: level 1

:: Mental processes ::

_________ is the state or quality of awareness or of being aware of an external object or something within oneself. It has been defined variously in terms of sentience, awareness, qualia, subjectivity, the ability to experience or to feel, wakefulness, having a sense of selfhood or soul, the fact that there is something "that it is like" to "have" or "be" it, and the executive control system of the mind. Despite the difficulty in definition, many philosophers believe that there is a broadly shared underlying intuition about what _________ is. As Max Velmans and Susan Schneider wrote in The Blackwell Companion to _________ : "Anything that we are aware of at a given moment forms part of our _________ , making conscious experience at once the most familiar and most mysterious aspect of our lives."

Exam Probability: **Low**

52. *Answer choices:*

(see index for correct answer)

- a. Neuronal memory allocation
- b. Consciousness
- c. Awareness
- d. Encoding

Guidance: level 1

:: Criminology ::

_________ is the act of a person repeating an undesirable behavior after they have either experienced negative consequences of that behavior, or have been trained to extinguish that behavior. It is also used to refer to the percentage of former prisoners who are rearrested for a similar offense.

Exam Probability: **Medium**

53. *Answer choices:*

(see index for correct answer)

- a. Lockdown
- b. Risk and actuarial criminology
- c. Recidivism
- d. Zero tolerance

Guidance: level 1

:: Sociological terminology ::

_________ is a concept developed by the anthropologist Ralph Linton denoting a social position that a person can acquire on the basis of merit; it is a position that is earned or chosen. It is the opposite of Ascribed status. It reflects personal skills, abilities, and efforts. Examples of _________ are being an Olympic athlete, being a criminal, or being a college professor.

Exam Probability: **High**

54. *Answer choices:*

(see index for correct answer)

- a. Traditional authority
- b. Reality distortion field
- c. Social system
- d. Enculturation

Guidance: level 1

:: Book of Daniel ::

_________ is a masculine given name and a surname of Hebrew origin. It means, "God is my judge", and derives from two early biblical figures, primary among them _________ from the Book of _________ . It is a common given name for males, and is also used as a surname. It is also the basis for various derived given names and surnames.

Exam Probability: **Low**

55. *Answer choices:*

(see index for correct answer)

- a. The writing on the wall
- b. The Burning Fiery Furnace
- c. Daniel
- d. Palmoni

Guidance: level 1

:: Adolescence ::

A secondary school is both an organization that provides secondary education and the building where this takes place. Some secondary schools can provide both lower secondary education and upper secondary education , but these can also be provided in separate schools, as in the American middle and _________ system.

Exam Probability: **High**

56. *Answer choices:*

(see index for correct answer)

- a. High school
- b. Amethyst Initiative
- c. Adolescent crystallization
- d. Reviving Ophelia

Guidance: level 1

:: Humanitarian aid ::

__________ is defined as living in housing that is below the minimum standard or lacks secure tenure. People can be categorized as homeless if they are: living on the streets ; moving between temporary shelters, including houses of friends, family and emergency accommodation ; living in private boarding houses without a private bathroom and/or security of tenure . The legal definition of homeless varies from country to country, or among different jurisdictions in the same country or region. According to the UK __________ charity Crisis, a home is not just a physical space: it also provides roots, identity, security, a sense of belonging and a place of emotional wellbeing. United States government homeless enumeration studies also include people who sleep in a public or private place not designed for use as a regular sleeping accommodation for human beings. People who are homeless are most often unable to acquire and maintain regular, safe, secure and adequate housing due to a lack of, or an unsteady income. __________ and poverty are interrelated.

Exam Probability: **Medium**

57. *Answer choices:*

(see index for correct answer)

- a. Sa Kaeo Refugee Camp
- b. Homelessness
- c. Helping Angels
- d. Central Asia Institute

Guidance: level 1

:: Ethnicity ::

An __________ , or an ethnicity, is a category of people who identify with each other based on similarities such as common ancestry, language, history, society, culture or nation. Ethnicity is usually an inherited status based on the society in which one lives. Membership of an __________ tends to be defined by a shared cultural heritage, ancestry, origin myth, history, homeland, language or dialect, symbolic systems such as religion, mythology and ritual, cuisine, dressing style, art or physical appearance.

Exam Probability: **High**

58. *Answer choices:*

(see index for correct answer)

- a. Ethnogenesis
- b. Ethnic group
- c. Geographical segregation
- d. Minority stress

Guidance: level 1

:: Political corruption ::

In ethics and governance, _________ is answerability, blameworthiness, liability, and the expectation of account-giving. As an aspect of governance, it has been central to discussions related to problems in the public sector, nonprofit and private and individual contexts. In leadership roles, _________ is the acknowledgment and assumption of responsibility for actions, products, decisions, and policies including the administration, governance, and implementation within the scope of the role or employment position and encompassing the obligation to report, explain and be answerable for resulting consequences.

Exam Probability: **High**

59. *Answer choices:*

(see index for correct answer)

- a. Accountability
- b. Rent-seeking
- c. Filibuster in the United States Senate
- d. Worldwide Governance Indicators

Guidance: level 1

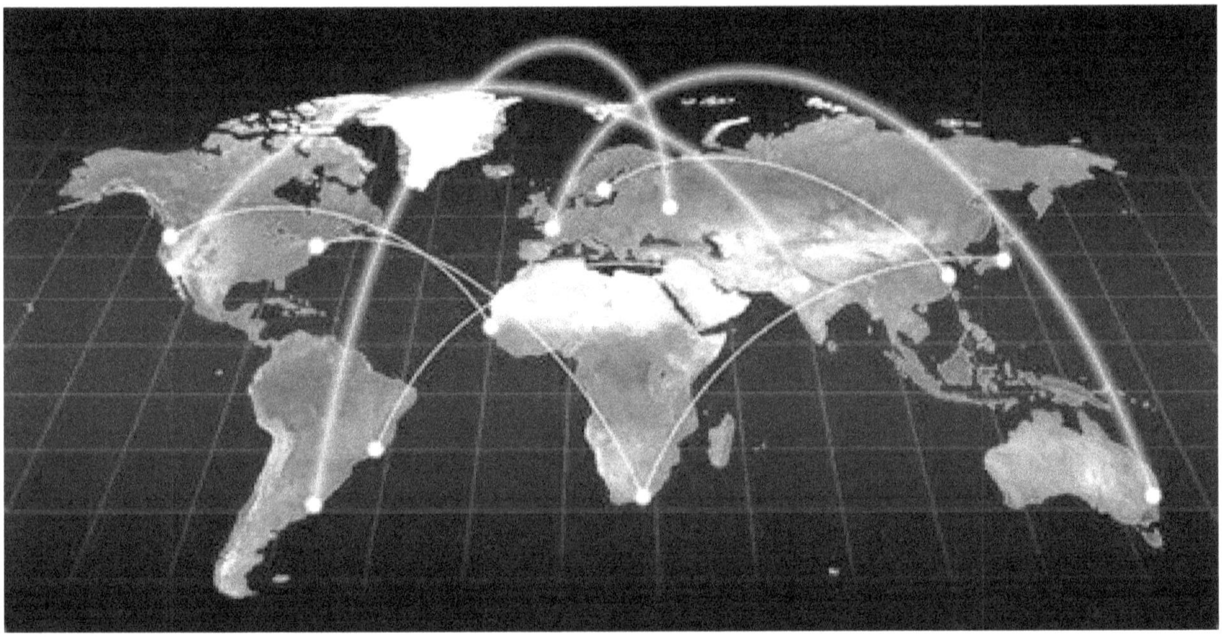

Globalization

Globalization is the process of interaction and integration among people, companies, and governments worldwide. Globalization has grown due to advances in transportation and communication technology. With increased global interactions comes the growth of international trade, ideas, and culture. Globalization is primarily an economic process of interaction and integration that's associated with social and cultural aspects. However, conflicts and diplomacy are also large parts of the history of globalization, and modern globalization.

:: Socialism ::

A __________ is the lowest remuneration that employers can legally pay their workers—the price floor below which workers may not sell their labor. Most countries had introduced __________ legislation by the end of the 20th century.

Exam Probability: **High**

1. *Answer choices:*

(see index for correct answer)

- a. Enterprises in the Soviet Union
- b. Nakaluluwag
- c. Minimum wage
- d. Market abolitionism

Guidance: level 1

:: War ::

World War II , also known as the Second World War, was a global war that lasted from 1939 to 1945. The vast majority of the world`s countries—including all the great powers—eventually formed two opposing military alliances: the Allies and the Axis. A state of total war emerged, directly involving more than 100 million people from over 30 countries. The major participants threw their entire economic, industrial, and scientific capabilities behind the war effort, blurring the distinction between civilian and military resources. __________ was the deadliest conflict in human history, marked by 50 to 85 million fatalities, most of whom were civilians in the Soviet Union and China. It included massacres, the genocide of the Holocaust, strategic bombing, premeditated death from starvation and disease, and the only use of nuclear weapons in war.

Exam Probability: **Medium**

2. *Answer choices:*

(see index for correct answer)

- a. Progressive war
- b. Military history
- c. Marketing of war
- d. Conflict resource

Guidance: level 1

:: Child care ::

__________ , otherwise known as day care, is the care and supervision of a child or multiple children at a time, whose ages range from six weeks to thirteen years. __________ is the action or skill of looking after children by a day-care center, nannies, babysitter, teachers or other providers. __________ is a broad topic that covers a wide spectrum of professionals, institutions, contexts, activities, and social and cultural conventions. Early __________ is an equally important and often overlooked component of child development. __________ providers can be children`s first teachers, and therefore play an integral role in systems of early childhood education. Quality care from a young age can have a substantial impact on the future successes of children. The main focus of childcare is on the development of the child, whether that be mental, social, or psychological.

Exam Probability: **Medium**

3. *Answer choices:*

(see index for correct answer)

- a. Baby bedding
- b. Anganwadi
- c. Primary caregiver
- d. Child care

Guidance: level 1

:: Liberalism ::

_________ is any effort to procure economic and social rights, political rights or equality, often for a specifically disenfranchised group, or more generally, in discussion of such matters. _________ stems from ex manus capere . Among others, Karl Marx discussed political _________ in his 1844 essay "On the Jewish Question", although often in addition to the term human _________ . Marx`s views of political _________ in this work were summarized by one writer as entailing "equal status of individual citizens in relation to the state, equality before the law, regardless of religion, property, or other `private` characteristics of individual people."

Exam Probability: **Medium**

4. *Answer choices:*

(see index for correct answer)

- a. Higher Superstition: The Academic Left and Its Quarrels with Science
- b. July Monarchy
- c. Constitutional liberalism
- d. Emancipation

Guidance: level 1

:: Social issues ::

__________ are a range of public services provided by the government, private, profit and non-profit organizations. These public services aim to create more effective organizations, build stronger communities, and promote equality and opportunity.

Exam Probability: **Low**

5. *Answer choices:*

(see index for correct answer)

- a. Privacy-enhancing technologies
- b. Youth exclusion
- c. Encyclopedia of World Problems and Human Potential
- d. Social services

Guidance: level 1

:: Religiously motivated violence in the United States ::

The __________ , commonly called the KKK or the Klan, is an American white supremacist hate group. The Klan has existed in three distinct eras at different points in time during the history of the United States. Each has advocated extremist reactionary positions such as white nationalism, anti-immigration and—especially in later iterations—Nordicism and anti-Catholicism. Historically, the Klan used terrorism—both physical assault and murder—against groups or individuals whom they opposed. All three movements have called for the "purification" of American society and all are considered right-wing extremist organizations. In each era, membership was secret and estimates of the total were highly exaggerated by both friends and enemies.

Exam Probability: **High**

6. *Answer choices:*

(see index for correct answer)

- a. Salem witch trials
- b. Philadelphia Nativist Riots
- c. 2009 Little Rock recruiting office shooting
- d. Philadelphia Election Riot

Guidance: level 1

:: Social justice ::

__________ is a concept of fair and just relations between the individual and society. This is measured by the explicit and tacit terms for the distribution of wealth, opportunities for personal activity, and social privileges. In Western as well as in older Asian cultures, the concept of __________ has often referred to the process of ensuring that individuals fulfill their societal roles and receive what was their due from society. In the current global grassroots movements for __________, the emphasis has been on the breaking of barriers for social mobility, the creation of safety nets and economic justice.

Exam Probability: **Medium**

7. *Answer choices:*

(see index for correct answer)

- a. Teaching for social justice
- b. Maintenance and Welfare of Parents and Senior Citizens Act, 2007
- c. Social justice
- d. Artists for Peace and Justice

Guidance: level 1

:: Globalization-related theories ::

Exam Probability: **High**

10. *Answer choices:*

(see index for correct answer)

- a. Global Social Change Research Project
- b. Social change
- c. SEED-SCALE
- d. New Marriage Law

Guidance: level 1

:: Drought-tolerant plants ::

__________ , also called summer lilac, butterfly-bush, or orange eye, is a species of flowering plant in the family Scrophulariaceae, native to Sichuan and Hubei provinces in central China, and also Japan. It is widely used as an ornamental plant, and many named varieties are in cultivation. B. davidii is named after the French missionary and explorer in China, Father Armand David, who was the first European to report the shrub. It was found near Ichang by Dr Augustine Henry about 1887 and sent to St Petersburg. Another botanist-missionary in China, Jean-André Soulié, sent seed to the French nursery Vilmorin, and B. davidii entered commerce in the 1890s.

Exam Probability: **High**

11. *Answer choices:*

(see index for correct answer)

- a. Ficus platypoda
- b. Trichostema lanatum
- c. Rosa sempervirens
- d. Buddleja davidii

Guidance: level 1

:: Egalitarianism ::

__________ is a state of fairness in which job applicants are treated similarly, unhampered by artificial barriers or prejudices or preferences, except when particular distinctions can be explicitly justified. According to this often complex and contested concept, the intent is that the important jobs in an organization should go to the people who are most qualified – persons most likely to perform ably in a given task – and not go to persons for reasons deemed arbitrary or irrelevant, such as circumstances of birth, upbringing, having well-connected relatives or friends, religion, sex, ethnicity, race, caste, or involuntary personal attributes such as disability, age, gender identity, or sexual orientation.

Exam Probability: **High**

12. *Answer choices:*

(see index for correct answer)

- a. Neohumanism
- b. Radical egalitarianism
- c. Social pension
- d. Equal opportunity

Guidance: level 1

:: Urban decay ::

De-industrialisation is a process of social and economic change caused by the removal or reduction of industrial capacity or activity in a country or region, especially of heavy industry or manufacturing industry. It is the opposite of industrialisation.

Exam Probability: **Low**

13. *Answer choices:*

(see index for correct answer)

- a. Punk house
- b. Redevelopment
- c. Planned shrinkage
- d. Deindustrialization

Guidance: level 1

:: Child poverty ::

_________ in the United States is a federal and state program that helps with medical costs for some people with limited income and resources. _________ also offers benefits not normally covered by Medicare, including nursing home care and personal care services. The Health Insurance Association of America describes _________ as "a government insurance program for persons of all ages whose income and resources are insufficient to pay for health care." _________ is the largest source of funding for medical and health-related services for people with low income in the United States, providing free health insurance to 74 million low-income and disabled people as of 2017. It is a means-tested program that is jointly funded by the state and federal governments and managed by the states, with each state currently having broad leeway to determine who is eligible for its implementation of the program. States are not required to participate in the program, although all have since 1982. _________ recipients must be U.S. citizens or qualified non-citizens, and may include low-income adults, their children, and people with certain disabilities. Poverty alone does not necessarily qualify someone for _________ .

Exam Probability: **High**

14. *Answer choices:*

(see index for correct answer)

- a. Compassion Suisse
- b. Asian Aid
- c. Child Poverty Act 2010
- d. Cycle of poverty

Guidance: level 1

:: Aftermath of war ::

The _________ was the period from 1863 to 1877 in American history. It was a significant chapter in the history of American civil rights.

Exam Probability: **High**

15. *Answer choices:*

(see index for correct answer)

- a. Reconstruction Era
- b. Aftermath of the Winter War
- c. Washington Agreement
- d. Nansen International Office for Refugees

Guidance: level 1

:: Personal financial problems ::

The causes of _________ are heavily debated. Classical economics, new classical economics, and the Austrian School of economics argued that market mechanisms are reliable means of resolving _________ . These theories argue against interventions imposed on the labor market from the outside, such as unionization, bureaucratic work rules, minimum wage laws, taxes, and other regulations that they claim discourage the hiring of workers. Keynesian economics emphasizes the cyclical nature of _________ and recommends government interventions in the economy that it claims will reduce _________ during recessions. This theory focuses on recurrent shocks that suddenly reduce aggregate demand for goods and services and thus reduce demand for workers. Keynesian models recommend government interventions designed to increase demand for workers; these can include financial stimuli, publicly funded job creation, and expansionist monetary policies. Its namesake economist, John Maynard Keynes, believed that the root cause of _________ is the desire of investors to receive more money rather than produce more products, which is not possible without public bodies producing new money. A third group of theories emphasize the need for a stable supply of capital and investment to maintain full employment. On this view, government should guarantee full employment through fiscal policy, monetary policy and trade policy as stated, for example, in the US Employment Act of 1946, by counteracting private sector or trade investment volatility, and reducing inequality.

Exam Probability: **Low**

16. *Answer choices:*

(see index for correct answer)

- a. Bankruptcy
- b. Insolvency law of Russia
- c. Unemployment
- d. Official Committee of Equity Security Holders

Guidance: level 1

:: Demography ::

_________ is the international movement of people into a destination country of which they are not natives or where they do not possess citizenship in order to settle or reside there, especially as permanent residents or naturalized citizens, or to take up employment as a migrant worker or temporarily as a foreign worker.

Exam Probability: **Medium**

17. *Answer choices:*

(see index for correct answer)

- a. Metapopulation
- b. Human behavioral ecology
- c. Fertility
- d. Immigration

Guidance: level 1

:: Sociology ::

__________ is the study of society, patterns of social relationships, social interaction, and culture of everyday life. It is a social science that uses various methods of empirical investigation and critical analysis to develop a body of knowledge about social order, acceptance, and change or social evolution. While some sociologists conduct research that may be applied directly to social policy and welfare, others focus primarily on refining the theoretical understanding of social processes. Subject matter ranges from the micro- __________ level of individual agency and interaction to the macro level of systems and the social structure.

Exam Probability: **Low**

18. *Answer choices:*

(see index for correct answer)

- a. Transformation in economics
- b. Sociology

Guidance: level 1

:: Systems ecology ::

__________ s are human-built, modified, or engineered niches of the Anthropocene. They exist in places that have been altered in structure and function by human agency. __________ s are part of the human environment and niche , they lack natural analogs, and they have extended an influence that has converted more than three-quarters of wild Earth. These anthropogenic biomes include technoecosystems that are fuelled by powerful energy sources including ecosystems populated with technodiversity, such as roads and unique combinations of soils called technosols. Vegetation associations on old buildings or along field boundary stone walls in old agricultural landscapes are examples of sites where research into __________ ecology is developing.

Exam Probability: **Low**

19. *Answer choices:*

(see index for correct answer)

- a. Macroecology
- b. Novel ecosystem
- c. Ecosystem model
- d. Synthetic ecosystems

Guidance: level 1

:: Social theories ::

Conservatism is a political and social philosophy promoting traditional social institutions in the context of culture and civilization. The central tenets of conservatism include tradition, human imperfection, organic society, hierarchy, authority, and property rights. _________ seek to preserve a range of institutions such as religion, parliamentary government, and property rights, with the aim of emphasizing social stability and continuity. The more traditional elements—reactionaries—oppose modernism and seek a return to "the way things were".

Exam Probability: **Low**

20. *Answer choices:*

(see index for correct answer)

- a. Conservatives
- b. Practice
- c. Analytical feminism
- d. Persianism

Guidance: level 1

:: Segregation ::

_________ is the process of ending the separation of two groups usually referring to races. This is most commonly used in reference to the United States. _________ was long a focus of the Civil Rights Movement, both before and after the United States Supreme Court`s decision in Brown v. Board of Education, particularly _________ of the school systems and the military . Racial integration of society was a closely related goal.

Exam Probability: **Medium**

21. *Answer choices:*

(see index for correct answer)

- a. Age-restricted community
- b. Segregation in Northern Ireland
- c. Age segregation
- d. Mainstreaming

Guidance: level 1

:: Types of marriage ::

_________ is a form of marriage outside a specific social group involving spouses who belong to different socially-defined races or racialized ethnicities. In the past, it was outlawed in the United States of America and in South Africa as miscegenation. It became legal in the entire United States in 1967 when the Supreme Court of the United States ruled in the case Loving v. Virginia that race-based restrictions on marriages violated the Equal Protection Clause of the United States Constitution.

Exam Probability: **High**

22. *Answer choices:*

(see index for correct answer)

- a. Endogamy

- b. Interracial marriage
- c. Jumping the broom
- d. Void marriage

Guidance: level 1

:: Public finance ::

A tax is a compulsory financial charge or some other type of levy imposed upon a taxpayer by a governmental organization in order to fund various public expenditures. A failure to pay, along with evasion of or resistance to taxation, is punishable by law. _________ consist of direct or indirect _________ and may be paid in money or as its labour equivalent.

Exam Probability: **High**

23. *Answer choices:*

(see index for correct answer)

- a. resource allocation
- b. Budgetary policy
- c. Taxes
- d. Tax revenue

Guidance: level 1

:: Sociological theories ::

_________ ism is a theory of knowledge in sociology and communication theory that examines the development of jointly constructed understandings of the world that form the basis for shared assumptions about reality. The theory centers on the notion that meanings are developed in coordination with others rather than separately within each individual.

Exam Probability: **Medium**

24. *Answer choices:*

(see index for correct answer)

- a. Amateur professionalism
- b. Social exchange theory
- c. Social construction
- d. Cultural selection theory

Guidance: level 1

:: Resource economics ::

_________ refers to the distribution of resources, including land, water, minerals, fuel and wealth in general among corresponding geographic entities . _________ refers to the geographic occurrence or spatial arrangement of resources on earth. In other words, where resources are located. Any particular place may be rich in the resources people desire and poor in others. ... As a result, resources are distributed unevenly across the globe.

Exam Probability: **Low**

25. *Answer choices:*

(see index for correct answer)

- a. Pooling
- b. Gordon-Schaefer Model
- c. Resource distribution
- d. Bioeconomics

Guidance: level 1

:: Sociological terminology ::

In the social sciences, _________ is the patterned social arrangements in society that are both emergent from and determinant of the actions of the individuals. On the macro scale, _________ is the system of socioeconomic stratification , social institutions, or, other patterned relations between large social groups. On the meso scale, it is the structure of social network ties between individuals or organizations. On the micro scale, it can be the way norms shape the behavior of individuals within the social system.

Exam Probability: **Medium**

26. *Answer choices:*

(see index for correct answer)

- a. Social structure

- b. Deindividuation
- c. State formation
- d. Social disintegration

Guidance: level 1

:: Distribution, retailing, and wholesaling ::

A _________ is a person who trades in commodities produced by other people. Historically, a _________ is anyone who is involved in business or trade. _________ s have operated for as long as industry, commerce, and trade have existed. During the 16th-century, in Europe, two different terms for _________ s emerged: One term, meerseniers, described local traders such as bakers, grocers, etc.; while a new term, koopman (Dutch: koopman, described _________ s who operated on a global stage, importing and exporting goods over vast distances, and offering added-value services such as credit and finance.

Exam Probability: **Medium**

27. *Answer choices:*

(see index for correct answer)

- a. Merchant
- b. Chicago Review Press
- c. Stock obsolescence
- d. Demand modeling

Guidance: level 1

:: Taxation ::

A _________ or government incentive is a form of financial aid or support extended to an economic sector generally with the aim of promoting economic and social policy. Although commonly extended from government, the term _________ can relate to any type of support – for example from NGOs or as implicit subsidies. Subsidies come in various forms including: direct and indirect .

Exam Probability: **High**

28. *Answer choices:*

(see index for correct answer)

- a. Depreciation
- b. tax credit
- c. Income tax

Guidance: level 1

:: Political theories ::

_________ is a condition of a person, nation, country, or state in which its residents and population, or some portion thereof, exercise self-government, and usually sovereignty, over the territory. The opposite of _________ is the status of a dependent territory.

Exam Probability: **High**

29. *Answer choices:*

(see index for correct answer)

- a. Political decay
- b. Antiestablishmentarianism
- c. Integral nationalism
- d. Independence

Guidance: level 1

:: Cultural assimilation ::

The _________ is a monocultural metaphor for a heterogeneous society becoming more homogeneous, the different elements "melting together" with a common culture or vice versa, for a homogeneous society becoming more heterogeneous through the influx of foreign elements with different cultural backgrounds, possessing the potential to create disharmony within the previous culture. Historically, it is often used to describe the assimilation of immigrants to the United States. The melting-together metaphor was in use by the 1780s. The exact term " _________ " came into general usage in the United States after it was used as a metaphor describing a fusion of nationalities, cultures and ethnicities in the 1908 play of the same name.

Exam Probability: **Low**

30. *Answer choices:*

(see index for correct answer)

- a. Sinhalisation
- b. Serbianisation
- c. Ananias of Adiabene
- d. Americanization

Guidance: level 1

:: Prejudice and discrimination ::

_________ , also known as reservation in India and Nepal, positive discrimination / action in the United Kingdom, and employment equity in Canada and South Africa, is the policy of promoting the education and employment of members of groups that are known to have previously suffered from discrimination. Historically and internationally, support for _________ has sought to achieve goals such as bridging inequalities in employment and pay, increasing access to education, promoting diversity, and redressing apparent past wrongs, harms, or hindrances.

Exam Probability: **Low**

31. *Answer choices:*

(see index for correct answer)

- a. Affirmative action
- b. Attribution-value model
- c. Article 153 of the Constitution of Malaysia
- d. Great Gypsy Round-up

Guidance: level 1

:: Identity politics ::

Civil and political rights are a class of rights that protect individuals` freedom from infringement by governments, social organizations, and private individuals. They ensure one`s entitlement to participate in the civil and political life of the society and state without discrimination or repression.

Exam Probability: **Low**

32. *Answer choices:*

(see index for correct answer)

- a. Civil Rights
- b. Cognitive liberty
- c. Black pride
- d. Consumer activism

Guidance: level 1

:: Social classes ::

The _________ is a class of people in the middle of a social hierarchy. The very definition of the term " _________ " is highly political and vigorously contested by various schools of political and economic philosophy. Modern social theorists - and especially economists - have defined and re-defined the term " _________ " in order to serve their particular political ends. The definitions of the term " _________ " therefore are the result of the more- or less-scientific methods used when delineating the parameters of what is and isn`t " _________ ".

Exam Probability: **Low**

33. *Answer choices:*

(see index for correct answer)

- a. Rayah
- b. Scholar-official
- c. Class struggle
- d. Middle class

Guidance: level 1

:: Welfare reform ::

The _________ was a series of programs, public work projects, financial reforms, and regulations enacted by President Franklin D. Roosevelt in the United States between 1933 and 1936. It responded to needs for relief, reform, and recovery from the Great Depression. Major federal programs included the Civilian Conservation Corps , the Civil Works Administration , the Farm Security Administration , the National Industrial Recovery Act of 1933 and the Social Security Administration . They provided support for farmers, the unemployed, youth and the elderly. The _________ included new constraints and safeguards on the banking industry and efforts to re-inflate the economy after prices had fallen sharply. _________ programs included both laws passed by Congress as well as presidential executive orders during the first term of the presidency of Franklin D. Roosevelt.

Exam Probability: **Low**

34. *Answer choices:*

(see index for correct answer)

- a. New Deal
- b. Welfare reform
- c. Workfare in the United Kingdom
- d. Self-Sufficiency Project

Guidance: level 1

:: Political theories ::

__________ is a political, social, and economic ideology and movement characterized by the promotion of the interests of a particular nation, especially with the aim of gaining and maintaining the nation`s sovereignty over its homeland. __________ holds that each nation should govern itself, free from outside interference , that a nation is a natural and ideal basis for a polity, and that the nation is the only rightful source of political power . It further aims to build and maintain a single national identity—based on shared social characteristics such as culture, language, religion, politics, and belief in a shared singular history—and to promote national unity or solidarity. __________ , therefore, seeks to preserve and foster a nation`s traditional culture, and cultural revivals have been associated with nationalist movements. It also encourages pride in national achievements, and is closely linked to patriotism. __________ is often combined with other ideologies, such as conservatism or socialism for example.

Exam Probability: **Low**

35. *Answer choices:*

(see index for correct answer)

- a. Juche
- b. Scientific socialism
- c. Nationalism
- d. Neo-Bonapartism

Guidance: level 1

:: Social inequality ::

__________ , in the abstract, is what belongs to or with something, whether as an attribute or as a component of said thing. In the context of this article, it is one or more components , whether physical or incorporeal, of a person's estate; or so belonging to, as in being owned by, a person or jointly a group of people or a legal entity like a corporation or even a society. Depending on the nature of the __________ , an owner of __________ has the right to consume, alter, share, redefine, rent, mortgage, pawn, sell, exchange, transfer, give away or destroy it, or to exclude others from doing these things, as well as to perhaps abandon it; whereas regardless of the nature of the __________ , the owner thereof has the right to properly use it , or at the very least exclusively keep it.

Exam Probability: **High**

36. *Answer choices:*

(see index for correct answer)

- a. Social deprivation
- b. Property
- c. Racial segregation
- d. Double standard

Guidance: level 1

:: Employment compensation ::

A __________ is monetary compensation paid by an employer to an employee in exchange for work done. Payment may be calculated as a fixed amount for each task completed , or at an hourly or daily rate , or based on an easily measured quantity of work done.

Exam Probability: **Medium**

37. *Answer choices:*

(see index for correct answer)

- a. Deferred compensation
- b. Wage
- c. Salary cap
- d. Dearness allowance

Guidance: level 1

:: Social inequality ::

__________ acknowledges that men and women are not equal and that gender affects an individual`s lived experience. These differences arise from distinctions in biology, psychology, and cultural norms. Some of these distinctions are empirically grounded while others appear to be socially constructed. Studies show the different lived experience of genders across many domains including education, life expectancy, personality, interests, family life, careers, and political affiliations. __________ is experienced differently across cultures.

Exam Probability: **Medium**

38. *Answer choices:*

(see index for correct answer)

- a. Kyriarchy
- b. Stereotype threat
- c. Gender inequality
- d. Petty bourgeoisie

Guidance: level 1

:: Scientific method ::

The __________ or method is a proposed description of scientific method. According to it, scientific inquiry proceeds by formulating a hypothesis in a form that can be falsifiable, using a test on observable data where the outcome is not yet known. A test outcome that could have and does run contrary to predictions of the hypothesis is taken as a falsification of the hypothesis. A test outcome that could have, but does not run contrary to the hypothesis corroborates the theory. It is then proposed to compare the explanatory value of competing hypotheses by testing how stringently they are corroborated by their predictions.

Exam Probability: **High**

39. *Answer choices:*

(see index for correct answer)

- a. Hypothetico-deductive model
- b. Theory
- c. study design
- d. Construct

Guidance: level 1

:: International relations theory ::

_________ is the policy of a nation seeking to extend or retain its authority over other people or territories, generally with the aim of opening trade opportunities. The colonising country seeks to benefit from the colonised country or land mass. In the process, colonisers imposed their religion, economics, and medicinal practices on the natives. Some argue this was a positive move toward modernisation, while other scholars counter that this is an intrinsically Eurocentric rationalisation, given that modernisation is itself a concept introduced by Europeans. _________ is largely regarded as a relationship of domination of an indigenous majority by a minority of foreign invaders where the latter rule in pursuit of its interests.

Exam Probability: **Medium**

40. *Answer choices:*

(see index for correct answer)

- a. Colonialism
- b. Foundationalism
- c. Postclassical realism
- d. North Atlantic triangle

Guidance: level 1

:: Urban decay in the United States ::

_________ was an extremely destructive and deadly Category 5 hurricane that made landfall on Florida and Louisiana in August 2005, causing catastrophic damage; particularly in the city of New Orleans and the surrounding areas. Subsequent flooding, caused largely as a result of fatal engineering flaws in the flood protection system known as levees around the city of New Orleans, precipitated most of the loss of lives. The storm was the third major hurricane of the record-breaking 2005 Atlantic hurricane season, as well as the fourth-most intense Atlantic hurricane on record to make landfall in the contiguous United States, behind only the 1935 Labor Day hurricane, Hurricane Camille in 1969, and Hurricane Michael in 2018.

Exam Probability: **Medium**

41. *Answer choices:*

(see index for correct answer)

- a. Hurricane Katrina
- b. English Avenue and Vine City
- c. Housing Act of 1937
- d. Baltimore riot of 1968

Guidance: level 1

:: Discrimination ::

In human social behavior, _________ is treatment or consideration of, or making a distinction towards, a person based on the group, class, or category to which the person is perceived to belong. These include age, colour, criminal record, height, disability, ethnicity, family status, gender identity, generation, genetic characteristics, marital status, nationality, race, religion, sex, and sexual orientation. _________ consists of treatment of an individual or group, based on their actual or perceived membership in a certain group or social category, "in a way that is worse than the way people are usually treated". It involves the group`s initial reaction or interaction going on to influence the individual`s actual behavior towards the group leader or the group, restricting members of one group from opportunities or privileges that are available to another group, leading to the exclusion of the individual or entities based on illogical or irrational decision making.

Exam Probability: **Medium**

42. *Answer choices:*

(see index for correct answer)

- a. Benevolent prejudice
- b. Hispanophobia
- c. Suspect classification
- d. Discrimination

Guidance: level 1

:: Social concepts ::

_________ is a concept of societal friendship and harmony in the absence of hostility and violence. In a social sense, _________ is commonly used to mean a lack of conflict and freedom from fear of violence between individuals or heterogeneous groups. Throughout history leaders have used _________ making and diplomacy to establish a certain type of behavioral restraint that has resulted in the establishment of regional _________ or economic growth through various forms of agreements or _________ treaties. Such behavioral restraint has often resulted in the reduction of conflicts, greater economic interactivity, and consequently substantial prosperity. The avoidance of war or violent hostility can be the result of thoughtful active listening and communication that enables greater genuine mutual understanding and therefore compromise. Leaders often benefit tremendously from the prestige of _________ talks and treaties that can result in substantially enhanced popularity.

Exam Probability: **Low**

43. *Answer choices:*

(see index for correct answer)

- a. Civic virtue
- b. Traditional values
- c. Peace
- d. Norm

Guidance: level 1

:: Globalization-related theories ::

Industrialisation is the period of social and economic change that transforms a human group from an agrarian society into an industrial society, involving the extensive re-organisation of an economy for the purpose of manufacturing.

Exam Probability: **High**

44. *Answer choices:*

(see index for correct answer)

- a. Industrialisation
- b. Industrialization
- c. Primitive accumulation
- d. post-industrial

Guidance: level 1

:: Environmental ethics ::

The __________ , also known as nature conservation, is a political, environmental, and social movement that seeks to protect natural resources including animal and plant species as well as their habitat for the future.

Exam Probability: **Medium**

45. *Answer choices:*

(see index for correct answer)

- a. J. Baird Callicott
- b. Conservation movement
- c. Mary Midgley
- d. Environmental virtue ethics

Guidance: level 1

:: Demography ::

In demographics, the _________ is the total number of humans currently living, and was estimated to have reached 7.7 billion people as of April 2019. It took over 200,000 years of human history for the world`s population to reach 1 billion; and only 200 years more to reach 7 billion.

Exam Probability: **High**

46. *Answer choices:*

(see index for correct answer)

- a. Isolation index
- b. Longevity claims
- c. World population
- d. Official statistics

Guidance: level 1

:: Sociological terminology ::

An ideology is a collection of normative beliefs and values that an individual or group holds for other than purely epistemic reasons. In other words, these rely on basic assumptions about reality that may or may not have any factual basis. The term is especially used to describe systems of ideas and ideals which form the basis of economic or political theories and resultant policies. In these there are tenuous causal links between policies and outcomes owing to the large numbers of variables available, so that many key assumptions have to be made. In political science the term is used in a descriptive sense to refer to political belief systems

Exam Probability: **Medium**

47. *Answer choices:*

(see index for correct answer)

- a. Collective action
- b. Social circle
- c. Ideologies
- d. Local community

Guidance: level 1

:: Systems ecology ::

_________ is the scientific discipline that involves the study of the chemical, physical, geological, and biological processes and reactions that govern the composition of the natural environment . In particular, _________ is the study of the cycles of chemical elements, such as carbon and nitrogen, and their interactions with and incorporation into living things transported through earth scale biological systems in space through time. The field focuses on chemical cycles which are either driven by or influence biological activity. Particular emphasis is placed on the study of carbon, nitrogen, sulfur, and phosphorus cycles. _________ is a systems science closely related to systems ecology.

Exam Probability: **Low**

48. *Answer choices:*

(see index for correct answer)

- a. Mineral cycle
- b. Biogeochemistry
- c. Effective evolutionary time
- d. Ecosystem engineer

Guidance: level 1

:: Resource economics ::

A __________ is a source or supply from which a benefit is produced and it has some utility. __________ s can broadly be classified upon their availability—they are classified into renewable and non-renewable __________ s.Examples of non renewable __________ s are coal ,crude oil natural gas nuclear energy etc. Examples of renewable __________ s are air,water,wind,solar energy etc. They can also be classified as actual and potential on the basis of level of development and use, on the basis of origin they can be classified as biotic and abiotic, and on the basis of their distribution, as ubiquitous and localized . An item becomes a __________ with time and developing technology. Typically, __________ s are materials, energy, services, staff, knowledge, or other assets that are transformed to produce benefit and in the process may be consumed or made unavailable. Benefits of __________ utilization may include increased wealth, proper functioning of a system, or enhanced well-being. From a human perspective a natural __________ is anything obtained from the environment to satisfy human needs and wants. From a broader biological or ecological perspective a __________ satisfies the needs of a living organism .

Exam Probability: **Medium**

49. *Answer choices:*

(see index for correct answer)

- a. Gordon-Schaefer Model
- b. Bioeconomics
- c. Resource distribution
- d. Backstop resources

Guidance: level 1

:: Social inequality ::

_________ is the systemic separation of people into racial or other ethnic groups in daily life. It may apply to activities such as eating in a restaurant, drinking from a water fountain, using a public toilet, attending school, going to the movies, riding on a bus, or in the rental or purchase of a home or of hotel rooms. Segregation is defined by the European Commission against Racism and Intolerance as "the act by which a person separates other persons on the basis of one of the enumerated grounds without an objective and reasonable justification, in conformity with the proposed definition of discrimination. As a result, the voluntary act of separating oneself from other people on the basis of one of the enumerated grounds does not constitute segregation". According to the UN Forum on Minority Issues, "The creation and development of classes and schools providing education in minority languages should not be considered impermissible segregation, if the assignment to such classes and schools is of a voluntary nature".

Exam Probability: **Medium**

50. *Answer choices:*

(see index for correct answer)

- a. Ruling class
- b. Distinction
- c. Social injustice
- d. Social deprivation

Guidance: level 1

:: Discrimination ::

__________ is the belief in the superiority of one race over another, which often results in discrimination and prejudice towards people based on their race or ethnicity. The use of the term " __________ " does not easily fall under a single definition.

Exam Probability: **High**

51. *Answer choices:*

(see index for correct answer)

- a. Anti-Filipino sentiment
- b. Anti-Catalanism
- c. Anti-Canadianism
- d. Racism

Guidance: level 1

:: Bonds (finance) ::

An indenture is a legal contract that reflects or covers a debt or purchase obligation. It specifically refers to two types of practices: in historical usage, an __________ servant status, and in modern usage, it is an instrument used for commercial debt or real estate transaction.

Exam Probability: **Medium**

52. *Answer choices:*

(see index for correct answer)

- a. 360-day calendar
- b. Bond option
- c. Bond Rider
- d. Indentured

Guidance: level 1

:: Cultural assimilation ::

In countries outside the United States of America, __________ or Americanisation, is the influence American culture and business have on other countries, such as their media, cuisine, business practices, popular culture, technology or political techniques. The term has been used since at least 1907. While not necessarily a pejorative term, it is most often used by critics in the target country who are against the influences. __________ has become more prevalent since the collapse of the Soviet Union in 1989–91, and especially since the advent of widespread high speed Internet use starting in the mid-2000s. In Europe, in recent years there is growing concern about __________ through Google, Facebook, Twitter, Amazon, Apple Inc. and Uber, among many other United States tech industry companies. European governments have increasingly expressed concern regarding privacy issues, as well as antitrust and taxation issues regarding the new American giants. The Wall Street Journal in 2015 reported "deep concerns in Europe's highest policy circles about the power of U.S. technology companies."

Exam Probability: **Medium**

53. *Answer choices:*

(see index for correct answer)

- a. Araucanization of Patagonia
- b. Persianization
- c. Cricket test
- d. Americanization

Guidance: level 1

:: Islam ::

__________ is an Abrahamic, monotheistic, universal religion teaching that there is only one God , and that Muhammad is the messenger of God. It is the world`s second-largest religion with over 1.8 billion followers or 24% of the world`s population, most commonly known as Muslims. Muslims make up a majority of the population in 50 countries. __________ teaches that God is merciful, all-powerful, and unique, and has guided humankind through prophets, revealed scriptures and natural signs. The primary scriptures of __________ are the Quran, viewed by Muslims as the verbatim word of God, and the teachings and normative examples of Muhammad .

Exam Probability: **High**

54. *Answer choices:*

(see index for correct answer)

- a. Muhammerah
- b. Islam
- c. Sunnah

- d. Quran belt

Guidance: level 1

:: Sociology of culture ::

The term _________ has a range of meanings within the contexts of sociology, of political philosophy, and of colloquial use. In sociology and in everyday usage, it is a synonym for "ethnic pluralism", with the two terms often used interchangeably, for example, a cultural pluralism in which various ethnic groups collaborate and enter into a dialogue with one another without having to sacrifice their particular identities. It can describe a mixed ethnic community area where multiple cultural traditions exist or a single country within which they do . Groups associated with an aboriginal or autochthonous ethnic group and foreigner ethnic groups are often the focus.

Exam Probability: **Low**

55. *Answer choices:*

(see index for correct answer)

- a. Collectivism
- b. Culture shock
- c. Multiculturalism
- d. Art world

Guidance: level 1

:: Crimes ::

_________ is intentional action to destroy a people in whole or in part. The hybrid word " _________ " is a combination of the Greek word and the Latin suffix -caedo . The term _________ was coined by Raphael Lemkin in his 1944 book Axis Rule in Occupied Europe;

Exam Probability: **Low**

56. *Answer choices:*

(see index for correct answer)

- a. Illegal emigration
- b. Obtaining pecuniary advantage by deception
- c. Reckless endangerment
- d. Theft

Guidance: level 1

:: Casual sex ::

_________ involves the transfer of goods or services from one person or entity to another, often in exchange for money. A system or network that allows _________ is called a market.

Exam Probability: **Medium**

57. *Answer choices:*

(see index for correct answer)

- a. One-night stand
- b. Threesome
- c. Slut
- d. Trade

Guidance: level 1

:: Adoption ::

__________ is a process whereby a person assumes the parenting of another, usually a child, from that person`s biological or legal parent or parents. Legal __________ s permanently transfers all rights and responsibilities, along with filiation, from the biological parent or parents.

Exam Probability: **Low**

58. *Answer choices:*

(see index for correct answer)

- a. Adoption
- b. Gotcha Day
- c. International adoption
- d. Adoption law

Guidance: level 1

:: Anarchist theory ::

The right of a people to _________ is a cardinal principle in modern international law , binding, as such, on the United Nations as authoritative interpretation of the Charter`s norms. It states that people, based on respect for the principle of equal rights and fair equality of opportunity, have the right to freely choose their sovereignty and international political status with no interference.

Exam Probability: **Low**

59. *Answer choices:*

(see index for correct answer)

- a. Especifismo
- b. Anarchy
- c. Swaraj
- d. Epistemological anarchism

Guidance: level 1

Ideologies

An ideology is a collection of normative beliefs and values that an individual or group holds for other than purely epistemic reasons. The term was coined by Antoine Destutt de Tracy in 1796, who conceived it as the "science of ideas". In contemporary philosophy it is narrower in scope than that original concept, or the ideas expressed in broad concepts such as worldview, imaginary and ontology.

:: Liberal democracies ::

__________ , officially the Italian Republic , is a European country consisting of a peninsula delimited by the Italian Alps and surrounded by several islands. Located in the middle of the Mediterranean sea and traversed along its length by the Apennines, __________ has a largely temperate seasonal climate. The country covers an area of 301,340 km2 and shares open land borders with France, Slovenia, Austria, Switzerland and the enclaved microstates of Vatican City and San Marino. __________ has a territorial exclave in Switzerland and a maritime exclave in the Tunisian sea . With around 60 million inhabitants, __________ is the fourth-most populous member state of the European Union.

Exam Probability: **High**

1. *Answer choices:*

(see index for correct answer)

- a. Republic of Ireland
- b. Italy
- c. Argentina
- d. Slovenia

Guidance: level 1

:: Social philosophy ::

The __________ is a research organization for sociology and continental philosophy, best known as the institutional home of the Frankfurt School and critical theory.

Exam Probability: **Low**

2. *Answer choices:*

(see index for correct answer)

- a. Golden Rule
- b. Formative context
- c. Hermeneutics
- d. Owenism

Guidance: level 1

:: Social philosophers ::

__________ was an American philosopher, sociologist and psychologist, primarily affiliated with the University of Chicago, where he was one of several distinguished pragmatists. He is regarded as one of the founders of symbolic interactionism and of what has come to be referred to as the Chicago sociological tradition.

Exam Probability: **Low**

3. *Answer choices:*

(see index for correct answer)

- a. Antoine Blanc de Saint-Bonnet
- b. Jeremy Bentham

- c. George Herbert Mead
- d. Gerald C. MacCallum, Jr

Guidance: level 1

:: International economics ::

_________ is a trade policy that does not restrict imports or exports; it can also be understood as the free market idea applied to international trade. In government, _________ is predominantly advocated by political parties that hold liberal economic positions while economically left-wing and nationalist political parties generally support protectionism, the opposite of _________ .

Exam Probability: **High**

4. *Answer choices:*

(see index for correct answer)

- a. Internationalization
- b. Multinational corporation
- c. Free trade
- d. Embargo

Guidance: level 1

:: Social justice ::

__________ is a concept of fair and just relations between the individual and society. This is measured by the explicit and tacit terms for the distribution of wealth, opportunities for personal activity, and social privileges. In Western as well as in older Asian cultures, the concept of __________ has often referred to the process of ensuring that individuals fulfill their societal roles and receive what was their due from society. In the current global grassroots movements for __________ , the emphasis has been on the breaking of barriers for social mobility, the creation of safety nets and economic justice.

Exam Probability: **Low**

5. *Answer choices:*

(see index for correct answer)

- a. Global justice
- b. Raging Grannies
- c. I Am a Man!
- d. Social justice

Guidance: level 1

:: Gender studies ::

_________ is a relationship of mutual affection between people. _________ is a stronger form of interpersonal bond than an association. _________ has been studied in academic fields such as communication, sociology, social psychology, anthropology, and philosophy. Various academic theories of _________ have been proposed, including social exchange theory, equity theory, relational dialectics, and attachment styles.

Exam Probability: **Medium**

6. *Answer choices:*

(see index for correct answer)

- a. Philogyny
- b. Friendship
- c. The Global Media Monitoring Project
- d. Depiction of women artists in art history

Guidance: level 1

:: Islamic republics ::

_________ , officially the Islamic Republic of _________ , is a landlocked country located in South-Central Asia. _________ is bordered by Pakistan in the south and east; Iran in the west; Turkmenistan, Uzbekistan, and Tajikistan in the north; and in the far northeast, China. Its territory covers 652,000 square kilometers and much of it is covered by the Hindu Kush mountain range, which experiences very cold winters. The north consists of fertile plains, while the south-west consists of deserts where temperatures can get very hot in summers. Kabul serves as the capital and its largest city.

Exam Probability: **Medium**

7. *Answer choices:*

(see index for correct answer)

- a. Afghanistan
- b. Pakistan
- c. Islamic republic

Guidance: level 1

:: Masterpieces of the Oral and Intangible Heritage of Humanity ::

A _________ is a human tower built traditionally at festivals in Catalonia, the Balearic islands and the Valencian Community, all in Spain. At these festivals, several colles _________ eres attempt to build and dismantle a tower's structure. On November 16, 2010, _________ s were declared by UNESCO to be amongst the Masterpieces of the Oral and Intangible Heritage of Humanity.

Exam Probability: **Low**

8. *Answer choices:*

(see index for correct answer)

- a. Guqin
- b. Day of the Dead
- c. Kashan rug

- d. Fujara

Guidance: level 1

:: Criminology ::

_________ is any sexual act or attempt to obtain a sexual act by violence or coercion, acts to traffic a person or acts directed against a person's sexuality, regardless of the relationship to the victim. It occurs in times of peace and armed conflict situations, is widespread and is considered to be one of the most traumatic, pervasive, and most common human rights violations.

Exam Probability: **Low**

9. *Answer choices:*

(see index for correct answer)

- a. Risk factor
- b. History of anthropometry
- c. Sexual violence
- d. Dark figure of crime

Guidance: level 1

:: Marxist theory ::

_________ is a concept frequently used by Karl Marx and Friedrich Engels in their theory of historical materialism and in Das Kapital. It is first explicitly used in Marx`s published book The Poverty of Philosophy, although Marx and Engels had already defined the term in The German Ideology.

Exam Probability: **Low**

10. *Answer choices:*

(see index for correct answer)

- a. Surplus value
- b. Relations of production
- c. Immiseration thesis
- d. Orthodox Trotskyism

Guidance: level 1

:: Religious occupations ::

A _________ or _________ ess is a religious leader authorized to perform the sacred rituals of a religion, especially as a mediatory agent between humans and one or more deities. They also have the authority or power to administer religious rites; in particular, rites of sacrifice to, and propitiation of, a deity or deities. Their office or position is the _________ hood, a term which also may apply to such persons collectively.

Exam Probability: **Medium**

11. *Answer choices:*

(see index for correct answer)

- a. Shugenja
- b. Priest
- c. Abbot
- d. Apostolic executor

Guidance: level 1

:: Feminism ::

A women-only space is an area where only women are allowed, thus providing a place where they do not have to interact with men. Historically and globally, many cultures had, and many still have, some form of female seclusion.

Exam Probability: **Medium**

12. *Answer choices:*

(see index for correct answer)

- a. Rosie Rosenzweig
- b. No gods, no masters
- c. Women only space
- d. Feminazi

Guidance: level 1

:: Critical theory ::

The __________ is an area in social life where individuals can come together to freely discuss and identify societal problems, and through that discussion influence political action. Such a discussion is called public debate and is defined as the expression of views on matters that are of concern to the public—often, but not always, with opposing or diverging views being expressed by participants in the discussion. Public debate takes place mostly through the mass media, but also at meetings or through social media, academic publications and government policy documents. The term was originally coined by German philosopher Jürgen Habermas who defined "the __________ as a virtual or imaginary community which does not necessarily exist in any identifiable space". Communication scholar Gerard A. Hauser defines it as "a discursive space in which individuals and groups associate to discuss matters of mutual interest and, where possible, to reach a common judgment about them". The __________ can be seen as "a theater in modern societies in which political participation is enacted through the medium of talk" and "a realm of social life in which public opinion can be formed".

Exam Probability: **Low**

13. *Answer choices:*

(see index for correct answer)

- a. Subversion and containment
- b. Critical technical practice
- c. Post-Marxism
- d. Public sphere

Guidance: level 1

:: Self ::

_________ is a conscious or subconscious process in which people attempt to influence the perceptions of other people about a person, object or event. They do so by regulating and controlling information in social interaction. It was first conceptualized by Erving Goffman in 1959 in The Presentation of Self in Everyday Life, and then was expanded upon in 1967. An example of _________ theory in play is in sports such as soccer. At an important game, a player would want to showcase themselves in the best light possible, because there are college recruiters watching. This person would have the flashiest pair of cleats and try and perform their best to show off their skills. Their main goal may be to impress the college recruiters in a way that maximizes their chances of being chosen for a college team rather than winning the game.

Exam Probability: **High**

14. *Answer choices:*

(see index for correct answer)

- a. Impression management
- b. sexual activity

Guidance: level 1

:: Philosophy of religion literature ::

_________ : Speeches to its Cultured Despisers is a book written by the German theologian Friedrich Schleiermacher . Originally published in 1799, two further editions were released in Schleiermacher`s lifetime in 1806 and 1831.

Exam Probability: **Low**

15. *Answer choices:*

(see index for correct answer)

- a. Philosophical Fragments
- b. The Doors of Perception
- c. On Religion
- d. Lectures on the Philosophy of Religion

Guidance: level 1

:: Social inequality ::

The _________ is the social class of a given society that decides upon and sets that society's political agenda.

Exam Probability: **Medium**

16. *Answer choices:*

(see index for correct answer)

- a. Gender inequality
- b. Social equity
- c. Ruling class
- d. Inequality in the workplace

Guidance: level 1

:: Futurology ::

Systems theory is the interdisciplinary study of systems. A system is a cohesive conglomeration of interrelated and interdependent parts that is either natural or man-made. Every system is delineated by its spatial and temporal boundaries, surrounded and influenced by its environment, described by its structure and purpose or nature and expressed in its functioning. In terms of its effects, a system can be more than the sum of its parts if it expresses synergy or emergent behavior. Changing one part of the system usually affects other parts and the whole system, with predictable patterns of behavior. For systems that are self-learning and self-adapting, the positive growth and adaptation depend upon how well the system is adjusted with its environment. Some systems function mainly to support other systems by aiding in the maintenance of the other system to prevent failure. The goal of systems theory is systematically discovering a system`s dynamics, constraints, conditions and elucidating principles that can be discerned and applied to systems at every level of nesting, and in every field for achieving optimized equifinality.

Exam Probability: **Low**

17. *Answer choices:*

(see index for correct answer)

- a. Citizen Cyborg
- b. Interdependence
- c. Personoid
- d. Dysgenics

Guidance: level 1

:: Religion ::

_________ is a cultural system of designated behaviors and practices, morals, worldviews, texts, sanctified places, prophecies, ethics, or organizations, that relates humanity to supernatural, transcendental, or spiritual elements. However, there is no scholarly consensus over what precisely constitutes a _________.

Exam Probability: **High**

18. *Answer choices:*

(see index for correct answer)

- a. Religion
- b. Tusholi
- c. Religious nationalism
- d. Secular morality

Guidance: level 1

:: Marxist theory ::

In classical logic, a _________ consists of a logical incompatibility between two or more propositions. It occurs when the propositions, taken together, yield two conclusions which form the logical, usually opposite inversions of each other. Illustrating a general tendency in applied logic, Aristotle`s law of non _________ states that "One cannot say of something that it is and that it is not in the same respect and at the same time."

Exam Probability: **High**

19. *Answer choices:*

(see index for correct answer)

- a. Contradiction
- b. Theory of historical trajectory
- c. Late capitalism
- d. Young Marx

Guidance: level 1

:: Social philosophers ::

_________ was a German philosopher, sociologist, psychologist and composer known for his critical theory of society.

Exam Probability: **Medium**

20. *Answer choices:*

(see index for correct answer)

- a. Norbert Leser
- b. Theodor W. Adorno
- c. Georg Wilhelm Friedrich Hegel
- d. Ernst Bloch

Guidance: level 1

:: Liberal democracies ::

_________ , officially the United Mexican States , is a country in the southern portion of North America. It is bordered to the north by the United States; to the south and west by the Pacific Ocean; to the southeast by Guatemala, Belize, and the Caribbean Sea; and to the east by the Gulf of _________ . Covering almost 2,000,000 square kilometres , the nation is the fifth largest country in the Americas by total area and the 13th largest independent state in the world. With an estimated population of over 120 million people, the country is the tenth most populous state and the most populous Spanish-speaking state in the world, while being the second most populous nation in Latin America after Brazil. _________ is a federation comprising 31 states and _________ City, a special federal entity that is also the capital city and its most populous city. Other metropolises in the state include Guadalajara, Monterrey, Puebla, Toluca, Tijuana and León.

Exam Probability: **Medium**

21. *Answer choices:*

(see index for correct answer)

- a. Botswana
- b. Mexico
- c. Dominican Republic
- d. Slovenia

Guidance: level 1

:: Political corruption ::

In ethics and governance, _________ is answerability, blameworthiness, liability, and the expectation of account-giving. As an aspect of governance, it has been central to discussions related to problems in the public sector, nonprofit and private and individual contexts. In leadership roles, _________ is the acknowledgment and assumption of responsibility for actions, products, decisions, and policies including the administration, governance, and implementation within the scope of the role or employment position and encompassing the obligation to report, explain and be answerable for resulting consequences.

Exam Probability: **Medium**

22. *Answer choices:*

(see index for correct answer)

- a. Patrimonialism
- b. Elite capture
- c. Naked official
- d. Clientelism

Guidance: level 1

:: Sociology of religion ::

__________ , also referred to as a civic religion, is the implicit religious values of a nation, as expressed through public rituals, symbols , and ceremonies on sacred days and at sacred places . It is distinct from churches, although church officials and ceremonies are sometimes incorporated into the practice of __________ . Countries described as having a __________ include France, South Korea, and the former Soviet Union. As a concept, it originated in French political thought and became a major topic for U.S. sociologists since its use by Robert Bellah in 1960.

Exam Probability: **Low**

23. *Answer choices:*

(see index for correct answer)

- a. Plausibility structure
- b. religious movement
- c. Civil religion
- d. Secondary conversion

Guidance: level 1

:: Islam ::

__________ is an Abrahamic, monotheistic, universal religion teaching that there is only one God , and that Muhammad is the messenger of God. It is the world`s second-largest religion with over 1.8 billion followers or 24% of the world`s population, most commonly known as Muslims. Muslims make up a majority of the population in 50 countries. __________ teaches that God is merciful, all-powerful, and unique, and has guided humankind through prophets, revealed scriptures and natural signs. The primary scriptures of __________ are the Quran, viewed by Muslims as the verbatim word of God, and the teachings and normative examples of Muhammad .

Exam Probability: **Medium**

24. *Answer choices:*

(see index for correct answer)

- a. Hajj
- b. Islam
- c. Sunnah
- d. When Strangers meet

Guidance: level 1

:: Consumer behaviour ::

__________ is an activity in which a customer browses the available goods or services presented by one or more retailers with the potential intent to purchase a suitable selection of them. A typology of shopper types has been developed by scholars which identifies one group of shoppers as recreational shoppers, that is, those who enjoy __________ and view it as a leisure activity.

Exam Probability: **High**

25. *Answer choices:*

(see index for correct answer)

- a. Consumption smoothing
- b. Shopping
- c. Buzz monitoring
- d. Abandonment rate

Guidance: level 1

:: Gender studies ::

A __________ , also known as a sex role, is a social role encompassing a range of behaviors and attitudes that are generally considered acceptable, appropriate, or desirable for people based on their actual or perceived sex. __________ s are usually centered on conceptions of femininity and masculinity, although there are exceptions and variations. The specifics regarding these gendered expectations may vary substantially among cultures, while other characteristics may be common throughout a range of cultures. There is ongoing debate as to what extent __________ s and their variations are biologically determined, and to what extent they are socially constructed.

Exam Probability: **Low**

26. *Answer choices:*

(see index for correct answer)

- a. Gender Park
- b. Gender role
- c. Rosalind Gill
- d. Separate Spheres

Guidance: level 1

:: Marxist theory ::

The __________ comprises those engaged in waged or salaried labour, especially in manual-labour occupations and industrial work. Working-class occupations include blue-collar jobs, some white-collar jobs, and most pink-collar jobs. Members of the __________ rely for their income exclusively upon their earnings from wage labour; thus, according to the more inclusive definitions, the category can include almost all of the working population of industrialized economies, as well as those employed in the urban areas of non-industrialized economies or in the rural workforce.

Exam Probability: **High**

27. *Answer choices:*

(see index for correct answer)

- a. Minimum programme
- b. Exchange value
- c. Working Class
- d. Value-form

Guidance: level 1

:: Single-party states ::

__________ , officially the People`s Republic of __________ , is a country in East Asia and the world`s most populous country, with a population of around 1.404 billion. Covering approximately 9,600,000 square kilometers , it is the third- or fourth-largest country by total area. Governed by the Communist Party of __________ , the state exercises jurisdiction over 22 provinces, five autonomous regions, four direct-controlled municipalities , and the special administrative regions of Hong Kong and Macau.

Exam Probability: **Medium**

28. *Answer choices:*

(see index for correct answer)

- a. Cuba
- b. Somali Democratic Republic
- c. China
- d. Sahrawi Arab Democratic Republic

Guidance: level 1

:: Liberal democracies ::

_________ , also known as the Republic of _________ , is a country in South Asia. It is the seventh largest country by area and with more than 1.3 billion people, it is the second most populous country as well as the most populous democracy in the world. Bounded by the _________ n Ocean on the south, the Arabian Sea on the southwest, and the Bay of Bengal on the southeast, it shares land borders with Pakistan to the west; China, Nepal, and Bhutan to the northeast; and Bangladesh and Myanmar to the east. In the _________ n Ocean, _________ is in the vicinity of Sri Lanka and the Maldives, while its Andaman and Nicobar Islands share a maritime border with Thailand and Indonesia.

Exam Probability: **Medium**

29. *Answer choices:*

(see index for correct answer)

- a. Belgium
- b. United Kingdom
- c. India
- d. Kiribati

Guidance: level 1

:: Sociological terminology ::

_________ is an awareness of shared interests, objectives, standards, and sympathies cause creating a psychological sense of unity of groups or classes. It refers to the ties in a society that bind people together as one. The term is generally employed in sociology, and the other social sciences as well as in philosophy, bioethics, and Catholic social teaching. It is also a significant concept in Catholic social teaching therefore it is a core concept in Christian democratic political ideology.

Exam Probability: **Low**

30. *Answer choices:*

(see index for correct answer)

- a. Social status
- b. Famous for being famous
- c. Social consciousness
- d. Public morality

Guidance: level 1

:: Liberal democracies ::

_________ is an island country in East Asia. Located in the Pacific Ocean, it lies off the eastern coast of the Asian continent and stretches from the Sea of Okhotsk in the north to the East China Sea and the Philippine Sea in the south.

Exam Probability: **Low**

31. *Answer choices:*

(see index for correct answer)

- a. Marshall Islands
- b. Norway
- c. Estonia
- d. Panama

Guidance: level 1

:: Anti-capitalism ::

In political and social sciences, _________ is the philosophical, social, political, and economic ideology and movement whose ultimate goal is the establishment of the communist society, which is a socioeconomic order structured upon the common ownership of the means of production and the absence of social classes, money, and the state.

Exam Probability: **High**

32. *Answer choices:*

(see index for correct answer)

- a. Industrial Workers of the World
- b. Industrial democracy
- c. Derrick Jensen
- d. Polynesian Panthers

Guidance: level 1

:: Sex-positive feminism ::

_________ , also known as pro-sex feminism, sex-radical feminism, or sexually liberal feminism, is a movement that began in the early 1980s centering on the idea that sexual freedom is an essential component of women`s freedom.

Exam Probability: **Low**

33. *Answer choices:*

(see index for correct answer)

- a. SlutWalk
- b. Dusk!
- c. Feminist Porn Award
- d. Feminist Sex Wars

Guidance: level 1

:: Discrimination ::

_________ is the belief in the superiority of one race over another, which often results in discrimination and prejudice towards people based on their race or ethnicity. The use of the term " _________ " does not easily fall under a single definition.

Exam Probability: **High**

34. *Answer choices:*

(see index for correct answer)

- a. Anti-Arabism
- b. Racism
- c. Days of Remembrance of the Victims of the Holocaust
- d. Discrimination

Guidance: level 1

:: Economic problems ::

_________ is the ability of an individual, family or some other group to improve their economic status—usually measured in income. _________ is often measured by movement between income quintiles. _________ may be considered a type of social mobility, which is often measured in change in income.

Exam Probability: **High**

35. *Answer choices:*

(see index for correct answer)

- a. Budget crisis
- b. The Nature of Mass Poverty
- c. Youth unemployment
- d. Panic selling

Guidance: level 1

:: Mental processes ::

__________ is the state or quality of awareness or of being aware of an external object or something within oneself. It has been defined variously in terms of sentience, awareness, qualia, subjectivity, the ability to experience or to feel, wakefulness, having a sense of selfhood or soul, the fact that there is something "that it is like" to "have" or "be" it, and the executive control system of the mind. Despite the difficulty in definition, many philosophers believe that there is a broadly shared underlying intuition about what __________ is. As Max Velmans and Susan Schneider wrote in The Blackwell Companion to __________ : "Anything that we are aware of at a given moment forms part of our __________ , making conscious experience at once the most familiar and most mysterious aspect of our lives."

Exam Probability: **Medium**

36. *Answer choices:*

(see index for correct answer)

- a. Conceptual blending
- b. Lucid dream
- c. Depth perception
- d. Set

Guidance: level 1

:: Social information processing ::

A _________ is a scattered population whose origin lies in a separate geographic locale. In particular, _________ has come to refer to involuntary mass dispersions of a population from its indigenous territories, most notably the expulsion of Jews from the Land of Israel and the fleeing of Greeks after the fall of Constantinople. Other examples are the African transatlantic slave trade, the southern Chinese or Indians during the coolie trade, the Irish during and after the Irish Famine, the Romani from India, the Italian _________ , the exile and deportation of Circassians, and the emigration of Anglo-Saxon warriors and their families after the Norman Conquest of England.

Exam Probability: **Low**

37. *Answer choices:*

(see index for correct answer)

- a. Diaspora
- b. Jeeran
- c. Wikirating
- d. Knowledge worker

Guidance: level 1

:: Activism ::

__________ is discrimination and social prejudice against people with disabilities or who are perceived to have disabilities. __________ characterizes persons as defined by their disabilities and as inferior to the non-disabled. On this basis, people are assigned or denied certain perceived abilities, skills, or character orientations.

Exam Probability: **Low**

38. *Answer choices:*

(see index for correct answer)

- a. The Sonja Haynes Stone Center for Black Culture and History
- b. Section 28
- c. Ableism
- d. Whirl-Mart

Guidance: level 1

:: Anarchist theory ::

_________ is a performative style of political activism, derived from mainstream cheerleading. _________ combines elements of peaceful protest or non-violent direct action with theatrical elements including costuming, spectacle and choreographed dance. _________ was created by sisters Cara Jennings,, Aimee Jennings and Coleen Jennings in Miami, Florida, USA in 1996. It grew to become an international movement with squads in the United States, Canada and Europe. Radical cheerleaders create and adapt cheers that promote feminism and left-wing ideals.

Exam Probability: **Low**

39. *Answer choices:*

(see index for correct answer)

- a. Polycentric law
- b. Radical cheerleading
- c. Property is theft!
- d. Anarchism and Friedrich Nietzsche

Guidance: level 1

:: Greek mythology ::

An _________ is a person who portrays a character in a performance . The _________ performs "in the flesh" in the traditional medium of the theatre or in modern media such as film, radio, and television. The analogous Greek term is pt , literally "one who answers". The _________ `s interpretation of their role—the art of acting—pertains to the role played, whether based on a real person or fictional character. Interpretation occurs even when the _________ is "playing themselves", as in some forms of experimental performance art.

Exam Probability: **Medium**

40. *Answer choices:*

(see index for correct answer)

- a. Calydnus
- b. Acanthis
- c. Actor
- d. Ceraon

Guidance: level 1

:: Sociological theories ::

_________ is a philosophical theory stating that certain knowledge is based on natural phenomena and their properties and relations. Thus, information derived from sensory experience, interpreted through reason and logic, forms the exclusive source of all certain knowledge. _________ holds that valid knowledge is found only in this a posteriori knowledge.

Exam Probability: **Low**

41. *Answer choices:*

(see index for correct answer)

- a. Social rule system theory
- b. Positivism
- c. Racial literacy
- d. Action theory

Guidance: level 1

:: Sociological theories ::

__________ is a collection of theories about how desirable change in society is best achieved. Such theories draw on a variety of social science disciplines and approaches. In this article, multiple theories are discussed, as are recent developments with regard to these theories. Depending on which theory that is being looked at, there are different explanations to the process of development and their inequalities

Exam Probability: **Medium**

42. *Answer choices:*

(see index for correct answer)

- a. Grand theory
- b. Three-component theory of stratification

- c. Social interface
- d. Social exchange theory

Guidance: level 1

:: Economic globalization ::

_________ is one of the three main dimensions of globalization commonly found in academic literature, with the two others being political globalization and cultural globalization, as well as the general term of globalization. _________ refers to the widespread international movement of goods, capital, services, technology and information. It is the increasing economic integration and interdependence of national, regional, and local economies across the world through an intensification of cross-border movement of goods, services, technologies and capital. _________ primarily comprises the globalization of production, finance, markets, technology, organizational regimes, institutions, corporations, and labour.

Exam Probability: **Low**

43. *Answer choices:*

(see index for correct answer)

- a. Systemically important financial institution
- b. Economic globalization
- c. global financial
- d. McWorld

Guidance: level 1

:: Marxist theory ::

_________ encompasses 20th-century approaches that amend or extend Marxism and Marxist theory, typically by incorporating elements from other intellectual traditions such as critical theory, psychoanalysis, or existentialism .

Exam Probability: **High**

44. *Answer choices:*

(see index for correct answer)

- a. Organic composition of capital
- b. Neo-Marxism
- c. General intellect
- d. Land reform

Guidance: level 1

:: LGBT rights ::

_________ is a system of attitudes, bias, and discrimination in favor of opposite-sex sexuality and relationships. It can include the presumption that other people are heterosexual or that opposite-sex attractions and relationships are the only norm and therefore superior.

Exam Probability: **High**

45. *Answer choices:*

(see index for correct answer)

- a. Identoba
- b. Declaration of Montreal
- c. Heterosexism
- d. International Human Rights Tribunal

Guidance: level 1

:: Marxist theory ::

__________ is the political, economic, or military predominance or control of one state over others. In ancient Greece , __________ denoted the politico-military dominance of a city-state over other city-states. The dominant state is known as the hegemon.

Exam Probability: **Low**

46. *Answer choices:*

(see index for correct answer)

- a. Hegemony
- b. Minimum programme
- c. Political Marxism

- d. Neo-Marxism

Guidance: level 1

:: Sociocultural evolution ::

_________ is a range of economic and social systems characterised by social ownership of the means of production and workers` self-management, as well as the political theories and movements associated with them. Social ownership can be public, collective or cooperative ownership, or citizen ownership of equity. There are many varieties of _________ and there is no single definition encapsulating all of them, with social ownership being the common element shared by its various forms.

Exam Probability: **Low**

47. *Answer choices:*

(see index for correct answer)

- a. Cultural history
- b. Thesis, antithesis, synthesis
- c. Information revolution
- d. Neolithic Revolution

Guidance: level 1

:: Antisemitism ::

National Socialism , more commonly known as _________ , is the ideology and practices associated with the Nazi Party—officially the National Socialist German Workers' Party —in Nazi Germany, and of other far-right groups with similar aims.

Exam Probability: **Medium**

48. *Answer choices:*

(see index for correct answer)

- a. Ottawa Protocol
- b. Syllabus against racism
- c. Pre-Adamite
- d. Kike

Guidance: level 1

:: Feudalism ::

_________ was a combination of legal and military customs in medieval Europe that flourished between the 9th and 15th centuries. Broadly defined, it was a way of structuring society around relationships derived from the holding of land in exchange for service or labour.Although derived from the Latin word feodum or feudum , then in use, the term _________ and the system it describes were not conceived of as a formal political system by the people living in the Middle Ages. In its classic definition, by François-Louis Ganshof , _________ describes a set of reciprocal legal and military obligations among the warrior nobility revolving around the three key concepts of lords, vassals and fiefs.

Exam Probability: **High**

49. *Answer choices:*

(see index for correct answer)

- a. Subinfeudation
- b. Polyptych
- c. Feudalism
- d. Feudal maintenance

Guidance: level 1

:: Social inequality ::

__________ occurs when resources in a given society are distributed unevenly, typically through norms of allocation, that engender specific patterns along lines of socially defined categories of persons. It is the differentiation preference of access of social goods in the society brought about by power, religion, kinship, prestige, race, ethnicity, gender, age, sexual orientation, and class. The social rights include labor market, the source of income, health care, and freedom of speech, education, political representation, and participation. __________ linked to economic inequality, usually described on the basis of the unequal distribution of income or wealth, is a frequently studied type of __________ . Though the disciplines of economics and sociology generally use different theoretical approaches to examine and explain economic inequality, both fields are actively involved in researching this inequality. However, social and natural resources other than purely economic resources are also unevenly distributed in most societies and may contribute to social status. Norms of allocation can also affect the distribution of rights and privileges, social power, access to public goods such as education or the judicial system, adequate housing, transportation, credit and financial services such as banking and other social goods and services.

Exam Probability: **Low**

50. *Answer choices:*

(see index for correct answer)

- a. Structural vulnerability
- b. Social inequality
- c. Double standard
- d. Social deprivation

Guidance: level 1

:: Sociological terminology ::

In the writings of Karl Marx and the Marxist theory of historical materialism, a _________ is a specific combination of the following.

Exam Probability: **Low**

51. *Answer choices:*

(see index for correct answer)

- a. Person of color
- b. Mode of production
- c. Semi-periphery countries
- d. Social structure

Guidance: level 1

:: Judaism terms ::

_________ is a Hebrew name which appears in the Book of Isaiah as a sign that God will protect the House of David.

Exam Probability: **Medium**

52. *Answer choices:*

(see index for correct answer)

- a. Shituf
- b. Klal Yisrael
- c. Live until 120
- d. Immanuel

Guidance: level 1

:: Sociocultural evolution ::

In sociology, __________ refers to a society characterized by an orientation to the past, not the future, with a predominant role for custom and habit. Such societies are marked by a lack of distinction between family and business, with the division of labor influenced primarily by age, gender, and status.

Exam Probability: **Low**

53. *Answer choices:*

(see index for correct answer)

- a. Information revolution
- b. Industrious Revolution
- c. Traditional society
- d. Human evolution

Guidance: level 1

:: Critical theory ::

__________ is a field of critical theory that emerged in the early 1990s out of the fields of queer studies and women`s studies. __________ is often cast in the dual role of method and method`s foil. __________ includes both queer readings of texts and the theorization of `queerness` itself. Heavily influenced by the work of Lauren Berlant, Leo Bersani, Judith Butler, Lee Edelman, Jack Halberstam, and Eve Kosofsky Sedgwick, __________ builds both upon feminist challenges to the idea that gender is part of the essential self and upon gay/lesbian studies` close examination of the socially constructed nature of sexual acts and identities. Whereas gay/lesbian studies focused its inquiries into natural and unnatural behavior with respect to homosexual behavior, __________ expands its focus to encompass any kind of sexual activity or identity that falls into normative and deviant categories. Italian feminist and film theorist Teresa de Lauretis coined the term __________ for a conference she organized at the University of California, Santa Cruz in 1990 and a special issue of Differences: A Journal of Feminist Cultural Studies she edited based on that conference.

Exam Probability: **Low**

54. *Answer choices:*

(see index for correct answer)

- a. Technocriticism
- b. Queer theory
- c. Ethan Kleinberg
- d. Aesthetic Journalism

Guidance: level 1

:: Islamic dress (female) ::

A __________ is an article of clothing or hanging cloth that is intended to cover some part of the head or face, or an object of some significance. __________ ing has a long history in European, Asian, and African societies. The practice has been prominent in different forms in Judaism, Christianity, and Islam. The practice of __________ ing is especially associated with women and sacred objects, though in some cultures it is men rather than women who are expected to wear a __________ . Besides its enduring religious significance, __________ ing continues to play a role in some modern secular contexts, such as wedding customs.

Exam Probability: **High**

55. *Answer choices:*

(see index for correct answer)

- a. Boshiya
- b. Veil
- c. Intimate parts
- d. Paranja

Guidance: level 1

:: Divorce ::

__________ , also known as dissolution of marriage, is the process of terminating a marriage or marital union. __________ usually entails the canceling or reorganizing of the legal duties and responsibilities of marriage, thus dissolving the bonds of matrimony between a married couple under the rule of law of the particular country or state. __________ laws vary considerably around the world, but in most countries __________ requires the sanction of a court or other authority in a legal process, which may involve issues of distribution of property, child custody, alimony , child visitation / access, parenting time, child support, and division of debt. In most countries, monogamy is required by law, so __________ allows each former partner to marry another person; where polygyny is legal but polyandry is not, __________ allows the woman to marry another person.

Exam Probability: **High**

56. *Answer choices:*

(see index for correct answer)

- a. Matthew 1:19
- b. Divorce Busting
- c. Divorce
- d. Grounds for divorce

Guidance: level 1

:: Social movements ::

The expression __________ was first used by Franklin Henry and employed later by Robert E. , Herbert , Ralph Turner and Lewis Killian , and Neil to refer to social processes and events which do not reflect existing social structure , but which emerge in a "spontaneous" way. Use of the term has been expanded to include reference to cells, social animals like birds and fish, and insects including ants. __________ takes many forms but generally violates societal norms . __________ can be tremendously destructive, as with riots or mob violence, silly, as with fads, or anywhere in between. __________ is always driven by group dynamics, encouraging people to engage in acts they might consider unthinkable under typical social circumstances .

Exam Probability: **Medium**

57. *Answer choices:*

(see index for correct answer)

- a. Minjung
- b. Mandela Park Backyarders
- c. Democratic transhumanism
- d. Collective behavior

Guidance: level 1

:: Social inequality ::

_________ acknowledges that men and women are not equal and that gender affects an individual's lived experience. These differences arise from distinctions in biology, psychology, and cultural norms. Some of these distinctions are empirically grounded while others appear to be socially constructed. Studies show the different lived experience of genders across many domains including education, life expectancy, personality, interests, family life, careers, and political affiliations. _________ is experienced differently across cultures.

Exam Probability: **Low**

58. *Answer choices:*

(see index for correct answer)

- a. Social equity
- b. Racial segregation
- c. Gender inequality
- d. Occupational inequality

Guidance: level 1

:: Anti-Protestantism ::

The __________ , also called the Catholic Reformation or the Catholic Revival, was the period of Catholic resurgence that was initiated in response to the Protestant Reformation. It began with the Council of Trent and largely ended with the 1781 Patent of Toleration, although smaller expulsions of Protestants continued into the 19th century. Initiated to preserve the power, influence and material wealth enjoyed by the Catholic Church and to present a theological and material challenge to Reformation, the __________ was a comprehensive effort composed of apologetic and polemical documents, ecclesiastical reconfiguration as decreed by the Council of Trent, a series of wars, political maneuvering including the efforts of Imperial Diets of the Holy Roman Empire, exiling of Protestant populations, confiscation of Protestant children for Catholic institutionalized upbringing, heresy trials and the Inquisition, anti-corruption efforts, spiritual movements, and the founding of new religious orders.

Exam Probability: **High**

59. *Answer choices:*

(see index for correct answer)

- a. Counter-Reformation
- b. Reginald Pole
- c. Bernardino de Mendoza
- d. Exsurge Domine

Guidance: level 1

Social Issues

A social issue is a problem that influences a considerable number of individuals within a society. It is often the consequence of factors extending beyond an individual's control, and is the source of a conflicting opinion on the grounds of what is perceived as a morally just personal life or societal order. Social issues are distinguished from economic issues; however, some issues (such as immigration) have both social and economic aspects. There are also issues that don't fall into either category, such as warfare.

:: Humanitarian aid ::

_________ is defined as living in housing that is below the minimum standard or lacks secure tenure. People can be categorized as homeless if they are: living on the streets ; moving between temporary shelters, including houses of friends, family and emergency accommodation ; living in private boarding houses without a private bathroom and/or security of tenure . The legal definition of homeless varies from country to country, or among different jurisdictions in the same country or region. According to the UK _________ charity Crisis, a home is not just a physical space: it also provides roots, identity, security, a sense of belonging and a place of emotional wellbeing. United States government homeless enumeration studies also include people who sleep in a public or private place not designed for use as a regular sleeping accommodation for human beings. People who are homeless are most often unable to acquire and maintain regular, safe, secure and adequate housing due to a lack of, or an unsteady income. _________ and poverty are interrelated.

Exam Probability: **Medium**

1. *Answer choices:*

(see index for correct answer)

- a. Homelessness
- b. Adventure capital
- c. Nong Chan Refugee Camp
- d. Target Poverty

Guidance: level 1

:: Debt ::

_________ is a situation in which a country is without the ability of paying back its government debt. When the expenditures of its government are more than its tax revenues for a prolonged period, a country may enter into a _________ .In any country,the government finances its expenditures primarily by raising money through taxation. When tax revenues are insufficient, the government can make up the difference by issuing debt.

Exam Probability: **Medium**

2. *Answer choices:*

(see index for correct answer)

- a. External debt
- b. Phantom debt
- c. Amortizing loan
- d. Debt crisis

Guidance: level 1

:: Sex industry ::

_________ is the business or practice of engaging in sexual activity in exchange for payment. _________ is sometimes described as sexual services, commercial sex or, colloquially, hooking. It is sometimes referred to euphemistically as "the world`s oldest profession" in the English-speaking world. A person who works in this field is called a prostitute, and is a type of sex worker.

Exam Probability: **High**

3. *Answer choices:*

(see index for correct answer)

- a. Fur massage
- b. Violence against women
- c. Call girl
- d. Prostitution

Guidance: level 1

:: Philosophy of sexuality ::

__________ is a motivational state and an interest in "sexual objects or activities, or as a wish, or drive to seek out sexual objects or to engage in sexual activities". Synonyms for __________ are libido, sexual attraction and lust. __________ is an aspect of a person's sexuality, which varies significantly from one person to another, and also varies depending on circumstances at a particular time. Not every person experiences __________ ; those who do not experience it may be labelled asexual.

Exam Probability: **Medium**

4. *Answer choices:*

(see index for correct answer)

- a. Sexual desire

- b. Marquis de Sade
- c. Jouissance
- d. Michael Uebel

Guidance: level 1

:: Sociological theories ::

_________ ism is a theory of knowledge in sociology and communication theory that examines the development of jointly constructed understandings of the world that form the basis for shared assumptions about reality. The theory centers on the notion that meanings are developed in coordination with others rather than separately within each individual.

Exam Probability: **High**

5. *Answer choices:*

(see index for correct answer)

- a. Generationism
- b. Social construction
- c. Neofunctionalism
- d. Anomie

Guidance: level 1

:: Social psychology ::

__________ is a process of communication by which one person reveals information about themself to another. The information can be descriptive or evaluative, and can include thoughts, feelings, aspirations, goals, failures, successes, fears, and dreams, as well as one`s likes, dislikes, and favorites.

Exam Probability: **High**

6. *Answer choices:*

(see index for correct answer)

- a. moral disengagement
- b. Self-disclosure
- c. contact theory
- d. Definition of the situation

Guidance: level 1

:: Behavioural syndromes associated with physiological disturbances and physical factors ::

An _________ is a mental disorder defined by abnormal eating habits that negatively affect a person`s physical or mental health. They include binge _________ where people eat a large amount in a short period of time, anorexia nervosa where people eat very little and thus have a low body weight, bulimia nervosa where people eat a lot and then try to rid themselves of the food, pica where people eat non-food items, rumination disorder where people regurgitate food, avoidant/restrictive food intake disorder where people have a lack of interest in food, and a group of other specified feeding or _________ s. Anxiety disorders, depression, and substance abuse are common among people with _________ s. These disorders do not include obesity.

Exam Probability: **Low**

7. *Answer choices:*

(see index for correct answer)

- a. Eating disorder
- b. Hypoactive sexual desire disorder
- c. Anorgasmia
- d. Sexual anhedonia

Guidance: level 1

:: Market research ::

Demography is the statistical study of populations, especially human beings. As a very general science, it can analyze any kind of dynamic living population, i.e., one that changes over time or space . Demography encompasses the study of the size, structure, and distribution of these populations, and spatial or temporal changes in them in response to birth, migration, aging, and death. Based on the demographic research of the earth, earth`s population up to the year 2050 and 2100 can be estimated by demographers. _________ are quantifiable characteristics of a given population.

Exam Probability: **Low**

8. *Answer choices:*

(see index for correct answer)

- a. Marketing research
- b. Fuld-Gilad-Herring Academy of Competitive Intelligence
- c. Demographics
- d. Shanghai Metals Market

Guidance: level 1

:: Crime data ::

The __________ , administered by the Bureau of Justice Statistics, is a national survey of approximately 49,000 to 77,400 households twice a year in the United States, on the frequency of crime victimization, as well as characteristics and consequences of victimization. The survey focuses on gathering information on the following crimes: assault, burglary, larceny, motor vehicle theft, rape, and robbery. The survey results are used for the purposes of building a crime index. It has been used in comparison with the Uniform Crime Reports and the National Incident-Based Reporting System to assess the dark figure of crime. The NCVS survey is comparable to the British Crime Survey conducted in the United Kingdom.

Exam Probability: **High**

9. *Answer choices:*

(see index for correct answer)

- a. ODIS
- b. National Crime Victimization Survey
- c. Uniform Crime Reports

Guidance: level 1

:: Dispute resolution ::

_________ is overt or covert, often harmful, social interaction with the intention of inflicting damage or other unpleasantness upon another individual. It may occur either reactively or without provocation. In humans, frustration due to blocked goals can cause _________ . Human _________ can be classified into direct and indirect _________ ; whilst the former is characterized by physical or verbal behavior intended to cause harm to someone, the latter is characterized by behavior intended to harm the social relations of an individual or group.

Exam Probability: **Low**

10. *Answer choices:*

(see index for correct answer)

- a. Suitable age and discretion
- b. Best alternative to a negotiated agreement
- c. Negotiation
- d. International Program in Conflict Resolution and Mediation

Guidance: level 1

:: Social issues ::

_________ is a composite color. In the CMYK color model used in printing or painting, _________ is made by combining red, black, and yellow, or red, yellow, and blue. In the RGB color model used to project colors onto television screens and computer monitors, _________ is made by combining red and green, in specific proportions. In painting, _________ is generally made by adding black to orange. Mixing red-green-blue pigments makes mud color. The _________ color is seen widely in nature, in wood, soil, human hair color, eye color and skin pigmentation. _________ is the color of dark wood or rich soil. According to public opinion surveys in Europe and the United States, _________ is the least favorite color of the public; the color is most often associated with plainness, the rustic and poverty.

Exam Probability: **High**

11. *Answer choices:*

(see index for correct answer)

- a. Mental health in East Africa
- b. Encyclopedia of World Problems and Human Potential
- c. Brown
- d. Gun Metal Grey

Guidance: level 1

:: Prenatal sex discernment ::

_________ is a medical procedure used in prenatal diagnosis of chromosomal abnormalities and fetal infections, and also for sex determination, in which a small amount of amniotic fluid, which contains fetal tissues, is sampled from the amniotic sac surrounding a developing fetus, and then the fetal DNA is examined for genetic abnormalities. The most common reason to have an "amnio" is to determine whether a fetus has certain genetic disorders or a chromosomal abnormality, such as Down syndrome. _________ can diagnose these problems in the womb. _________ is performed when a woman is between 14 and 16 weeks gestation. Women who choose to have this test are primarily those at increased risk for genetic and chromosomal problems, in part because the test is invasive and carries a small risk of miscarriage. This process can be used for prenatal sex discernment and hence this procedure has legal restrictions in some countries.

Exam Probability: **Medium**

12. *Answer choices:*

(see index for correct answer)

- a. Prenatal sex discernment
- b. Obstetric ultrasonography
- c. Chorionic villus sampling
- d. Baby Gender Mentor

Guidance: level 1

:: Offender profiling ::

__________ is the act of suspecting or targeting a person of a certain race on the basis of observed or assumed characteristics or behavior of a racial or ethnic group, rather than on individual suspicion. __________ , however, isn`t limited only to an individual`s race, but can also be based on the individual`s ethnicity, religion, or national origin. More commonly in the United States, __________ is referred to regarding its use by law enforcement at the local, state, and federal levels, and its use leading to discrimination against people in the African American, Native American, Asian, Latino, Arab, and Muslim communities of the U.S. In European countries, the term "ethnic profiling" is also used instead of __________ .

Exam Probability: **Medium**

13. *Answer choices:*

(see index for correct answer)

- a. Analysis of the Personality of Adolph Hitler
- b. Residential Burglary Expert System
- c. Richard Walter
- d. Dayle Hinman

Guidance: level 1

:: Gender studies ::

_________ is romantic attraction, sexual attraction or sexual behavior between members of the same sex or gender. As a sexual orientation, _________ is "an enduring pattern of emotional, romantic, and/or sexual attractions" to people of the same sex. It "also refers to a person`s sense of identity based on those attractions, related behaviors, and membership in a community of others who share those attractions."

Exam Probability: **High**

14. *Answer choices:*

(see index for correct answer)

- a. Homosexuality
- b. Friendship
- c. Josette Rey-Debove
- d. The Modern Amazons: Warrior Women On-Screen

Guidance: level 1

:: Child sexual abuse ::

_________ is human sexual activity between family members or close relatives. This typically includes sexual activity between people in consanguinity , and sometimes those related by affinity , adoption, clan, or lineage.

Exam Probability: **High**

15. *Answer choices:*

(see index for correct answer)

- a. Kippumjo
- b. Nayoung Case
- c. Convention for the Suppression of the Traffic in Persons and of the Exploitation of the Prostitution of Others
- d. Little Children are Sacred

Guidance: level 1

:: Dynamical systems ::

A _________ is any event that is going to lead to an unstable and dangerous situation affecting an individual, group, community, or whole society. Crises are deemed to be negative changes in the security, economic, political, societal, or environmental affairs, especially when they occur abruptly, with little or no warning. More loosely, it is a term meaning "a testing time" or an "emergency event".

Exam Probability: **Medium**

16. *Answer choices:*

(see index for correct answer)

- a. Valentin Afraimovich
- b. Lorenz-95
- c. Shift space

- d. Projected dynamical system

Guidance: level 1

:: Teratogens ::

Fetal alcohol spectrum disorders are a group of conditions that can occur in a person whose mother drank alcohol during pregnancy. Problems may include an abnormal appearance, short height, low body weight, small head size, poor coordination, low intelligence, behavior problems, and problems with hearing or seeing. Those affected are more likely to have trouble in school, legal problems, participate in high-risk behaviors, and have trouble with alcohol or other drugs. The most severe form of the condition is known as _________ . Other types include partial _________ , alcohol-related neurodevelopmental disorder and alcohol-related birth defects . Some accept only FAS as a diagnosis, seeing the evidence as inconclusive with respect to other types.

Exam Probability: **High**

17. *Answer choices:*

(see index for correct answer)

- a. Valproic acid
- b. Ethyl methanesulfonate
- c. Fetal Alcohol Syndrome
- d. Warfarin

Guidance: level 1

:: Ethnicity ::

An __________ , or an ethnicity, is a category of people who identify with each other based on similarities such as common ancestry, language, history, society, culture or nation. Ethnicity is usually an inherited status based on the society in which one lives. Membership of an __________ tends to be defined by a shared cultural heritage, ancestry, origin myth, history, homeland, language or dialect, symbolic systems such as religion, mythology and ritual, cuisine, dressing style, art or physical appearance.

Exam Probability: **High**

18. *Answer choices:*

(see index for correct answer)

- a. Ethnic group
- b. Ethnic option
- c. Race impersonation
- d. Native American identity in the United States

Guidance: level 1

:: Bullying ::

A __________ is a communicated intent to inflict harm or loss on another person. A __________ is considered an act of coercion. __________ s are widely observed in animal behavior, particularly in a ritualized form, chiefly in order to avoid the unnecessary physical violence that can lead to physical damage or the death of both conflicting parties.

Exam Probability: **Medium**

19. *Answer choices:*

(see index for correct answer)

- a. Threat
- b. Happy slapping
- c. Emotional abuse
- d. Innuendo

Guidance: level 1

:: Criminology ::

__________ is an approach to justice in which the response to a crime is to organize a meeting between the victim and the offender, sometimes with representatives of the wider community. The goal is for them to share their experience of what happened, to discuss who was harmed by the crime and how, and to create a consensus for what the offender can do to repair the harm from the offense. This may include a payment of money given from the offender to the victim, apologies and other amends, and other actions to compensate those affected and to prevent the offender from causing future harm.

Exam Probability: **Low**

20. *Answer choices:*

(see index for correct answer)

- a. Constitutive criminology
- b. Restorative justice
- c. Physiognomy
- d. Social Network Analysis

Guidance: level 1

:: Native American religion ::

_________ is a product prepared from the leaves of the _________ plant by curing them. The plant is part of the genus Nicotiana and of the Solanaceae family. While more than 70 species of _________ are known, the chief commercial crop is N. tabacum. The more potent variant N. rustica is also used around the world.

Exam Probability: **Medium**

21. *Answer choices:*

(see index for correct answer)

- a. Seven fires prophecy
- b. Southeastern Ceremonial Complex

- c. Tobacco
- d. Willamette Meteorite

Guidance: level 1

:: Social problems ::

A __________ is a particular abnormal condition that negatively affects the structure or function of part or all of an organism, and that is not due to any external injury. __________ s are often construed as medical conditions that are associated with specific symptoms and signs. A __________ may be caused by external factors such as pathogens or by internal dysfunctions. For example, internal dysfunctions of the immune system can produce a variety of different __________ s, including various forms of immunodeficiency, hypersensitivity, allergies and autoimmune disorders.

Exam Probability: **Low**

22. *Answer choices:*

(see index for correct answer)

- a. drug addiction
- b. Sedentism
- c. Substance dependence
- d. Disease

Guidance: level 1

:: Childhood ::

_________ entails the biological, psychological and emotional changes that occur in human beings between birth and the conclusion of adolescence, as the individual progresses from dependency to increasing autonomy. It is a continuous process with a predictable sequence, yet having a unique course for every child. It does not progress at the same rate and each stage is affected by the preceding developmental experiences. Because these developmental changes may be strongly influenced by genetic factors and events during prenatal life, genetics and prenatal development are usually included as part of the study of _________ . Related terms include developmental psychology, referring to development throughout the lifespan, and pediatrics, the branch of medicine relating to the care of children. Developmental change may occur as a result of genetically-controlled processes known as maturation, or as a result of environmental factors and learning, but most commonly involves an interaction between the two. It may also occur as a result of human nature and our ability to learn from our environment.

Exam Probability: **Medium**

23. *Answer choices:*

(see index for correct answer)

- a. Cubby-hole
- b. Child development
- c. Crib talk
- d. Global Movement for Children

Guidance: level 1

:: Futurology ::

Systems theory is the interdisciplinary study of systems. A system is a cohesive conglomeration of interrelated and interdependent parts that is either natural or man-made. Every system is delineated by its spatial and temporal boundaries, surrounded and influenced by its environment, described by its structure and purpose or nature and expressed in its functioning. In terms of its effects, a system can be more than the sum of its parts if it expresses synergy or emergent behavior. Changing one part of the system usually affects other parts and the whole system, with predictable patterns of behavior. For systems that are self-learning and self-adapting, the positive growth and adaptation depend upon how well the system is adjusted with its environment. Some systems function mainly to support other systems by aiding in the maintenance of the other system to prevent failure. The goal of systems theory is systematically discovering a system`s dynamics, constraints, conditions and elucidating principles that can be discerned and applied to systems at every level of nesting, and in every field for achieving optimized equifinality.

Exam Probability: **Low**

24. *Answer choices:*

(see index for correct answer)

- a. Redshift
- b. Interdependence
- c. Cyborg Foundation
- d. System analysis

Guidance: level 1

:: Bullying ::

_________ is when a person forcefully criticizes, insults, or denounces someone else. Characterized by underlying anger and hostility, it is a destructive form of communication intended to harm the self-concept of the other person and produce negative emotions. _________ is a maladaptive mechanism that anyone can display occasionally, such as during times of high stress or physical discomfort. For some people, it is a pattern of behaviors used intentionally to control or manipulate others or to get revenge.

Exam Probability: **High**

25. *Answer choices:*

(see index for correct answer)

- a. Verbal abuse
- b. Scapegoating
- c. Psychological manipulation
- d. Self-esteem

Guidance: level 1

:: Stress ::

A __________ is a chemical or biological agent, environmental condition, external stimulus or an event that causes stress to an organism. Psychologically speaking, a __________ can be events or environments that an individual would consider demanding, challenging, and or threaten the individual`s safety.

Exam Probability: **Medium**

26. *Answer choices:*

(see index for correct answer)

- a. Tend and befriend
- b. Effects of stress on memory
- c. Biotic stress
- d. CernySmith Assessment

Guidance: level 1

:: Oligarchy ::

The __________ in modern societies is the social class composed of people who hold the highest social status, usually are the wealthiest members of society, and wield the greatest political power. According to this view, the __________ is generally distinguished by immense wealth which is passed on from generation to generation. Prior to the 20th century, the emphasis was on aristocracy, which emphasized generations of inherited noble status, not just recent wealth.

Exam Probability: **High**

27. *Answer choices:*

(see index for correct answer)

- a. Despotism
- b. Ethnarch
- c. Upper class
- d. Upper ten thousand

Guidance: level 1

:: Taxation ::

An _________ is a tax imposed on individuals or entities that varies with respective income or profits . _________ generally is computed as the product of a tax rate times taxable income. Taxation rates may vary by type or characteristics of the taxpayer.

Exam Probability: **Medium**

28. *Answer choices:*

(see index for correct answer)

- a. subsidy
- b. Income Tax
- c. Depreciation

Guidance: level 1

:: Human rights abuses ::

A __________ is an authoritarian form of government, characterized by a single leader or group of leaders with either no party or a weak party, little mass mobilization, and limited political pluralism. According to other definitions, democracies are regimes in which "those who govern are selected through contested elections"; therefore __________ s are "not democracies". With the advent of the 19th and 20th centuries, __________ s and constitutional democracies emerged as the world's two major forms of government, gradually eliminating monarchies, one of the traditional widespread forms of government of the time. Typically, in a dictatorial regime, the leader of the country is identified with the title of dictator, although their formal title may more closely resemble something similar to "leader". A common aspect that characterized dictators is taking advantage of their strong personality, usually by suppressing freedom of thought and speech of the masses, in order to maintain complete political and social supremacy and stability. __________ s and totalitarian societies generally employ political propaganda to decrease the influence of proponents of alternative governing systems.

Exam Probability: **Medium**

29. *Answer choices:*

(see index for correct answer)

- a. Cristero War
- b. Periodic Report of the United States of America to the United Nations Committee Against Torture
- c. School of the Americas Watch
- d. Dictatorship

Guidance: level 1

:: Social philosophy ::

_________ is the social status a person is assigned at birth or assumed involuntarily later in life. It is a position that is neither earned nor chosen but assigned. These rigid social designators remain fixed throughout an individual`s life and are inseparable from the positive or negative stereotypes that are linked with one`s _________ es.

Exam Probability: **Medium**

30. *Answer choices:*

(see index for correct answer)

- a. Ascribed status
- b. The Values Exchange All Schools Project
- c. Evil
- d. Computer ethics

Guidance: level 1

:: Columbine High School massacre ::

__________ is a public high school in Columbine, Colorado, United States, in the Denver metropolitan area. It is part of the Jefferson County Public Schools district. It is known for being the site of the 1999 massacre.

Exam Probability: **High**

31. *Answer choices:*

(see index for correct answer)

- a. Columbine High School massacre
- b. Eric Harris and Dylan Klebold
- c. Columbine High School

Guidance: level 1

:: Wealth ::

__________ is the abundance of valuable financial assets or physical possessions which can be converted into a form that can be used for transactions. This includes the core meaning as held in the originating old English word weal, which is from an Indo-European word stem. A community, region or country that possesses an abundance of such possessions or resources to the benefit of the common good is known as __________ y.

Exam Probability: **High**

32. *Answer choices:*

(see index for correct answer)

- a. Wealth
- b. Wealth effect
- c. High-net-worth individual
- d. Millionaire

Guidance: level 1

:: Debt ::

Student loans can be used to meet financial need of students in higher education when personal resources, grants, and scholarships are insufficient for the cost of attendance. A student`s debt is the sum of the amount borrowed and interest accrued minus payments made. While several countries offer students loans of various types, the size of _________ in the US has increased to what some consider crisis levels.

Exam Probability: **Medium**

33. *Answer choices:*

(see index for correct answer)

- a. Phantom debt
- b. Student debt
- c. Household debt
- d. Debt restructuring

Guidance: level 1

:: Sustainability organisations ::

_________ is a non-governmental environmental organization with offices in over 39 countries and an international coordinating body in Amsterdam, the Netherlands. _________ was founded in 1971 by Irving Stowe, and Dorothy Stowe, Canadian and US ex-pat environmental activists. _________ states its goal is to "ensure the ability of the Earth to nurture life in all its diversity" and focuses its campaigning on worldwide issues such as climate change, deforestation, overfishing, commercial whaling, genetic engineering, and anti-nuclear issues. It uses direct action, lobbying, research, and ecotage to achieve its goals. The global organization does not accept funding from governments, corporations, or political parties, relying on three million individual supporters and foundation grants. _________ has a general consultative status with the United Nations Economic and Social Council and is a founding member of the INGO Accountability Charter, an international non-governmental organization that intends to foster accountability and transparency of non-governmental organizations.

Exam Probability: **Medium**

34. *Answer choices:*

(see index for correct answer)

- a. Tools for Self Reliance
- b. Ancient Forest Alliance
- c. Sean R. Garner
- d. CODEFF

Guidance: level 1

:: Employment compensation ::

A _________ is monetary compensation paid by an employer to an employee in exchange for work done. Payment may be calculated as a fixed amount for each task completed , or at an hourly or daily rate , or based on an easily measured quantity of work done.

Exam Probability: **Low**

35. *Answer choices:*

(see index for correct answer)

- a. Employee stock purchase plan
- b. Employees%27 Compensation Appeals Board
- c. Australian Fair Pay Commission
- d. Wage

Guidance: level 1

:: Social enterprises ::

_________ is a microfinance organisation and community development bank founded in Bangladesh. It makes small loans to the impoverished without requiring collateral.

Exam Probability: **Medium**

36. *Answer choices:*

(see index for correct answer)

- a. Patients Know Best
- b. Ecoisland Partnership CIC
- c. Grameen Bank
- d. VolunteerMatch

Guidance: level 1

:: Prejudices ::

_________ is the fear, hatred of, or prejudice against the Islamic religion or Muslims generally, especially when seen as a geopolitical force or the source of terrorism.

Exam Probability: **High**

37. *Answer choices:*

(see index for correct answer)

- a. Cocolo
- b. Xenophobia
- c. Idola fori
- d. Islamophobia

Guidance: level 1

:: Criminology ::

_________ is the scientific study of the nature, extent, management, causes, control, consequences, and prevention of criminal behavior, both on individual and social levels. _________ is an interdisciplinary field in both the behavioral and social sciences, which draws primarily upon the research of sociologists, psychologists, philosophers, psychiatrists, biologists, social anthropologists, as well as scholars of law.

Exam Probability: **Low**

38. *Answer choices:*

(see index for correct answer)

- a. Sadistic personality disorder
- b. Criminology
- c. Crime prevention
- d. Insanity defense

Guidance: level 1

:: Poverty ::

The _________ is the segment of the population that occupies the lowest possible position in a class hierarchy, below the core body of the working class.

Exam Probability: **Medium**

39. *Answer choices:*

(see index for correct answer)

- a. The SING Campaign
- b. Underclass
- c. Fish fur
- d. Brooks World Poverty Institute

Guidance: level 1

:: Casual sex ::

_________ is sexual activity that takes places outside a romantic relationship and implies an absence of commitment, emotional attachment, or familiarity between sexual partners. Examples are sexual activity while casually dating, one-night stands, extramarital sex, prostitution, or swinging.

Exam Probability: **Low**

40. *Answer choices:*

(see index for correct answer)

- a. Glory hole
- b. Sexual addiction
- c. Sex club
- d. Cruising for sex

Guidance: level 1

:: Patient safety ::

In medicine, an _________ is an undesired harmful effect resulting from a medication or other intervention such as surgery.

Exam Probability: **Low**

41. *Answer choices:*

(see index for correct answer)

- a. Rapid response system
- b. ECRI Institute
- c. Clinical handover
- d. To Err is Human

Guidance: level 1

:: Social inequality ::

A __________ is the application of different sets of principles for situations that are, in principle, the same, and is often used to describe freedom that is given to one party over another. A double-standard arises when two or more people, circumstances, or events are treated differently even though they should be treated the same way. Margaret Eichler, author of The __________ : A Feminist Critique of Feminist Social Science, explains that a __________ "implies that two things which are the same are measured by different standards".

Exam Probability: **High**

42. *Answer choices:*

(see index for correct answer)

- a. Social injustice
- b. Occupational inequality
- c. Double standard
- d. Petty bourgeoisie

Guidance: level 1

:: Androgyny ::

__________ is the combination of masculine and feminine characteristics into an ambiguous form. __________ may be expressed with regard to biological sex, gender identity, gender expression, or sexual identity. The different meanings of __________ point to the complex interrelationship between aspects of sex, gender, and sexuality.

Exam Probability: **Medium**

43. *Answer choices:*

(see index for correct answer)

- a. Ardhanarishvara
- b. Mana
- c. Peter
- d. Androgyny

Guidance: level 1

:: Branches of sociology (interdisciplinary) ::

_________ is the interdisciplinary study of systems. A system is a cohesive conglomeration of interrelated and interdependent parts that is either natural or man-made. Every system is delineated by its spatial and temporal boundaries, surrounded and influenced by its environment, described by its structure and purpose or nature and expressed in its functioning. In terms of its effects, a system can be more than the sum of its parts if it expresses synergy or emergent behavior. Changing one part of the system usually affects other parts and the whole system, with predictable patterns of behavior. For systems that are self-learning and self-adapting, the positive growth and adaptation depend upon how well the system is adjusted with its environment. Some systems function mainly to support other systems by aiding in the maintenance of the other system to prevent failure. The goal of _________ is systematically discovering a system`s dynamics, constraints, conditions and elucidating principles that can be discerned and applied to systems at every level of nesting, and in every field for achieving optimized equifinality.

Exam Probability: **Medium**

44. *Answer choices:*

(see index for correct answer)

- a. Geodemography
- b. Ethogenics
- c. Social epistemology
- d. Systems theory

Guidance: level 1

:: Globalization-related theories ::

Industrialisation is the period of social and economic change that transforms a human group from an agrarian society into an industrial society, involving the extensive re-organisation of an economy for the purpose of manufacturing.

Exam Probability: **Medium**

45. *Answer choices:*

(see index for correct answer)

- a. Primitive accumulation
- b. Economic development
- c. Industrialization
- d. Postmodernism

Guidance: level 1

:: Psychiatric diagnosis ::

Substance abuse, also known as _________ , is a patterned use of a drug in which the user consumes the substance in amounts or with methods which are harmful to themselves or others, and is a form of substance-related disorder. Widely differing definitions of _________ are used in public health, medical and criminal justice contexts. In some cases criminal or anti-social behaviour occurs when the person is under the influence of a drug, and long term personality changes in individuals may occur as well. In addition to possible physical, social, and psychological harm, use of some drugs may also lead to criminal penalties, although these vary widely depending on the local jurisdiction.

Exam Probability: **Medium**

46. *Answer choices:*

(see index for correct answer)

- a. Drug Abuse
- b. Borderline personality disorder
- c. Body dysmorphic disorder
- d. Schizoid personality disorder

Guidance: level 1

:: Marriage ::

_________ is the marriage of two persons of the same sex or gender, entered into in a civil or religious ceremony.

Exam Probability: **Medium**

47. *Answer choices:*

(see index for correct answer)

- a. Levirate marriage
- b. War bride
- c. Same-sex marriage
- d. Sexual ritual

Guidance: level 1

:: Statistical terminology ::

The _________ is the value separating the higher half from the lower half of a data sample . For a data set, it may be thought of as the "middle" value. For example, in the data set , the _________ is 6, the fourth largest, and also the fourth smallest, number in the sample. For a continuous probability distribution, the _________ is the value such that a number is equally likely to fall above or below it.

Exam Probability: **High**

48. *Answer choices:*

(see index for correct answer)

- a. Median
- b. confounding
- c. Control group
- d. Univariate

Guidance: level 1

:: Human pregnancy ::

_________ development includes the development of the embryo and of the fetus during a viviparous animal`s gestation. _________ development starts with fertilization, in the germinal stage of embryonic development, and continues in fetal development until birth.

Exam Probability: **High**

49. *Answer choices:*

(see index for correct answer)

- a. Prenatal
- b. Prenatal development
- c. Maternity clothing
- d. Spiegelberg criteria

Guidance: level 1

:: Cultural generations ::

According to the Pew Research Center, just over one of every eight Americans aged 40 to 70 is both raising a child and caring for a parent, in addition to between seven and ten million adults caring for their aging parents from a long distance. US Census Bureau statistics indicate that the number of older Americans aged 65 or older will double by the year 2030, to over 70 million. In Australia, the term `sandwich carer` relevant to the 2.6 million unpaid caregivers. A Carers UK report in 2012 said that approximately 2.4 million people are combining childcare with caring for older or disabled relatives.

Exam Probability: **Medium**

50. *Answer choices:*

(see index for correct answer)

- a. Burnt Generation
- b. Me generation
- c. Sandwich generation
- d. 1837 generation

Guidance: level 1

:: Labeling theory ::

In social psychology, a _________ is an over-generalized belief about a particular category of people. _________ s are generalized because one assumes that the _________ is true for each individual person in the category. While such generalizations may be useful when making quick decisions, they may be erroneous when applied to particular individuals. _________ s encourage prejudice and may arise for a number of reasons.

Exam Probability: **Medium**

51. *Answer choices:*

(see index for correct answer)

- a. Stig-9
- b. Stereotype

- c. Frank Tannenbaum

Guidance: level 1

:: Feminism and social class ::

__________ is a phenomenon referring to the widening gap between women and men caught in a sequence of economic deprivation and scarcity. This phenomenon is not only a consequence of lack of income, but is also the result of the deprivation of capabilities and gender biases present in both societies and governments. It covers the poverty of choices and opportunities such as the ability to lead a long, healthy, and creative life, and enjoying basic rights like freedom, respect, and dignity.

Exam Probability: **Medium**

52. *Answer choices:*

(see index for correct answer)

- a. Contingent workforce
- b. Feminisation of the workplace
- c. Triple oppression
- d. Feminization of poverty

Guidance: level 1

:: Casual sex ::

_________ involves the transfer of goods or services from one person or entity to another, often in exchange for money. A system or network that allows _________ is called a market.

Exam Probability: **High**

53. *Answer choices:*

(see index for correct answer)

- a. Trade
- b. Gay bathhouse
- c. Hypersexuality
- d. 2010 Duke University faux sex thesis controversy

Guidance: level 1

:: Sexuality and society ::

_________ is the instruction of issues relating to human sexuality, including emotional relations and responsibilities, human sexual anatomy, sexual activity, sexual reproduction, age of consent, reproductive health, reproductive rights, safe sex, birth control and sexual abstinence. _________ that covers all of these aspects is known as comprehensive _________ . Common avenues for _________ are parents or caregivers, formal school programs, and public health campaigns.

Exam Probability: **High**

54. *Answer choices:*

(see index for correct answer)

- a. Strip game
- b. House party
- c. $pread
- d. Sexual opportunism

Guidance: level 1

:: Activism ::

_________ consists of efforts to promote, impede, direct, or intervene in social, political, economic, or environmental reform with the desire to make changes in society. Forms of _________ range from mandate building in the community , petitioning elected officials, running or contributing to a political campaign, preferential patronage of businesses, and demonstrative forms of _________ like rallies, street marches, strikes, sit-ins, or hunger strikes.

Exam Probability: **Low**

55. *Answer choices:*

(see index for correct answer)

- a. Government-organized demonstration
- b. 1984 Network Liberty Alliance
- c. Antimilitarism

- d. The Sonja Haynes Stone Center for Black Culture and History

Guidance: level 1

:: Torture ::

_________ is any action, surgical, chemical, or otherwise, by which an individual loses use of the testicles: the male reproductive gland or gonad. Surgical _________ is bilateral orchiectomy , and chemical _________ uses pharmaceutical drugs to deactivate the testes. _________ causes sterilization ; it also greatly reduces the production of certain hormones, such as testosterone. Surgical _________ in animals is often called neutering.

Exam Probability: **High**

56. *Answer choices:*

(see index for correct answer)

- a. Torture Abolition and Survivors Support Coalition
- b. Castration
- c. Torture chamber
- d. Long-term complications of standing

Guidance: level 1

:: War ::

World War II , also known as the Second World War, was a global war that lasted from 1939 to 1945. The vast majority of the world`s countries—including all the great powers—eventually formed two opposing military alliances: the Allies and the Axis. A state of total war emerged, directly involving more than 100 million people from over 30 countries. The major participants threw their entire economic, industrial, and scientific capabilities behind the war effort, blurring the distinction between civilian and military resources. _________ was the deadliest conflict in human history, marked by 50 to 85 million fatalities, most of whom were civilians in the Soviet Union and China. It included massacres, the genocide of the Holocaust, strategic bombing, premeditated death from starvation and disease, and the only use of nuclear weapons in war.

Exam Probability: **High**

57. *Answer choices:*

(see index for correct answer)

- a. World War II
- b. Progressive war
- c. Civilian casualty ratio
- d. Philosophy of war

Guidance: level 1

:: Drug rehabilitation ::

__________ are judicially supervised court dockets that provide a sentencing alternative of treatment combined with supervision for people living with serious substance use and mental health disorders. __________ are problem-solving courts that take a public health approach using a specialized model in which the judiciary, prosecution, defense bar, probation, law enforcement, mental health, social service, and treatment communities work together to help addicted offenders into long-term recovery.

Exam Probability: **Medium**

58. *Answer choices:*

(see index for correct answer)

- a. Memory erasure
- b. Kent/Sussex Detoxification Center
- c. Sober companion
- d. The Cabin Chiang Mai

Guidance: level 1

:: Individualism ::

In developmental psychology and moral, political, and bioethical philosophy, _________ is the capacity to make an informed, uncoerced decision. Autonomous organizations or institutions are independent or self-governing. _________ can also be defined from a human resources perspective, where it denotes a level of discretion granted to an employee in his or her work. In such cases, _________ is known to generally increase job satisfaction. _________ is a term that is also widely used in the field of medicine — personal _________ is greatly recognized and valued in health care.

Exam Probability: **Medium**

59. *Answer choices:*

(see index for correct answer)

- a. Autonomy
- b. Psychological egoism
- c. existential
- d. Individualism Old and New

Guidance: level 1

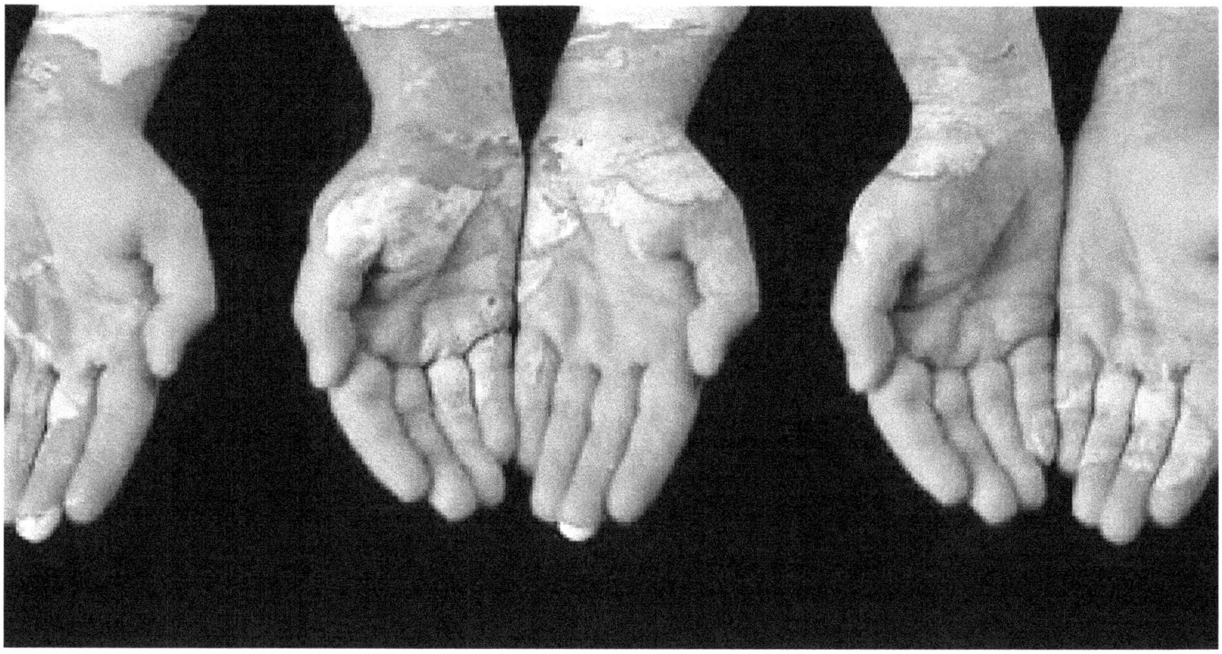

Social change

Social change is an alteration in the social order of a society. Social change may include changes in nature, social institutions, social behaviours, or social relations.

Social change usually refers to the notion of social progress or sociocultural evolution, the philosophical idea that society moves forward by evolutionary means. It may refer to a paradigmatic change in the socio-economic structure, for instance a shift away from feudalism and towards capitalism. Accordingly, it may also refer to social revolution, such as the Socialist revolution presented in, or to other social movements, such as Women's suffrage or the Civil rights movement. Social change may be driven by cultural, religious, economic, scientific or technological forces.

:: Ethical principles ::

_________ , in its broadest context, includes both the attainment of that which is just and the philosophical discussion of that which is just. The concept of _________ is based on numerous fields, and many differing viewpoints and perspectives including the concepts of moral correctness based on ethics, rationality, law, religion, equity and fairness. Often, the general discussion of _________ is divided into the realm of social _________ as found in philosophy, theology and religion, and, procedural _________ as found in the study and application of the law.

Exam Probability: **High**

1. *Answer choices:*

(see index for correct answer)

- a. Misplaced loyalty
- b. JUSTICE
- c. Two wrongs make a right
- d. Loyalty

Guidance: level 1

:: Hostage taking ::

The _________ is a Native American advocacy group in the United States, founded in July 1968 in Minneapolis, Minnesota. AIM was initially formed to address Native American affirmation, treaty issues, spirituality, and leadership while simultaneously addressing incidents of police harassment and racism against Natives forced to move away from reservations and tribal culture by the Indian Termination Policies. AIM`s paramount objective is to create "real economic independence for the Indians".

Exam Probability: **Medium**

2. *Answer choices:*

(see index for correct answer)

- a. American Indian Movement
- b. Foreign hostages in Pakistan
- c. United Nations Security Council Resolution 638
- d. Norrmalmstorg robbery

Guidance: level 1

:: Sociology of culture ::

_________ is the act of judging another culture based on preconceptions that are found in the values and standards of one`s own culture – especially regarding language, behavior, customs, and religion. These aspects or categories are distinctions that define each ethnicity`s unique cultural identity.

Exam Probability: **High**

3. *Answer choices:*

(see index for correct answer)

- a. Symbolic boundaries
- b. Culture shock
- c. Ethnocentrism
- d. The Invention of Art: A Cultural History

Guidance: level 1

:: Welfare ::

_________ is often used to describe a government`s bestowal of money grants, tax breaks, or other special favorable treatment for corporations. It highlights how wealthy corporations are less in need of such treatment than the poor.

Exam Probability: **Low**

4. *Answer choices:*

(see index for correct answer)

- a. Telecare
- b. Welfare capitalism
- c. Anti-austerity protests

- d. Welfare technology

Guidance: level 1

:: Child abuse ::

_________ or child maltreatment is physical, sexual, and/or psychological maltreatment or neglect of a child or children, especially by a parent or a caregiver. _________ may include any act or failure to act by a parent or a caregiver that results in actual or potential harm to a child, and can occur in a child's home, or in the organizations, schools or communities the child interacts with.

Exam Probability: **Low**

5. *Answer choices:*

(see index for correct answer)

- a. Lost children of Francoism
- b. Child abuse
- c. Covert incest
- d. Outcry witness

Guidance: level 1

:: Child sexual abuse ::

In common law jurisdictions, _________ is nonforcible sexual activity in which one of the individuals is below the age of consent . Although it usually refers to adults engaging in sexual contact with minors under the age of consent, it is a generic term, and very few jurisdictions use the actual term _________ in the language of statutes.

Exam Probability: **Low**

6. *Answer choices:*

(see index for correct answer)

- a. Duplessis Orphans
- b. Statutory rape
- c. North Wales child abuse scandal
- d. Harish Iyer

Guidance: level 1

:: Non-governmental organizations ::

_________ is a major international humanitarian agency delivering emergency relief and long-term international development projects. Founded in 1945, _________ is nonsectarian, impartial, and non-governmental. It is one of the largest and oldest humanitarian aid organizations focused on fighting global poverty. In 2016, _________ reported working in 94 countries, supporting 962 poverty-fighting projects and humanitarian aid projects, and reaching over 80 million people and 256 million people indirectly.

Exam Probability: **Medium**

7. *Answer choices:*

(see index for correct answer)

- a. Atlantic Council of Serbia
- b. CARE
- c. Eurodad
- d. Olivaint Conference of Belgium

Guidance: level 1

:: Social issues ::

__________ is a category of transnational, national, or local groupings of highly centralized enterprises run by criminals who intend to engage in illegal activity, most commonly for profit. Some criminal organizations, such as terrorist groups, are politically motivated. Sometimes criminal organizations force people to do business with them, such as when a gang extorts money from shopkeepers for "protection". Gangs may become disciplined enough to be considered organized. A criminal organization or gang can also be referred to as a mafia, mob, or crime syndicate; the network, subculture and community of criminals may be referred to as the underworld. European sociologists define the mafia as a type of __________ group that specializes in the supply of extra-legal protection and quasi law enforcement. Gambetta`s classic work on the Sicilian Mafia generates an economic study of the mafia, which exerts great influence on studies of the Russian Mafia, the Chinese Mafia, Hong Kong Triads and the Japanese Yakuza.

Exam Probability: **Medium**

8. *Answer choices:*

(see index for correct answer)

- a. Mental health in East Africa
- b. Social innovation
- c. The Globalized City
- d. Organized crime

Guidance: level 1

:: Antinatalism ::

A _________ is a person who selectively breeds carefully selected mates, normally of the same breed to sexually reproduce offspring with specific, consistently replicable qualities and characteristics. This might be as a farmer, agriculturalist, or hobbyist, and can be practiced on a large or small scale, for food, fun, or profit.

Exam Probability: **High**

9. *Answer choices:*

(see index for correct answer)

- a. National Alliance for Optional Parenthood
- b. Fernando Vallejo
- c. David Benatar
- d. No Kidding!

Guidance: level 1

:: Critical race theory ::

__________ is a theoretical framework in the social sciences that uses critical theory to examine society and culture as they relate to categorizations of race, law, and power. It began as a theoretical movement within American law schools in the mid- to late 1980s as a reworking of critical legal studies on race issues and is loosely unified by two common themes: First, CRT proposes that white supremacy and racial power are maintained over time, and in particular, that the law may play a role in this process. Second, CRT work has investigated the possibility of transforming the relationship between law and racial power, and more broadly, pursues a project of achieving racial emancipation and anti-subordination. Scholars important to the theory include Derrick Bell, Patricia Williams, Richard Delgado, Kimberlé Williams Crenshaw, Camara Phyllis Jones, and Mari Matsuda. By 2002, over 20 American law schools and at least three law schools in other countries offered __________ courses or classes which covered the issue centrally. __________ is taught and innovated in the fields of education, law, political science, women`s studies, ethnic studies, communication, and American studies.

Exam Probability: **Low**

10. *Answer choices:*

(see index for correct answer)

- a. Elizabeth V. Spelman
- b. Bell hooks
- c. Kiarina Kordela
- d. Peggy McIntosh

Guidance: level 1

:: Crimes against humanity ::

_________ is the act of deliberately inflicting severe physical or psychological suffering on someone by another as a punishment or in order to fulfill some desire of the _________ r or force some action from the victim. _________ , by definition, is a knowing and intentional act; deeds which unknowingly or negligently inflict suffering or pain, without a specific intent to do so, are not typically considered _________ .

Exam Probability: **Medium**

11. *Answer choices:*

(see index for correct answer)

- a. Holodomor
- b. Death march
- c. Human rights violations during the Military government of Chile
- d. Torture

Guidance: level 1

:: Demography ::

__________ or resettlement is the movement of a large group of people from one region to another, often a form of forced migration imposed by state policy or international authority and most frequently on the basis of ethnicity or religion but also due to economic development. Banishment or exile is a similar process, but is forcibly applied to individuals and groups.

Exam Probability: **High**

12. *Answer choices:*

(see index for correct answer)

- a. Dependency ratio
- b. Population transfer
- c. Gestational age
- d. Household

Guidance: level 1

:: Internments ::

__________ is the imprisonment of people, commonly in large groups, without charges or intent to file charges, and thus no trial. The term is especially used for the confinement "of enemy citizens in wartime or of terrorism suspects". Thus, while it can simply mean imprisonment, it tends to refer to preventive confinement, rather than confinement after having been convicted of some crime. Use of these terms is subject to debate and political sensitivities.

Exam Probability: **High**

13. *Answer choices:*

(see index for correct answer)

- a. Internment
- b. Defence Regulation 18B
- c. Japanese Canadian internment
- d. Frank Riethmuller

Guidance: level 1

:: Demography ::

__________ , sometimes abbreviated ZPG , is a condition of demographic balance where the number of people in a specified population neither grows nor declines, considered as a social aim by some.According to some, __________ , perhaps after stabilizing at some optimum population, is the ideal towards which countries and the whole world should aspire in the interests of accomplishing long-term environmental sustainability. What it means by 'the number of people neither grows nor declines' is that births plus in-migrants equal deaths plus out-migrants.

Exam Probability: **Low**

14. *Answer choices:*

(see index for correct answer)

- a. Decrement table
- b. Population transfer
- c. Longevity claims
- d. Urban density

Guidance: level 1

:: Itinerant living ::

A _________ is a migrant worker or homeless vagrant, especially one who is impoverished. The term originated in the Western—probably Northwestern—United States around 1890. Unlike a "tramp", who works only when forced to, and a "bum", who does not work at all, a " _________ " is a traveling worker.

Exam Probability: **High**

15. *Answer choices:*

(see index for correct answer)

- a. Bohemianism
- b. Post-vacation blues
- c. Hobo
- d. Lifestyle travelling

Guidance: level 1

:: Positive psychology ::

_________ is both a research area and a practical skill encompassing the ability of an individual or organization to "lead" or guide other individuals, teams, or entire organizations. Specialist literature debates various viewpoints, contrasting Eastern and Western approaches to _________ , and also United States versus European approaches. U.S. academic environments define _________ as "a process of social influence in which a person can enlist the aid and support of others in the accomplishment of a common task".

Exam Probability: **Low**

16. *Answer choices:*

(see index for correct answer)

- a. Leadership
- b. Character Strengths and Virtues
- c. Project Happiness
- d. Tal Ben-Shahar

Guidance: level 1

:: Bullying ::

_________ is a type of harassment technique that relates to a sexual nature and the unwelcome or inappropriate promise of rewards in exchange for sexual favors. _________ includes a range of actions from mild transgressions to sexual abuse or assault. Harassment can occur in many different social settings such as the workplace, the home, school, churches, etc. Harassers or victims may be of any gender.

Exam Probability: **Low**

17. *Answer choices:*

(see index for correct answer)

- a. Dehumanization
- b. Embarrassment
- c. Physical abuse
- d. Mobbing

Guidance: level 1

:: Ageing ::

A _________ is a person who has lived to the age of 100 years. Because life expectancies worldwide are below 100 years, the term is invariably associated with longevity. In 2012, the United Nations estimated that there were 316,600 living _________ s worldwide.

Exam Probability: **High**

18. *Answer choices:*

(see index for correct answer)

- a. How to Live Forever
- b. Jean Carper
- c. Age and Ageing
- d. Centenarian

Guidance: level 1

:: Hate crime ::

_________ is a statement intended to demean and brutalize another, or the use of cruel and derogatory language or gestures on the basis of real or alleged membership in a social group. _________ is speech that attacks a person or a group on the basis of protected attributes such as race, religion, ethnic origin, national origin, sex, disability, sexual orientation, or gender identity. The laws of some countries describe _________ as speech, gestures, conduct, writing, or displays that incite violence or prejudicial actions against a group or individuals on the basis of their membership in the group, or disparages or intimidates a group, or individuals on the basis of their membership in the group. The law may identify a group based on certain characteristics. In some countries, _________ is not a legal term. Additionally in some countries, including the United States, _________ is constitutionally protected.

Exam Probability: **Low**

19. *Answer choices:*

(see index for correct answer)

- a. Gendercide
- b. Melissa King assault case
- c. Price tag policy
- d. Hate speech

Guidance: level 1

:: Criminology ::

__________ is the attempt to reduce and deter crime and criminals. It is applied specifically to efforts made by governments to reduce crime, enforce the law, and maintain criminal justice.

Exam Probability: **Low**

20. *Answer choices:*

(see index for correct answer)

- a. Social control theory
- b. Marxist criminology
- c. Differential identification
- d. Crime prevention

Guidance: level 1

:: Social movements ::

_________ is the process of getting resources from resource provider, using different mechanisms, to implement an organization`s pre-determined goals.

Exam Probability: **High**

21. *Answer choices:*

(see index for correct answer)

- a. Men Against Rape and Discrimination
- b. Chain Reaction 2008
- c. Opportunity structures
- d. Clean living movement

Guidance: level 1

:: Courts ::

A _________ is any person or institution with authority to judge or adjudicate, often as a government institution, with the authority to adjudicate legal disputes between parties and carry out the administration of justice in civil, criminal, and administrative matters in accordance with the rule of law. In both common law and civil law legal systems, _________ s are the central means for dispute resolution, and it is generally understood that all people have an ability to bring their claims before a _________ . Similarly, the rights of those accused of a crime include the right to present a defense before a _________ .

Exam Probability: **High**

22. *Answer choices:*

(see index for correct answer)

- a. Chief Justice of Kenya
- b. Mental health court
- c. Oyer and terminer
- d. Supreme Court of Palau

Guidance: level 1

:: Cultural assimilation ::

_________ is the process during which newcomers or minorities are incorporated into the social structure of the host society.

Exam Probability: **High**

23. *Answer choices:*

(see index for correct answer)

- a. Anglicisation
- b. Castilianization
- c. Social integration
- d. Cultural leveling

Guidance: level 1

:: Social justice ::

__________ is a concept of fair and just relations between the individual and society. This is measured by the explicit and tacit terms for the distribution of wealth, opportunities for personal activity, and social privileges. In Western as well as in older Asian cultures, the concept of __________ has often referred to the process of ensuring that individuals fulfill their societal roles and receive what was their due from society. In the current global grassroots movements for __________ , the emphasis has been on the breaking of barriers for social mobility, the creation of safety nets and economic justice.

Exam Probability: **Low**

24. *Answer choices:*

(see index for correct answer)

- a. Social Justice Party
- b. Education, Citizenship and Social Justice
- c. Social justice
- d. International Society for Justice Research

Guidance: level 1

:: White supremacy in the United States ::

_________ is an anti-Semitic, neo-Nazi, white supremacist terrorist organization that was originally based in Hayden, Idaho. Richard Girnt Butler founded the group in the 1970s, as an arm of the Christian Identity organization Church of Jesus Christ–Christian. The Federal Bureau of Investigation has called _________ a "terrorist threat", and the RAND Corporation has called it the "first truly nationwide terrorist network" in the United States.

Exam Probability: **Medium**

25. *Answer choices:*

(see index for correct answer)

- a. The Turner Diaries
- b. From Swastika to Jim Crow
- c. Colfax massacre
- d. Robert Wilson Shufeldt

Guidance: level 1

:: Internet censorship ::

The __________ was a series of anti-government protests, uprisings, and armed rebellions that spread across the Middle East in late 2010. It began in response to oppressive regimes and a low standard of living, beginning with protests in Tunisia . In the news, social media has been heralded as the driving force behind the swift spread of revolution throughout the world, as new protests appear in response to success stories shared from those taking place in other countries . In many countries, the governments have also recognized the importance of social media for organizing and have shut down certain sites or blocked Internet service entirely, especially in the times preceding a major rally . Governments have also scrutinized or suppressed discussion in those forums through accusing content creators of unrelated crimes or shutting down communication on specific sites or groups, such as through Facebook .

Exam Probability: **Low**

26. *Answer choices:*

(see index for correct answer)

- a. Internet kill switch
- b. Arab Spring
- c. Consent of the Networked
- d. Ultrasurf

Guidance: level 1

:: AIDS activism ::

_________ is a French community-based non-profit organisation that was founded in 1984 by Daniel Defert, following the death of his partner Michel Foucault. The name is a word play on _________ and AIDS hence the name.

Exam Probability: **High**

27. *Answer choices:*

(see index for correct answer)

- a. Chi Heng Foundation
- b. Elizabeth Glaser Pediatric AIDS Foundation
- c. Red Hot Organization
- d. AIDES

Guidance: level 1

:: Marriage ::

_________ is the marriage of two persons of the same sex or gender, entered into in a civil or religious ceremony.

Exam Probability: **Medium**

28. *Answer choices:*

(see index for correct answer)

- a. Trailing spouse
- b. Shotgun wedding
- c. Marriage and health
- d. Same-sex marriage

Guidance: level 1

:: Imagination ::

_________ is the ability to produce and simulate novel objects, peoples and ideas in the mind without any immediate input of the senses. It is also described as the forming of experiences in the mind, which can be recreations of past experiences such as vivid memories with imagined changes or that they are completely invented. _________ helps make knowledge applicable in solving problems and is fundamental to integrating experience and the learning process. A basic training for _________ is listening to storytelling , in which the exactness of the chosen words is the fundamental factor to "evoke worlds".

Exam Probability: **Low**

29. *Answer choices:*

(see index for correct answer)

- a. Worldbuilding
- b. Hypotheticals
- c. Imagination
- d. Hypothesis

Guidance: level 1

:: Demography ::

_________ , also known as contraception and fertility control, is a method or device used to prevent pregnancy. _________ has been used since ancient times, but effective and safe methods of _________ only became available in the 20th century. Planning, making available, and using _________ is called family planning. Some cultures limit or discourage access to _________ because they consider it to be morally, religiously, or politically undesirable.

Exam Probability: **Medium**

30. *Answer choices:*

(see index for correct answer)

- a. Birth control
- b. Sex ratio
- c. Population momentum
- d. Preston curve

Guidance: level 1

:: Public housing ::

Social housing is an umbrella term referring to rental housing which may be owned and managed by the state, by non-profit organizations, or by a combination of the two, usually with the aim of providing affordable housing. Social housing can also be seen as a potential remedy to housing inequality.

Exam Probability: **High**

31. *Answer choices:*

(see index for correct answer)

- a. American Radburn design for public housing
- b. Million Programme
- c. Gemeindebau
- d. Public housing

Guidance: level 1

:: Bullying ::

__________ covers a wide range of behaviors of an offensive nature. It is commonly understood as behavior that demeans, humiliates or embarrasses a person, and it is characteristically identified by its unlikelihood in terms of social and moral reasonableness. In the legal sense, these are behaviors that appear to be disturbing, upsetting or threatening. They evolve from discriminatory grounds, and have an effect of nullifying or impairing a person from benefiting their rights. When these behaviors become repetitive they are defined as bullying.

Exam Probability: **Medium**

32. *Answer choices:*

(see index for correct answer)

- a. Harassment
- b. Sexual bullying
- c. SuEllen Fried
- d. Wedgie

Guidance: level 1

:: Feminist theory ::

_________ is a perspective within feminism that calls for a radical reordering of society in which male supremacy is eliminated in all social and economic contexts.

Exam Probability: **High**

33. *Answer choices:*

(see index for correct answer)

- a. Chicana feminism
- b. Compulsory Heterosexuality and Lesbian Existence
- c. When We Dead Awaken: Writing as Re-Vision
- d. Hegemonic masculinity

Guidance: level 1

:: Communism-based civil wars ::

The _________ , also known as the Second Indochina War, and in Vietnam as the Resistance War Against America or simply the American War, was an undeclared war in Vietnam, Laos, and Cambodia from 1 November 1955 to the fall of Saigon on 30 April 1975. It was the second of the Indochina Wars and was officially fought between North Vietnam and South Vietnam. North Vietnam was supported by the Soviet Union, China, and other communist allies; South Vietnam was supported by the United States, South Korea, the Philippines, Australia, Thailand and other anti-communist allies. The war is considered a Cold War-era proxy war from some US perspectives. It lasted some 19 years with direct U.S. involvement ending in 1973 following the Paris Peace Accords, and included the Laotian Civil War and the Cambodian Civil War, resulting in all three countries becoming communist states in 1975.

Exam Probability: **Low**

34. *Answer choices:*

(see index for correct answer)

- a. Insurgency of the Communist Party of the Philippines
- b. Vietnam War
- c. Cambodian Civil War

Guidance: level 1

:: Recessions ::

In economics, a __________ is a business cycle contraction when there is a general decline in economic activity. Macroeconomic indicators such as GDP , investment spending, capacity utilization, household income, business profits, and inflation fall, while bankruptcies and the unemployment rate rise. In the United Kingdom, it is defined as a negative economic growth for two consecutive quarters.

Exam Probability: **High**

35. *Answer choices:*

(see index for correct answer)

- a. Early 2000s recession
- b. Recession shapes
- c. Recession of 1949
- d. Rolling recession

Guidance: level 1

:: Philanthropy ::

__________ s, nongovernmental organizations, or nongovernment organizations, commonly referred to as NGOs, are usually non-profit and sometimes international organizations independent of governments and international governmental organizations that are active in humanitarian, educational, health care, public policy, social, human rights, environmental, and other areas to effect changes according to their objectives. They are thus a subgroup of all organizations founded by citizens, which include clubs and other associations that provide services, benefits, and premises only to members. Sometimes the term is used as a synonym of "civil society organization" to refer to any association founded by citizens, but this is not how the term is normally used in the media or everyday language, as recorded by major dictionaries. The explanation of the term by NGO.org is ambivalent. It first says an NGO is any non-profit, voluntary citizens` group which is organized on a local, national or international level, but then goes on to restrict the meaning in the sense used by most English speakers and the media: Task-oriented and driven by people with a common interest, NGOs perform a variety of service and humanitarian functions, bring citizen concerns to Governments, advocate and monitor policies and encourage political participation through provision of information.

Exam Probability: **Low**

36. *Answer choices:*

(see index for correct answer)

- a. Non-governmental organization
- b. Democratized transactional giving
- c. The Gospel of Wealth
- d. The Ripple Effect

Guidance: level 1

:: Female foeticide and infanticide ::

_________ is the deliberate killing of newborn female children. In countries with a history of _________ , the modern practice of sex-selective abortion is often discussed as a closely related issue. _________ is a major cause of concern in several nations such as China and India. It has been argued that the low status in which women are viewed in patriarchal societies creates a bias against females.

Exam Probability: **Low**

37. *Answer choices:*

(see index for correct answer)

- a. Sex-selective abortion
- b. Female infanticide
- c. Missing women of Asia
- d. Female infanticide in Pakistan

Guidance: level 1

:: Criminology ::

In criminology, _________ refers to crimes committed either by a corporation , or by individuals acting on behalf of a corporation or other business entity . For the worst _________ s, corporations may face judicial dissolution, sometimes called the "corporate death penalty", which is a legal procedure in which a corporation is forced to dissolve or cease to exist.

Exam Probability: **Low**

38. *Answer choices:*

(see index for correct answer)

- a. Symbolic interactionism
- b. Corporate crime
- c. Crime science
- d. Inquest

Guidance: level 1

:: Criminology ::

__________ is the scientific study of the nature, extent, management, causes, control, consequences, and prevention of criminal behavior, both on individual and social levels. __________ is an interdisciplinary field in both the behavioral and social sciences, which draws primarily upon the research of sociologists, psychologists, philosophers, psychiatrists, biologists, social anthropologists, as well as scholars of law.

Exam Probability: **Medium**

39. *Answer choices:*

(see index for correct answer)

- a. Predictive policing
- b. Minneapolis Domestic Violence Experiment

- c. Criminology
- d. Chicago school

Guidance: level 1

:: Femininity ::

_________ is a set of attributes, behaviors, and roles generally associated with girls and women. _________ is socially constructed, but made up of both socially-defined and biologically-created factors. This makes it distinct from the definition of the biological female sex, as both males and females can exhibit feminine traits.

Exam Probability: **Medium**

40. *Answer choices:*

(see index for correct answer)

- a. Femininity
- b. Whipping Girl
- c. Femminiello
- d. Marianismo

Guidance: level 1

:: Resource economics ::

_________ is the consumption of a resource faster than it can be replenished. Natural resources are commonly divided between renewable resources and non-renewable resources . Use of either of these forms of resources beyond their rate of replacement is considered to be _________ . The value of a resource is a direct result of its availability in nature and the cost of extracting the resource, the more a resource is depleted the more the value of the resource increases. There are several types of _________ the most known being; Aquifer depletion, deforestation, mining for fossil fuels and minerals, pollution or contamination of resources, slash-and-burn agricultural practices, Soil erosion, and overconsumption, excessive or unnecessary use of resources.

Exam Probability: **Medium**

41. *Answer choices:*

(see index for correct answer)

- a. Crude
- b. Resource depletion
- c. World Resources Forum
- d. Backstop resources

Guidance: level 1

:: Commercial crimes ::

__________ is the act of withholding assets for the purpose of conversion of such assets, by one or more persons to whom the assets were entrusted, either to be held or to be used for specific purposes. __________ is a type of financial fraud. For example, a lawyer might embezzle funds from the trust accounts of their clients; a financial advisor might embezzle the funds of investors; and a husband or a wife might embezzle funds from a bank account jointly held with the spouse.

Exam Probability: **High**

42. *Answer choices:*

(see index for correct answer)

- a. Late trading
- b. Fence
- c. Global Forum on Transparency and Exchange of Information for Tax Purposes
- d. Clearstream

Guidance: level 1

:: Economic globalization ::

__________ is one of the three main dimensions of globalization commonly found in academic literature, with the two others being political globalization and cultural globalization, as well as the general term of globalization. __________ refers to the widespread international movement of goods, capital, services, technology and information. It is the increasing economic integration and interdependence of national, regional, and local economies across the world through an intensification of cross-border movement of goods, services, technologies and capital. __________ primarily comprises the globalization of production, finance, markets, technology, organizational regimes, institutions, corporations, and labour.

Exam Probability: **Low**

43. *Answer choices:*

(see index for correct answer)

- a. Development economics
- b. World currency
- c. global financial
- d. Economic globalization

Guidance: level 1

:: Development economics ::

_________ programs consist of loans provided by the International Monetary Fund and the World Bank to countries that experienced economic crises. The two Bretton Woods Institutions require borrowing countries to implement certain policies in order to obtain new loans . The conditionality clauses attached to the loans have been criticized because of their effects on the social sector.

Exam Probability: **Low**

44. *Answer choices:*

(see index for correct answer)

- a. Dependency theory
- b. Structural adjustment
- c. Critical minimum effort theory
- d. United Nations Economic Commission for Latin America and the Caribbean

Guidance: level 1

:: Crimes against humanity ::

__________ is any system in which principles of property law are applied to people, allowing individuals to own, buy and sell other individuals, as a de jure form of property. A slave is unable to withdraw unilaterally from such an arrangement and works without remuneration. Many scholars now use the term chattel __________ to refer to this specific sense of legalised, de jure __________. In a broader sense, however, the word __________ may also refer to any situation in which an individual is de facto forced to work against their own will. Scholars also use the more generic terms such as unfree labour or forced labour to refer to such situations. However, and especially under __________ in broader senses of the word, slaves may have some rights and protections according to laws or customs.

Exam Probability: **High**

45. *Answer choices:*

(see index for correct answer)

- a. Rome Statute of the International Criminal Court
- b. Slavery
- c. Nuremberg principles
- d. Tokyo Charter

Guidance: level 1

:: Discrimination ::

_________ is the belief in the superiority of one race over another, which often results in discrimination and prejudice towards people based on their race or ethnicity. The use of the term " _________ " does not easily fall under a single definition.

Exam Probability: **Low**

46. *Answer choices:*

(see index for correct answer)

- a. Legacy preferences
- b. Racial quota
- c. Sexism
- d. Racism

Guidance: level 1

:: Nonviolent resistance movements ::

_________ was a left-wing protest movement that began on September 17, 2011, in Zuccotti Park, located in New York City`s Wall Street financial district, against economic inequality.

Exam Probability: **Low**

47. *Answer choices:*

(see index for correct answer)

- a. Roscommon Hospital Action Committee
- b. Guruvayur Satyagraha
- c. Free Trade Unions of the Coast
- d. Budrus

Guidance: level 1

:: Social movements ::

Civil _________ is the active, professed refusal of a citizen to obey certain laws, demands, orders or commands of a government or occupying international power. Civil _________ is sometimes defined as having to be nonviolent to be called civil _________ . Civil _________ is sometimes, therefore, equated with nonviolent resistance.

Exam Probability: **High**

48. *Answer choices:*

(see index for correct answer)

- a. Disobedience
- b. YaLa-Young Leaders
- c. Lomaland
- d. Leaderless resistance

Guidance: level 1

:: Sexual abuse ::

_________ , also referred to as molestation, is usually undesired sexual behavior by one person upon another. It is often perpetrated using force or by taking advantage of another. When force is immediate, of short duration, or infrequent, it is called sexual assault. The offender is referred to as a _________ r or molester. The term also covers any behavior by an adult or older adolescent towards a child to stimulate any of the involved sexually. The use of a child, or other individuals younger than the age of consent, for sexual stimulation is referred to as child _________ or statutory rape.

Exam Probability: **High**

49. *Answer choices:*

(see index for correct answer)

- a. National Sexual Violence Resource Center
- b. Sexual abuse
- c. Agraphobia
- d. Rind et al. controversy

Guidance: level 1

:: Pornography ::

_________ is the portrayal of sexual subject matter for the exclusive purpose of sexual arousal. _________ may be presented in a variety of media, including books, magazines, postcards, photographs, sculpture, drawing, painting, animation, sound recording, phone calls, writing, film, video, and video games. The term applies to the depiction of the act rather than the act itself, and so does not include live exhibitions like sex shows and striptease. The primary subjects of present-day pornographic depictions are pornographic models, who pose for still photographs, and pornographic actors or "porn stars", who perform in pornographic films. If dramatic skills are not involved, a performer in pornographic media may also be called a model.

Exam Probability: **Medium**

50. *Answer choices:*

(see index for correct answer)

- a. Ethnic pornography
- b. Religious views on pornography
- c. Pornography
- d. War porn

Guidance: level 1

:: Bioethics ::

_________ —also known as germ warfare—is the use of biological toxins or infectious agents such as bacteria, viruses, and fungi with the intent to kill or incapacitate humans, animals or plants as an act of war. Biological weapons are living organisms or replicating entities that reproduce or replicate within their host victims. Entomological warfare is also considered a type of biological weapon. This type of warfare is distinct from nuclear warfare and chemical warfare, which together with _________ make up NBC, the military initialism for nuclear, biological, and chemical warfare using weapons of mass destruction . None of these are considered conventional weapons, which are deployed primarily for their explosive, kinetic, or incendiary potential.

Exam Probability: **Low**

51. *Answer choices:*

(see index for correct answer)

- a. Operation Cauldron
- b. Gene therapy
- c. Metabolic supermice
- d. Biological warfare

Guidance: level 1

:: Discrimination ::

__________ is stereotyping and discrimination against individuals or groups on the basis of their age. This may be casual or systematic. The term was coined in 1969 by Robert Neil Butler to describe discrimination against seniors, and patterned on sexism and racism. Butler defined " __________ " as a combination of three connected elements. Among them were prejudicial attitudes towards older people, old age, and the aging process; discriminatory practices against older people; and institutional practices and policies that perpetuate stereotypes about elderly people.

Exam Probability: **Low**

52. *Answer choices:*

(see index for correct answer)

- a. Dalit
- b. Surdophobia
- c. Sex Discrimination Commissioner
- d. Reverse racism

Guidance: level 1

:: Demography ::

The existence of some kind of _________ is widely accepted in the social sciences because of the well-established historical correlation linking dropping fertility to social and economic development. Scholars debate whether industrialization and higher incomes lead to lower population, or whether lower populations lead to industrialization and higher incomes. Scholars also debate to what extent various proposed and sometimes inter-related factors such as higher per capita income, higher female income, lower mortality, old-age security, and rise of demand for human capital are involved.

Exam Probability: **Low**

53. *Answer choices:*

(see index for correct answer)

- a. Family planning
- b. Human sex ratio
- c. Chain migration
- d. Demographic transition

Guidance: level 1

:: Sex industry ::

__________ , also known as gender-based violence and sexual and gender-based violence are violent acts primarily or exclusively committed against women and girls. Often considered a form of hate crime, this type of violence is gender-based, meaning that the acts of violence are committed against women and girls expressly because they are female. The UN Declaration on the Elimination of __________ states, " __________ is a manifestation of historically unequal power relations between men and women" and " __________ is one of the crucial social mechanisms by which women are forced into a subordinate position compared with men."

Exam Probability: **Medium**

54. *Answer choices:*

(see index for correct answer)

- a. Violence against women
- b. Prostitution
- c. Forced prostitution
- d. Sexual economy

Guidance: level 1

:: Human migration ::

A " __________ " is a person who either migrates within their home country or outside it to pursue work such as seasonal work. __________ s usually do not have an intention to stay permanently in the country or region in which they work.

Exam Probability: **Low**

55. *Answer choices:*

(see index for correct answer)

- a. Emigration from Ecuador
- b. Vietnamese boat people
- c. Migrant worker
- d. Emigration from Colombia

Guidance: level 1

:: Prejudice and discrimination ::

__________ , also known as reservation in India and Nepal, positive discrimination / action in the United Kingdom, and employment equity in Canada and South Africa, is the policy of promoting the education and employment of members of groups that are known to have previously suffered from discrimination. Historically and internationally, support for __________ has sought to achieve goals such as bridging inequalities in employment and pay, increasing access to education, promoting diversity, and redressing apparent past wrongs, harms, or hindrances.

Exam Probability: **High**

56. *Answer choices:*

(see index for correct answer)

- a. Article 153 of the Constitution of Malaysia
- b. Affirmative action
- c. Albinism in popular culture
- d. Great Gypsy Round-up

Guidance: level 1

:: Wealth ::

__________ is the abundance of valuable financial assets or physical possessions which can be converted into a form that can be used for transactions. This includes the core meaning as held in the originating old English word weal, which is from an Indo-European word stem. A community, region or country that possesses an abundance of such possessions or resources to the benefit of the common good is known as __________ y.

Exam Probability: **Low**

57. *Answer choices:*

(see index for correct answer)

- a. Paper wealth
- b. Private bank
- c. Wealth
- d. Net worth

Guidance: level 1

:: Religion and children ::

_________ is the removal of the foreskin from the human penis. In the most common procedure, the foreskin is opened, adhesions are removed, and the foreskin is separated from the glans. After that, a _________ device may be placed, and then the foreskin is cut off. Topical or locally injected anesthesia is sometimes used to reduce pain and physiologic stress. For adults and children, general anesthesia is an option, and the procedure may be performed without a specialized _________ device. The procedure is most often an elective surgery performed on babies and children, for religious or cultural reasons. In other cases it may be done as a treatment for certain medical conditions or for preventative reasons. Medically it is a treatment option for problematic cases of phimosis, balanoposthitis that does not resolve with other treatments, and chronic urinary tract infections . It is contraindicated in cases of certain genital structure abnormalities or poor general health.

Exam Probability: **Medium**

58. *Answer choices:*

(see index for correct answer)

- a. Child Jesus
- b. Daugherty v. Vanguard
- c. Sunday school
- d. Kumari

Guidance: level 1

:: Teratogens ::

Fetal alcohol spectrum disorders are a group of conditions that can occur in a person whose mother drank alcohol during pregnancy. Problems may include an abnormal appearance, short height, low body weight, small head size, poor coordination, low intelligence, behavior problems, and problems with hearing or seeing. Those affected are more likely to have trouble in school, legal problems, participate in high-risk behaviors, and have trouble with alcohol or other drugs. The most severe form of the condition is known as _________ . Other types include partial _________ , alcohol-related neurodevelopmental disorder and alcohol-related birth defects . Some accept only FAS as a diagnosis, seeing the evidence as inconclusive with respect to other types.

Exam Probability: **Low**

59. *Answer choices:*

(see index for correct answer)

- a. Dacarbazine
- b. Ethanol
- c. N-Nitroso-N-methylurea
- d. Fetal alcohol syndrome

Guidance: level 1

Criminology

Criminology is the scientific study of the nature, extent, management, causes, control, consequences, and prevention of criminal behavior, both on individual and social levels. Criminology is an interdisciplinary field in both the behavioral and social sciences, drawing especially upon the research of sociologists, psychologists, philosophers, psychiatrists, biologists, social anthropologists, as well as scholars of law.

:: Criminology ::

_________ is a process often applied to what are regarded, by those working towards _________ , as victimless crimes, of which one example is the consumption of illegal drugs .

Exam Probability: **Medium**

1. *Answer choices:*

(see index for correct answer)

- a. Radical criminology
- b. Sex differences in crime
- c. Legalization
- d. Reassurance policing

Guidance: level 1

:: Criminology ::

The __________ , also known as the CSI syndrome and the CSI infection, is any of several ways in which the exaggerated portrayal of forensic science on crime television shows such as CSI: Crime Scene Investigation influences public perception. The term was first reported in a 2004 USA Today article describing the effect being made on trial jurors by television programs featuring forensic science. It most often refers to the belief that jurors have come to demand more forensic evidence in criminal trials, thereby raising the effective standard of proof for prosecutors. While this belief is widely held among American legal professionals, some studies have suggested that crime shows are unlikely to cause such an effect, although frequent CSI viewers may place a lower value on circumstantial evidence. As technology improves and becomes more prevalent throughout society, people may also develop higher expectations for the capabilities of forensic technology.

Exam Probability: **Low**

2. *Answer choices:*

(see index for correct answer)

- a. Crime prevention
- b. CSI effect
- c. Electronic harassment
- d. Crime analysis

Guidance: level 1

:: Sociological terminology ::

In the social sciences, _________ is the patterned social arrangements in society that are both emergent from and determinant of the actions of the individuals. On the macro scale, _________ is the system of socioeconomic stratification , social institutions, or, other patterned relations between large social groups. On the meso scale, it is the structure of social network ties between individuals or organizations. On the micro scale, it can be the way norms shape the behavior of individuals within the social system.

Exam Probability: **Low**

3. *Answer choices:*

(see index for correct answer)

- a. Iron cage
- b. Reality distortion field
- c. Social structure

- d. Analytic frame

Guidance: level 1

:: Smuggling ::

_________ is the illegal transportation of objects, substances, information or people, such as out of a house or buildings, into a prison, or across an international border, in violation of applicable laws or other regulations.

Exam Probability: **Low**

4. *Answer choices:*

(see index for correct answer)

- a. Go-fast boat
- b. Rum-running
- c. Illegal drug trade
- d. Smuggling

Guidance: level 1

:: Mental and behavioural disorders ::

A _________ , also called a mental illness or psychiatric disorder, is a behavioral or mental pattern that causes significant distress or impairment of personal functioning. Such features may be persistent, relapsing and remitting, or occur as a single episode. Many disorders have been described, with signs and symptoms that vary widely between specific disorders. Such disorders may be diagnosed by a mental health professional.

Exam Probability: **High**

5. *Answer choices:*

(see index for correct answer)

- a. Sexual maturation disorder
- b. Hallucinogen persisting perception disorder
- c. Acedia
- d. Mental disorder

Guidance: level 1

:: Penal imprisonment ::

A _________ , or for-profit prison, is a place where people are imprisoned by third party that is contracted by a government agency. _________ companies typically enter into contractual agreements with governments that commit prisoners and then pay a per diem or monthly rate, either for each prisoner in the facility, or for each place available, whether occupied or not. Such contracts may be for the operation only of a facility, or for design, construction and operation.

Exam Probability: **Low**

6. *Answer choices:*

(see index for correct answer)

- a. Private prison
- b. European Prison Rules
- c. Prison rape
- d. Maze Prison escape

Guidance: level 1

:: Criminal defenses ::

Coercion is the practice of forcing another party to act in an involuntary manner by use of threats or force. It involves a set of various types of forceful actions that violate the free will of an individual to induce a desired response, for example: a bully demanding lunch money from a student or the student gets beaten. These actions may include extortion, blackmail, torture, threats to induce favors, or even sexual assault. In law, coercion is codified as a _________ crime. Such actions are used as leverage, to force the victim to act in a way contrary to their own interests. Coercion may involve the actual infliction of physical pain/injury or psychological harm in order to enhance the credibility of a threat. The threat of further harm may lead to the cooperation or obedience of the person being coerced.

Exam Probability: **Medium**

7. *Answer choices:*

(see index for correct answer)

- a. Excuse
- b. Twinkie defense
- c. Diminished responsibility
- d. Duress

Guidance: level 1

:: Sex industry ::

_________ is the business or practice of engaging in sexual activity in exchange for payment. _________ is sometimes described as sexual services, commercial sex or, colloquially, hooking. It is sometimes referred to euphemistically as "the world`s oldest profession" in the English-speaking world. A person who works in this field is called a prostitute, and is a type of sex worker.

Exam Probability: **Medium**

8. *Answer choices:*

(see index for correct answer)

- a. Ping pong show
- b. Prostitution
- c. Strip club
- d. Sexual slavery

Guidance: level 1

:: Crimes ::

An _________ is the act of inflicting physical harm or unwanted physical contact upon a person or, in some specific legal definitions, a threat or attempt to commit such an action. It is both a crime and a tort and, therefore, may result in either criminal and/or civil liability. Generally, the common law definition is the same in criminal and tort law.

Exam Probability: **High**

9. *Answer choices:*

(see index for correct answer)

- a. Aggravated battery
- b. Assault
- c. Failure to appear
- d. Coney-catching

Guidance: level 1

:: Terrorism ::

_________ is, in the broadest sense, the use of intentionally indiscriminate violence as a means to create terror among masses of people; or fear to achieve a religious or political aim. It is used in this regard primarily to refer to violence during peacetime or in war against non-combatants . The terms "terrorist" and " _________ " originated during the French Revolution of the late 18th century but gained mainstream popularity in the 1970s in news reports and books covering the conflicts in Northern Ireland, the Basque Country and Palestine. The increased use of suicide attacks from the 1980s onwards was typified by the September 11 attacks in New York City and Washington, D.C. in 2001.

Exam Probability: **Medium**

10. *Answer choices:*

(see index for correct answer)

- a. Terrorism insurance
- b. Radicalization
- c. Terrorism
- d. Lone wolf

Guidance: level 1

:: Criminal law legal terms ::

A _________ is a person accused of committing a crime in criminal prosecution or a person against whom some type of civil relief is being sought in a civil case.

Exam Probability: **Medium**

11. *Answer choices:*

(see index for correct answer)

- a. misdemeanors
- b. Defendant

Guidance: level 1

:: Economic theories ::

_________ theory, also known as choice theory or rational action theory, is a framework for understanding and often formally modeling social and economic behavior. The basic premise of _________ theory is that aggregate social behavior results from the behavior of individual actors, each of whom is making their individual decisions. The theory also focuses on the determinants of the individual choices .

Exam Probability: **High**

12. *Answer choices:*

(see index for correct answer)

- a. Rahn curve
- b. Technological dualism
- c. Shareholder ownership value
- d. Lipstick effect

Guidance: level 1

:: Sociobiology ::

_________ is a social system in which men hold primary power and predominate in roles of political leadership, moral authority, social privilege and control of property. Some patriarchal societies are also patrilineal, meaning that property and title are inherited by the male lineage.

Exam Probability: **Low**

13. *Answer choices:*

(see index for correct answer)

- a. Alate
- b. Queen ant
- c. Patriarchy
- d. Human Universals

Guidance: level 1

:: Political science ::

_________ is the principled guide to action taken by the administrative executive branches of the state with regard to a class of issues, in a manner consistent with law and institutional customs.

Exam Probability: **High**

14. *Answer choices:*

(see index for correct answer)

- a. Policy analysis
- b. Jose Azel
- c. Ernesto Laclau
- d. Deep politics

Guidance: level 1

:: Critical theory ::

An _________ is a collection of normative beliefs and values that an individual or group holds for other than purely epistemic reasons. In other words, these rely on basic assumptions about reality that may or may not have any factual basis. The term is especially used to describe systems of ideas and ideals which form the basis of economic or political theories and resultant policies. In these there are tenuous causal links between policies and outcomes owing to the large numbers of variables available, so that many key assumptions have to be made. In political science the term is used in a descriptive sense to refer to political belief systems

Exam Probability: **Medium**

15. *Answer choices:*

(see index for correct answer)

- a. Critical Psychiatry Network
- b. Post-Critical
- c. Ideology
- d. Critical Practice

Guidance: level 1

:: Crime data ::

The __________ , administered by the Bureau of Justice Statistics, is a national survey of approximately 49,000 to 77,400 households twice a year in the United States, on the frequency of crime victimization, as well as characteristics and consequences of victimization. The survey focuses on gathering information on the following crimes: assault, burglary, larceny, motor vehicle theft, rape, and robbery. The survey results are used for the purposes of building a crime index. It has been used in comparison with the Uniform Crime Reports and the National Incident-Based Reporting System to assess the dark figure of crime. The NCVS survey is comparable to the British Crime Survey conducted in the United Kingdom.

Exam Probability: **Low**

16. *Answer choices:*

(see index for correct answer)

- a. Uniform Crime Reports
- b. ODIS
- c. Crime Survey for England and Wales

Guidance: level 1

:: Social problems ::

In general, _________ is a form of dishonesty or criminal activity undertaken by a person or organization entrusted with a position of authority, often to acquire illicit benefit. _________ may include many activities including bribery and embezzlement, though it may also involve practices that are legal in many countries. Political _________ occurs when an office-holder or other governmental employee acts in an official capacity for personal gain. _________ is most commonplace in kleptocracies, oligarchies, narco-states and mafia states.

Exam Probability: **High**

17. *Answer choices:*

(see index for correct answer)

- a. Sedentism
- b. Corruption
- c. Disease
- d. Debt

Guidance: level 1

:: International law ::

__________ is the practical authority granted to a legal body to administer justice within a defined field of responsibility, e.g., Michigan tax law. In federations like the United States, areas of __________ apply to local, state, and federal levels; e.g. the court has __________ to apply federal law.

Exam Probability: **Medium**

18. *Answer choices:*

(see index for correct answer)

- a. Pacific blockade
- b. The Hague Justice Portal
- c. Terra nullius
- d. De jure belli ac pacis

Guidance: level 1

:: Child pornography ::

_________ is pornography that exploits children for sexual stimulation. It may be produced with the direct involvement or sexual assault of a child or it may be simulated _________ . Abuse of the child occurs during the sexual acts or lascivious exhibitions of genitals or pubic areas which are recorded in the production of _________ . _________ may use a variety of media, including writings, magazines, photos, sculpture, drawing, cartoon, painting, animation, sound recording, film, video, and video games.

Exam Probability: **High**

19. *Answer choices:*

(see index for correct answer)

- a. Operation Cathedral
- b. Child pornography
- c. Sean Paul Lockhart
- d. R v. Fellows; R v. Arnold

Guidance: level 1

:: Ethnicity in politics ::

A _________ is the concept of the place where a: cultural, national, or racial identity had formed . The definition can also mean simply one`s country of birth. When used as a proper noun, the _________ , as well as its equivalents in other languages, often have ethnic nationalist connotations. A _________ may also be referred to as a fatherland, a motherland, or a mother country, depending on the culture and language of the nationality in question.

Exam Probability: **Low**

20. *Answer choices:*

(see index for correct answer)

- a. Homeland
- b. Partition of Belgium
- c. Diaspora politics
- d. Ethnic interest groups in the United States

Guidance: level 1

:: Economic sociology ::

__________ is the social science that studies how economic activity affects and is shaped by social processes. In general it analyzes how societies progress, stagnate, or regress because of their local or regional economy, or the global economy. Societies are divided into 3 groups: social, cultural and economic.

Exam Probability: **High**

21. *Answer choices:*

(see index for correct answer)

- a. social support
- b. The Philosophy of Money

- c. Economic sociology
- d. Socioeconomics

Guidance: level 1

:: Social philosophy ::

__________ is an economic system based on the private ownership of the means of production and their operation for profit. Characteristics central to __________ include private property, capital accumulation, wage labor, voluntary exchange, a price system, and competitive markets. In a capitalist market economy, decision-making and investment are determined by every owner of wealth, property or production ability in financial and capital markets, whereas prices and the distribution of goods and services are mainly determined by competition in goods and services markets.

Exam Probability: **Medium**

22. *Answer choices:*

(see index for correct answer)

- a. Philosophy of love
- b. Capitalism
- c. Deliberation
- d. Dasein

Guidance: level 1

:: Punishments ::

_________ in criminal law is a period of supervision over an offender, ordered by the court instead of serving time in prison.

Exam Probability: **High**

23. *Answer choices:*

(see index for correct answer)

- a. Probation
- b. Amende honorable
- c. Castigation
- d. Symbole

Guidance: level 1

:: Surveillance ::

In physics, _________ is the change in position of an object with respect to its surroundings in a given interval of time. _________ is mathematically described in terms of displacement, distance, velocity, acceleration, and speed. _________ of a body is observed by attaching a frame of reference to an observer and measuring the change in position of the body relative to that frame.

Exam Probability: **High**

24. *Answer choices:*

(see index for correct answer)

- a. Pedestrian detection
- b. Talking CCTV
- c. Library Awareness Program
- d. Motion

Guidance: level 1

:: Penal imprisonment ::

The __________ is a penal method of the 19th century in which persons worked during the day in groups and were kept in solitary confinement at night, with enforced silence at all times. The silent system evolved during the 1820s at Auburn Prison in Auburn, New York, as an alternative to and modification of the Pennsylvania system of solitary confinement, which it gradually replaced in the United States. Whigs favored this system because it promised to rehabilitate criminals by teaching them personal discipline and respect for work, property, and other people.

Exam Probability: **Low**

25. *Answer choices:*

(see index for correct answer)

- a. Compassionate release
- b. correctional officer

- c. Auburn system
- d. Pay jail

Guidance: level 1

:: Criminal procedure ::

A _________ is a jury – a group of citizens – empowered by law to conduct legal proceedings and investigate potential criminal conduct, and determine whether criminal charges should be brought. A _________ may subpoena physical evidence or a person to testify. A _________ is separate from the courts, which do not preside over its functioning.

Exam Probability: **High**

26. *Answer choices:*

(see index for correct answer)

- a. Exoneration
- b. Duty solicitor
- c. Grand jury
- d. Stay of proceedings

Guidance: level 1

:: Commercial crimes ::

_________ is the act of withholding assets for the purpose of conversion of such assets, by one or more persons to whom the assets were entrusted, either to be held or to be used for specific purposes. _________ is a type of financial fraud. For example, a lawyer might embezzle funds from the trust accounts of their clients; a financial advisor might embezzle the funds of investors; and a husband or a wife might embezzle funds from a bank account jointly held with the spouse.

Exam Probability: **Low**

27. *Answer choices:*

(see index for correct answer)

- a. Copyfraud
- b. National White Collar Crime Center
- c. False advertising
- d. Embezzlement

Guidance: level 1

:: Workplace bullying ::

_________ or occupational violence refers to violence, usually in the form of physical abuse or threat, that creates a risk to the health and safety of an employee or multiple employees. The National Institute for Occupational Safety and Health defines worker on worker, personal relationship, customer/client, and criminal intent all as categories of violence in the workplace. These four categories are further broken down into three levels: Level one displays early warning signs of violence, Level two is slightly more violent, and level three is significantly violent. Many workplaces have initiated programs and protocols to protect their workers as the Occupational Health Act of 1970 states that employers must provide an environment in which employees are free of harm or harmful conditions.

Exam Probability: **Medium**

28. *Answer choices:*

(see index for correct answer)

- a. Micromanagement
- b. Workplace politics
- c. Workplace violence
- d. Sarah J. Tracy

Guidance: level 1

:: Criminal justice ::

__________ is the delivery of justice to those who have committed crimes. The __________ system is a series of government agencies and institutions whose goals are to identify and catch unlawful individuals to inflict a form of punishment on them. Other goals include the rehabilitation of offenders, preventing other crimes, and moral support for victims. The primary institutions of the __________ system are the police, prosecution and defense lawyers, the courts and prisons.

Exam Probability: **Low**

29. *Answer choices:*

(see index for correct answer)

- a. Alphonse Bertillon
- b. Errors of impunity
- c. Office of Juvenile Justice and Delinquency Prevention
- d. Criminal justice

Guidance: level 1

:: Computer crimes ::

__________ is the fraudulent attempt to obtain sensitive information such as usernames, passwords and credit card details by disguising oneself as a trustworthy entity in an electronic communication. Typically carried out by email spoofing or instant messaging, it often directs users to enter personal information at a fake website which matches the look and feel of the legitimate site.

Exam Probability: **High**

30. *Answer choices:*

(see index for correct answer)

- a. Stop Cyberbullying Day
- b. Phishing
- c. Text spy
- d. Cyberstalking

Guidance: level 1

:: United States federal criminal legislation ::

The _________ of 1994, H.R. 3355, Pub.L. 103–322 is an Act of Congress dealing with crime and law enforcement; it became law in 1994. It is the largest crime bill in the history of the United States and consisted of 356 pages that provided for 100,000 new police officers, $9.7 billion in funding for prisons and $6.1 billion in funding for prevention programs, which were designed with significant input from experienced police officers. Sponsored by Representative Jack Brooks of Texas, the bill was originally written by Senator Joe Biden of Delaware and then was passed by Congress and signed into law by President Bill Clinton.

Exam Probability: **High**

31. *Answer choices:*

(see index for correct answer)

- a. Clery Act
- b. Volstead Act
- c. Speedy Trial Act
- d. Violent Crime Control and Law Enforcement Act

Guidance: level 1

:: Behavior ::

__________ is a theory of learning process and social behavior which proposes that new behaviors can be acquired by observing and imitating others. It states that learning is a cognitive process that takes place in a social context and can occur purely through observation or direct instruction, even in the absence of motor reproduction or direct reinforcement. In addition to the observation of behavior, learning also occurs through the observation of rewards and punishments, a process known as vicarious reinforcement. When a particular behavior is rewarded regularly, it will most likely persist; conversely, if a particular behavior is constantly punished, it will most likely desist. The theory expands on traditional behavioral theories, in which behavior is governed solely by reinforcements, by placing emphasis on the important roles of various internal processes in the learning individual.

Exam Probability: **Low**

32. *Answer choices:*

(see index for correct answer)

- a. Behavior change
- b. Bad habit

- c. Reality mining
- d. Social learning theory

Guidance: level 1

:: Criminology ::

In criminology, the _________ theory emphasizes the importance of shame in criminal punishment. The theory holds that punishments should focus on the offender`s behavior rather than characteristics of the offender. It was developed by Australian criminologist John Braithwaite at Australian National University in 1989. It is related to the emerging perspective of positive criminology, developed by the Israeli criminologist Natti Ronel and his research team.

Exam Probability: **Medium**

33. *Answer choices:*

(see index for correct answer)

- a. Copycat crime
- b. Rational choice theory
- c. Differential association
- d. White IT

Guidance: level 1

:: Individualism ::

_________ or personal freedoms are personal guarantees and freedoms that the government cannot abridge, either by law or by judicial interpretation, without due process. Though the scope of the term differs between countries, _________ may include the freedom of conscience, freedom of press, freedom of religion, freedom of expression, freedom of assembly, the right to security and liberty, freedom of speech, the right to privacy, the right to equal treatment under the law and due process, the right to a fair trial, and the right to life. Other _________ include the right to own property, the right to defend oneself, and the right to bodily integrity. Within the distinctions between _________ and other types of liberty, distinctions exist between positive liberty/positive rights and negative liberty/negative rights.

Exam Probability: **Low**

34. *Answer choices:*

(see index for correct answer)

- a. Civil liberties
- b. Anticonformism
- c. Anti-individualism
- d. Autonomy

Guidance: level 1

:: Courts ::

A _________ in the criminal justice system is a form of sentence in which the criminal offender joins a rehabilitation program, which will help remedy the behavior leading to the original arrest, and avoid conviction and a criminal record. The programs are often run by a police department, court, a district attorney's office, or outside agency. Problem-solving courts typically include a diversion component as part of their program. The purposes of diversion are generally thought to include relief to the courts, police department and probation office, better outcomes compared to direct involvement of the court system, and an opportunity for the offender to avoid prosecution by completing various requirements for the program. These requirements may include.

Exam Probability: **Medium**

35. *Answer choices:*

(see index for correct answer)

- a. Land court
- b. Diversion program
- c. Law and motion calendar
- d. High Court of Justice

Guidance: level 1

:: Criminology ::

_________ is a personality disorder characterized by a long term pattern of disregard for, or violation of, the rights of others. A low moral sense or conscience is often apparent, as well as a history of crime, legal problems, or impulsive and aggressive behavior.

Exam Probability: **Low**

36. *Answer choices:*

(see index for correct answer)

- a. Radical criminology
- b. Broken windows theory
- c. Postmodernist school
- d. Antisocial personality disorder

Guidance: level 1

:: Social philosophy ::

A _________ is a place of work and residence where a great number of similarly situated people, cut off from the wider community for a considerable time, together lead an enclosed, formally administered round of life. The concept is mostly associated with the work of sociologist Erving Goffman.

Exam Probability: **High**

37. *Answer choices:*

(see index for correct answer)

- a. Ethics
- b. Philosophy of law
- c. Subjective expected utility
- d. Societal attitudes towards abortion

Guidance: level 1

:: Criminal procedure ::

_________ is a legal phrase for a variety of procedures connected with jury trials. It originally referred to an oath taken by jurors to tell the truth , i.e., to say what is true, what is objectively accurate or subjectively honest, or both.

Exam Probability: **High**

38. *Answer choices:*

(see index for correct answer)

- a. R v Thomas
- b. Allocution
- c. Court of public opinion
- d. Voir dire

Guidance: level 1

:: Criminal law ::

__________ is a term from Western jurisprudence referring to the principle that a crime must be proved to have occurred before a person can be convicted of committing that crime.

Exam Probability: **Low**

39. *Answer choices:*

(see index for correct answer)

- a. Resisting arrest
- b. Unindicted co-conspirator
- c. Eurodac
- d. Brawling

Guidance: level 1

:: School bullying ::

_________ encompasses physical violence, including student-on-student fighting and corporal punishment; psychological violence, including verbal abuse; sexual violence, including rape and sexual harassment; many forms of bullying, including cyberbullying; and carrying weapons in school. It is widely held to have become a serious problem in recent decades in many countries, especially where weapons such as guns or knives are involved. It includes violence between school students as well as physical attacks by students on school staff.

Exam Probability: **Low**

40. *Answer choices:*

(see index for correct answer)

- a. School bullying
- b. Raumanmeri school shooting
- c. SchoolTipline
- d. New Jersey Anti-Bullying Bill of Rights Act

Guidance: level 1

:: Juvenile law ::

_________ is the legal process by which an arbiter or judge reviews evidence and argumentation, including legal reasoning set forth by opposing parties or litigants to come to a decision which determines rights and obligations between the parties involved.

Exam Probability: **Medium**

41. *Answer choices:*

(see index for correct answer)

- a. Reform school
- b. School leaving age
- c. Jena Six
- d. Adjudication

Guidance: level 1

:: Crime ::

_________ is a white-collar crime that generally refers to the false making or material alteration of a legal instrument with the specific intent to defraud anyone . Tampering with a certain legal instrument may be forbidden by law in some jurisdictions but such an offense is not related to _________ unless the tampered legal instrument was actually used in the course of the crime to defraud another person or entity. Copies, studio replicas, and reproductions are not considered forgeries, though they may later become forgeries through knowing and willful misrepresentations.

Exam Probability: **Medium**

42. *Answer choices:*

(see index for correct answer)

- a. Confiscation
- b. False lien
- c. Gun violence
- d. Intimate partner violence

Guidance: level 1

:: Social concepts ::

_________ are legal, social, or ethical principles of freedom or entitlement; that is, _________ are the fundamental normative rules about what is allowed of people or owed to people, according to some legal system, social convention, or ethical theory. _________ are of essential importance in such disciplines as law and ethics, especially theories of justice and deontology.

Exam Probability: **Medium**

43. *Answer choices:*

(see index for correct answer)

- a. Rights
- b. Habitus
- c. Social relation
- d. Peace

Guidance: level 1

:: Domestic violence ::

__________ is violence or other abuse by one person against another in a domestic setting, such as in marriage or cohabitation. It may be termed intimate partner violence when committed by a spouse or partner in an intimate relationship against the other spouse or partner, and can take place in heterosexual or same-sex relationships, or between former spouses or partners. __________ can also involve violence against children, parents, or the elderly. It takes a number of forms, including physical, verbal, emotional, economic, religious, reproductive, and sexual abuse, which can range from subtle, coercive forms to marital rape and to violent physical abuse such as choking, beating, female genital mutilation, and acid throwing that results in disfigurement or death. Domestic murders include stoning, bride burning, honor killings, and dowry deaths.

Exam Probability: **High**

44. *Answer choices:*

(see index for correct answer)

- a. Management of domestic violence
- b. Power and control in abusive relationships
- c. Godelina
- d. George Puttenham

Guidance: level 1

:: Globalization-related theories ::

_________ is a broad movement that developed in the mid- to late 20th century across philosophy, the arts, architecture, and criticism and that marked a departure from modernism. The term has also more generally been applied to the historical era following modernity and the tendencies of this era.

Exam Probability: **Low**

45. *Answer choices:*

(see index for correct answer)

- a. Postmodernism
- b. Primitive accumulation
- c. Primitive accumulation of capital
- d. post-industrial

Guidance: level 1

:: Criminal law ::

A _________ is a legal representative of the prosecution in countries with either the common law adversarial system, or the civil law inquisitorial system. The prosecution is the legal party responsible for presenting the case in a criminal trial against an individual accused of breaking the law. Typically, the _________ represents the government in the case brought against the accused person.

Exam Probability: **High**

46. *Answer choices:*

(see index for correct answer)

- a. Conviction
- b. Due diligence
- c. Common purpose
- d. Prosecutor

Guidance: level 1

:: Law enforcement ::

__________ is freedom from, or resilience against, potential harm caused by others. Beneficiaries of __________ may be of persons and social groups, objects and institutions, ecosystems or any other entity or phenomenon vulnerable to unwanted change by its environment.

Exam Probability: **Medium**

47. *Answer choices:*

(see index for correct answer)

- a. Security
- b. Police and Security News
- c. Ticket quota
- d. Military aid to the civil power

Guidance: level 1

:: Islamic terrorism ::

_________ is a militant Sunni Islamist multi-national organization founded in 1988 by Osama bin Laden, Abdullah Azzam, and several other Arab volunteers during the Soviet invasion of Afghanistan.

Exam Probability: **Medium**

48. *Answer choices:*

(see index for correct answer)

- a. 2011 Al Hillah bombing
- b. 19 March 2013 Iraq attacks
- c. History of Iraq
- d. Al-Qaeda

Guidance: level 1

:: Crimes ::

_________ is intentional action to destroy a people in whole or in part. The hybrid word " _________ " is a combination of the Greek word and the Latin suffix -caedo . The term _________ was coined by Raphael Lemkin in his 1944 book Axis Rule in Occupied Europe;

Exam Probability: **Medium**

49. *Answer choices:*

(see index for correct answer)

- a. Solicitation
- b. Genocide
- c. Assault
- d. Endangerment

Guidance: level 1

:: Gun politics ::

_________ is the set of laws or policies that regulate the manufacture, sale, transfer, possession, modification, or use of firearms by civilians.

Exam Probability: **Low**

50. *Answer choices:*

(see index for correct answer)

- a. Toy weapon
- b. Escopetarra
- c. To extend the Undetectable Firearms Act of 1988 for 10 years
- d. Gun Control

Guidance: level 1

:: United States federal criminal legislation ::

The _________ of 1970 , was an Act of Congress sponsored by Democratic Senator John L. McClellan and signed into law by U.S. President Richard Nixon.

Exam Probability: **High**

51. *Answer choices:*

(see index for correct answer)

- a. Child Protection and Obscenity Enforcement Act
- b. Organized Crime Control Act
- c. Foreign Narcotics Kingpin Designation Act
- d. Violent Crime Control and Law Enforcement Act

Guidance: level 1

:: Criminal law ::

A __________ is an action that is prohibited only to a certain class of people, and most often applied only to offenses committed by minors.

Exam Probability: **Low**

52. *Answer choices:*

(see index for correct answer)

- a. Nulla poena sine lege
- b. Juvenile offender
- c. Status offense
- d. Additional Protocol to the Convention on Cybercrime

Guidance: level 1

:: Concepts in ethics ::

A __________ is a proposition or value that is a guide for behavior or evaluation. In law, it is a rule that has to be or usually is to be followed, or can be desirably followed, or is an inevitable consequence of something, such as the laws observed in nature or the way that a system is constructed. The __________ s of such a system are understood by its users as the essential characteristics of the system, or reflecting system`s designed purpose, and the effective operation or use of which would be impossible if any one of the __________ s was to be ignored. A system may be explicitly based on and implemented from a document of __________ s as was done in IBM`s 360/370 __________ s of Operation.

Exam Probability: **Low**

53. *Answer choices:*

(see index for correct answer)

- a. Principle
- b. Open-question argument
- c. Mental reservation
- d. Greed

Guidance: level 1

:: Criminal procedure ::

In common law jurisdictions, an __________ certifies that the accused is free from the charge of an offense, as far as the criminal law is concerned. This is so even where the prosecution is simply abandoned by the prosecution. The finality of an __________ is dependent on the jurisdiction. In some countries, such as the United States, an __________ operates to bar the retrial of the accused for the same offense, even if new evidence surfaces that further implicates the accused. The effect of an __________ on criminal proceedings is the same whether it results from a jury verdict or results from the operation of some other rule that discharges the accused. In other countries, the prosecuting authority may appeal an __________ similar to how a defendant may appeal a conviction.

Exam Probability: **High**

54. *Answer choices:*

(see index for correct answer)

- a. Double jeopardy
- b. Mug shot
- c. Cross-examination
- d. Acquittal

Guidance: level 1

:: Uniforms ::

A _________ is a type of clothing worn by members of an organization while participating in that organization`s activity. Modern _________ s are most often worn by armed forces and paramilitary organizations such as police, emergency services, security guards, in some workplaces and schools and by inmates in prisons. In some countries, some other officials also wear _________ s in their duties; such is the case of the Commissioned Corps of the United States Public Health Service or the French prefects. For some organizations, such as police, it may be illegal for non members to wear the _________ .

Exam Probability: **Medium**

55. *Answer choices:*

(see index for correct answer)

- a. Uniform
- b. Burka
- c. Papakha

- d. Japanese school uniform

Guidance: level 1

:: Legal terms ::

__________ is the legal requirement that the state must respect all legal rights that are owed to a person. __________ balances the power of law of the land and protects the individual person from it. When a government harms a person without following the exact course of the law, this constitutes a __________ violation, which offends the rule of law.

Exam Probability: **Low**

56. *Answer choices:*

(see index for correct answer)

- a. Felon
- b. respondent
- c. redirect examination
- d. Due process

Guidance: level 1

:: Crimes ::

__________ is the crime of taking or attempting to take anything of value by force, threat of force, or by putting the victim in fear. According to common law, __________ is defined as taking the property of another, with the intent to permanently deprive the person of that property, by means of force or fear; that is, it is a larceny or theft accomplished by an assault. Precise definitions of the offence may vary between jurisdictions. __________ is differentiated from other forms of theft by its inherently violent nature ; whereas many lesser forms of theft are punished as misdemeanors, __________ is always a felony in jurisdictions that distinguish between the two. Under English law, most forms of theft are triable either way, whereas __________ is triable only on indictment. The word "rob" came via French from Late Latin words of Germanic origin, from Common Germanic raub -- "theft".

Exam Probability: **Medium**

57. *Answer choices:*

(see index for correct answer)

- a. Cruelty to animals
- b. Robbery
- c. Identity cleansing
- d. Malfeasance in office

Guidance: level 1

:: Penal imprisonment ::

A _________ is a panel of people who decide whether an offender should be released from prison on parole after serving at least a minimum portion of their sentence as prescribed by the sentencing judge. _________ s are used in many jurisdictions, including the United Kingdom, the United States, and New Zealand. A related concept is the board of pardons and paroles, which may deal with pardons and commutations as well as paroles.

Exam Probability: **High**

58. *Answer choices:*

(see index for correct answer)

- a. Parole board
- b. Prisoner security categories in the United Kingdom
- c. Penal harm
- d. Prison ring

Guidance: level 1

:: Murder ::

_________ is the act of murdering a number of people, typically simultaneously or over a relatively short period of time and in close geographic proximity. The FBI defines _________ as murdering four or more people during an event with no "cooling-off period" between the murders. A _________ typically occurs in a single location where one or more people kill several others.

Exam Probability: **Medium**

59. *Answer choices:*

(see index for correct answer)

- a. Dragging death
- b. Horse murders
- c. Premeditated murder
- d. Crime of passion

Guidance: level 1

Social work

Social work is an academic discipline and profession that concerns itself with individuals, families, groups and communities in an effort to enhance health and social functioning and overall well-being. Social functioning refers to the way in which people perform their social roles, and the structural institutions that are provided to sustain them. Social work applies social sciences, such as sociology, psychology, political science, public health, community development, law, and economics, to engage with client systems, conduct assessments, and develop interventions to solve social and personal problems; and create social change. Social work practice is often divided into micro-work, which involves working directly with individuals or small groups; and macro-work, which involves working communities, and within social policy, to create change on a larger scale.

:: Child abuse ::

_________ or child maltreatment is physical, sexual, and/or psychological maltreatment or neglect of a child or children, especially by a parent or a caregiver. _________ may include any act or failure to act by a parent or a caregiver that results in actual or potential harm to a child, and can occur in a child`s home, or in the organizations, schools or communities the child interacts with.

Exam Probability: **High**

1. *Answer choices:*

(see index for correct answer)

- a. Shaken baby syndrome
- b. Andrew John Yellowbear Jr.
- c. Psychohistory
- d. Child abuse

Guidance: level 1

:: Diagnostic and Statistical Manual of Mental Disorders ::

The __________ , published by the American Psychiatric Association , offers a common language and standard criteria for the classification of mental disorders. It is used, or relied upon, by clinicians, researchers, psychiatric drug regulation agencies, health insurance companies, pharmaceutical companies, the legal system, and policy makers together with alternatives such as the ICD-10 Classification of Mental and Behavioural Disorders, produced by the WHO.

Exam Probability: **Low**

2. *Answer choices:*

(see index for correct answer)

- a. DSM-5
- b. Structured Clinical Interview for DSM-IV
- c. Diagnostic and Statistical Manual of Mental Disorders
- d. Global Assessment of Functioning

Guidance: level 1

:: Methods in sociology ::

__________ , field studies, or fieldwork is the collection of raw data outside a laboratory, library, or workplace setting. The approaches and methods used in __________ vary across disciplines. For example, biologists who conduct __________ may simply observe animals interacting with their environments, whereas social scientists conducting __________ may interview or observe people in their natural environments to learn their languages, folklore, and social structures.

Exam Probability: **Medium**

3. *Answer choices:*

(see index for correct answer)

- a. Content analysis
- b. Participant observation
- c. Autoethnography
- d. KH Coder

Guidance: level 1

:: Hazard analysis ::

Broadly speaking, a __________ is the combined effort of 1. identifying and analyzing potential events that may negatively impact individuals, assets, and/or the environment ; and 2. making judgments "on the tolerability of the risk on the basis of a risk analysis" while considering influencing factors . Put in simpler terms, a __________ analyzes what can go wrong, how likely it is to happen, what the potential consequences are, and how tolerable the identified risk is. As part of this process, the resulting determination of risk may be expressed in a quantitative or qualitative fashion. The __________ is an inherent part of an overall risk management strategy, which attempts to, after a __________ , "introduce control measures to eliminate or reduce" any potential risk-related consequences.

Exam Probability: **Low**

4. *Answer choices:*

(see index for correct answer)

- a. Hazard
- b. Risk assessment
- c. Hazardous Materials Identification System
- d. Hazard identification

Guidance: level 1

:: Bullying ::

__________ covers a wide range of behaviors of an offensive nature. It is commonly understood as behavior that demeans, humiliates or embarrasses a person, and it is characteristically identified by its unlikelihood in terms of social and moral reasonableness. In the legal sense, these are behaviors that appear to be disturbing, upsetting or threatening. They evolve from discriminatory grounds, and have an effect of nullifying or impairing a person from benefiting their rights. When these behaviors become repetitive they are defined as bullying.

Exam Probability: **Medium**

5. *Answer choices:*

(see index for correct answer)

- a. Harassment
- b. Threat
- c. Mind games

- d. Social undermining

Guidance: level 1

:: United States federal child welfare legislation ::

The __________ of 1988 was a federal law that amended Title IV of the Social Security Act to revise the Aid to Families with Dependent Children program to emphasize work, child support and family benefits, as well as on withholding the wages of absentee parents. The Job Opportunities and Basic Skills Training program was a welfare-to-work program created by the __________ to replace the Work Incentive program .

Exam Probability: **Medium**

6. *Answer choices:*

(see index for correct answer)

- a. Virginia Graeme Baker Pool and Spa Safety Act
- b. Juvenile Justice and Delinquency Prevention Act
- c. National Child Search Assistance Act
- d. Family Support Act

Guidance: level 1

:: Child abuse ::

_________ was a pediatrician and the first in the medical community to identify and recognize child abuse.

Exam Probability: **High**

7. *Answer choices:*

(see index for correct answer)

- a. Court Appointed Special Advocates
- b. Covert incest
- c. C. Henry Kempe
- d. Deadbeat parent

Guidance: level 1

:: Psychometrics ::

A descriptive statistic is a summary statistic that quantitatively describes or summarizes features of a collection of information, while _________ in the mass noun sense is the process of using and analyzing those statistics. _________ is distinguished from inferential statistics , in that _________ aims to summarize a sample, rather than use the data to learn about the population that the sample of data is thought to represent. This generally means that _________ , unlike inferential statistics, is not developed on the basis of probability theory, and are frequently nonparametric statistics. Even when a data analysis draws its main conclusions using inferential statistics, _________ are generally also presented. For example, in papers reporting on human subjects, typically a table is included giving the overall sample size, sample sizes in important subgroups , and demographic or clinical characteristics such as the average age, the proportion of subjects of each sex, the proportion of subjects with related comorbidities, etc.

Exam Probability: **Medium**

8. *Answer choices:*

(see index for correct answer)

- a. Common-method variance
- b. Descriptive statistics
- c. Questionnaire construction
- d. Polynomial conjoint measurement

Guidance: level 1

:: Sectarianism ::

_________ ism is a form of prejudice, discrimination, or hatred arising from attaching relations of inferiority and superiority to differences between subdivisions within a group. Common examples are denominations of a religion, ethnic identity, class, or region for citizens of a state and factions of a political movement.

Exam Probability: **High**

9. *Answer choices:*

(see index for correct answer)

- a. My Sweet Lord
- b. Nil by Mouth
- c. White Lotus
- d. Sectarian

Guidance: level 1

:: Special education ::

In the U.S. the Individuals with Disabilities Education Act is a special education law that mandates regulation for students with disabilities in order to protect their rights as students and the rights of their parents. Under this act it is required that all students receive a Free and Appropriate Public Education , and that these students should be educated in the _________ . The _________ clause states that students with disabilities should be educated with students without disabilities to the maximum appropriate extent. If a student should require supplementary aids and services necessary to achieve educational goals while being placed in a classroom with students without disabilities, they should be provided as needed.

Exam Probability: **Low**

10. *Answer choices:*

(see index for correct answer)

- a. Orton-Gillingham
- b. Developmental disability
- c. Least restrictive environment
- d. ACE Centre

Guidance: level 1

:: Social philosophy ::

_________ or moral philosophy is a branch of philosophy that involves systematizing, defending, and recommending concepts of right and wrong conduct. The field of _________ , along with aesthetics, concerns matters of value, and thus comprises the branch of philosophy called axiology.

Exam Probability: **High**

11. *Answer choices:*

(see index for correct answer)

- a. Pirate utopia
- b. Communicative rationality
- c. Ethics
- d. Julian Gumperz

Guidance: level 1

:: Socialism ::

A _________ is a set of subjectively defined concepts in the social sciences and political theory centered on models of social stratification in which people are grouped into a set of hierarchical social categories, the most common being the upper, middle and lower classes.

Exam Probability: **High**

12. *Answer choices:*

(see index for correct answer)

- a. Venceremos Brigade
- b. Third World Socialism
- c. Nationalist Clubs

- d. Social class

Guidance: level 1

:: Life skills ::

__________ , emotional leadership , emotional quotient and __________ quotient , is the capability of individuals to recognize their own emotions and those of others, discern between different feelings and label them appropriately, use emotional information to guide thinking and behavior, and manage and/or adjust emotions to adapt to environments or achieve one`s goal.

Exam Probability: **High**

13. *Answer choices:*

(see index for correct answer)

- a. Emotional intelligence
- b. Emotional Intelligence 2.0
- c. Interpersonal emotion regulation
- d. Social intelligence

Guidance: level 1

:: Ethics ::

_________ is a process for getting permission before conducting a healthcare intervention on a person, or for disclosing personal information. A health care provider may ask a patient to consent to receive therapy before providing it, or a clinical researcher may ask a research participant before enrolling that person into a clinical trial. _________ is collected according to guidelines from the fields of medical ethics and research ethics.

Exam Probability: **Low**

14. *Answer choices:*

(see index for correct answer)

- a. Ultimate end
- b. Research ethics consultation
- c. Dignity
- d. Ethical calculus

Guidance: level 1

:: Mental processes ::

_________ is the means to see, hear, or become aware of something or someone through our fundamental senses. The term _________ derives from the Latin word perceptio, and is the organization, identification, and interpretation of sensory information in order to represent and understand the presented information, or the environment.

Exam Probability: **Medium**

15. *Answer choices:*

(see index for correct answer)

- a. Perception
- b. Strong reciprocity
- c. Adaptive memory
- d. Set

Guidance: level 1

:: Conceptions of self ::

One`s _________ is a collection of beliefs about oneself. Generally, _________ embodies the answer to "Who am I".

Exam Probability: **Medium**

16. *Answer choices:*

(see index for correct answer)

- a. Looking-glass self
- b. Self-concept

Guidance: level 1

:: VMAT inhibitors ::

_________ is a potent central nervous system stimulant that is mainly used as a recreational drug and less commonly as a second-line treatment for attention deficit hyperactivity disorder and obesity. _________ was discovered in 1893 and exists as two enantiomers: levo- _________ and dextro- _________ . _________ properly refers to a specific chemical, the racemic free base, which is an equal mixture of levo _________ and dextro _________ in their pure amine forms. It is rarely prescribed over concerns involving human neurotoxicity and potential for recreational use as an aphrodisiac and euphoriant, among other concerns, as well as the availability of safer substitute drugs with comparable treatment efficacy. Dextro _________ is a much stronger CNS stimulant than levo _________ .

Exam Probability: **Low**

17. *Answer choices:*

(see index for correct answer)

- a. Amphetamine
- b. Adderall
- c. Methamphetamine

Guidance: level 1

:: Sociological terminology ::

__________ , a topic in the humanities and social sciences, is both a historical period , as well as the ensemble of particular socio-cultural norms, attitudes and practices that arose in the wake of the Renaissance—in the "Age of Reason" of 17th-century thought and the 18th-century "Enlightenment". Some commentators consider the era of __________ to have ended by 1930, with World War II in 1945, or the 1980s or 1990s; the following era is called post __________ . The term "contemporary history" is also used to refer to the post-1945 timeframe, without assigning it to either the modern or postmodern era.

Exam Probability: **Medium**

18. *Answer choices:*

(see index for correct answer)

- a. Modernity
- b. Primary socialisation
- c. Intergroup anxiety
- d. Contested ideological terrain

Guidance: level 1

:: Bias ::

In statistics, _________ is a bias in which a sample is collected in such a way that some members of the intended population are less likely to be included than others. It results in a biased sample, a non-random sample of a population in which all individuals, or instances, were not equally likely to have been selected. If this is not accounted for, results can be erroneously attributed to the phenomenon under study rather than to the method of sampling.

Exam Probability: **Medium**

19. *Answer choices:*

(see index for correct answer)

- a. Affordance
- b. Modifiable areal unit problem
- c. Sampling bias
- d. Social desirability bias

Guidance: level 1

:: Sociology of knowledge ::

_________ is a theory of knowledge in sociology and communication theory that examines the development of jointly constructed understandings of the world that form the basis for shared assumptions about reality. The theory centers on the notion that meanings are developed in coordination with others rather than separately within each individual.

Exam Probability: **Low**

20. *Answer choices:*

(see index for correct answer)

- a. Problems of a Sociology of Knowledge
- b. Sociology of Knowledge Approach to Discourse
- c. Social constructionism
- d. Folksonomy

Guidance: level 1

:: Meetings ::

A __________ is someone who engages in facilitation—any activity that makes a social process easy or easier. A __________ often helps a group of people to understand their common objectives and assists them to plan how to achieve these objectives; in doing so, the __________ remains "neutral", meaning he/she does not take a particular position in the discussion. Some __________ tools will try to assist the group in achieving a consensus on any disagreements that preexist or emerge in the meeting so that it has a strong basis for future action.

Exam Probability: **High**

21. *Answer choices:*

(see index for correct answer)

- a. Agenda
- b. Facilitator

- c. Skeptics in the Pub
- d. Meeting

Guidance: level 1

:: Organizations ::

An _________ or organisation is an entity comprising multiple people, such as an institution or an association, that has a particular purpose.

Exam Probability: **Medium**

22. *Answer choices:*

(see index for correct answer)

- a. Alliance for Affordable Internet
- b. World Association for Disaster and Emergency Medicine
- c. DOHaD China
- d. Organization

Guidance: level 1

:: Political theories ::

__________ is the mixed or compound mode of government, combining a general government with regional governments in a single political system. Its distinctive feature, exemplified in the founding example of modern __________ by the United States under the Constitution of 1787, is a relationship of parity between the two levels of government established. It can thus be defined as a form of government in which there is a division of powers between two levels of government of equal status.

Exam Probability: **High**

23. *Answer choices:*

(see index for correct answer)

- a. Eco-socialism
- b. Neo-Nazism
- c. Hellenoturkism
- d. Sadaejuui

Guidance: level 1

:: Historically African-American Christian denominations ::

Biserica Neagra or __________ (German: Schwarze Kirche; Romanian: Biserica Neagra; is a church in Braov, a city in south-eastern Transylvania, Romania. It was built by the German community of the city and stands as the main Gothic style monument in the country, as well as being the largest and one of the most important Lutheran places of worship in the region.

Exam Probability: **Medium**

24. *Answer choices:*

(see index for correct answer)

- a. Black Church
- b. African Methodist Episcopal Zion Church
- c. Original Church of God or Sanctified Church
- d. United American Free Will Baptist Church

Guidance: level 1

:: Native American religion ::

_________ is a product prepared from the leaves of the _________ plant by curing them. The plant is part of the genus Nicotiana and of the Solanaceae family. While more than 70 species of _________ are known, the chief commercial crop is N. tabacum. The more potent variant N. rustica is also used around the world.

Exam Probability: **High**

25. *Answer choices:*

(see index for correct answer)

- a. Teachings of the Seven Grandfathers
- b. Peyote

- c. Underwater panther
- d. Traditional narratives

Guidance: level 1

:: Welfare reform ::

The __________ was a series of programs, public work projects, financial reforms, and regulations enacted by President Franklin D. Roosevelt in the United States between 1933 and 1936. It responded to needs for relief, reform, and recovery from the Great Depression. Major federal programs included the Civilian Conservation Corps , the Civil Works Administration , the Farm Security Administration , the National Industrial Recovery Act of 1933 and the Social Security Administration . They provided support for farmers, the unemployed, youth and the elderly. The __________ included new constraints and safeguards on the banking industry and efforts to re-inflate the economy after prices had fallen sharply. __________ programs included both laws passed by Congress as well as presidential executive orders during the first term of the presidency of Franklin D. Roosevelt.

Exam Probability: **High**

26. *Answer choices:*

(see index for correct answer)

- a. Workfare
- b. Self-Sufficiency Project
- c. Work for the Dole
- d. Get Britain Working

Guidance: level 1

:: Ethics ::

An _________ or ethical paradox is a decision-making problem between two possible moral imperatives, neither of which is unambiguously acceptable or preferable. The complexity arises out of the situational conflict in which obeying one would result in transgressing another. Sometimes called ethical paradoxes in moral philosophy, _________ s may be invoked to refute an ethical system or moral code, or to improve it so as to resolve the paradox.

Exam Probability: **Medium**

27. *Answer choices:*

(see index for correct answer)

- a. Universal code
- b. Visual ethics
- c. Inequitable conduct
- d. Ethical dilemma

Guidance: level 1

:: bad_topic ::

A _________ , from the Latin _________ ia meaning wakefulness , is a period of purposeful sleeplessness, an occasion for devotional watching, or an observance. The Italian word _________ ia has become generalized in this sense and means "eve" .

Exam Probability: **Medium**

28. *Answer choices:*

(see index for correct answer)

- a. economics education
- b. Swope
- c. Michael Hannan
- d. Vigil

Guidance: level 1

:: Child abuse ::

_________ is a form of child abuse, and is a deficit in meeting a child`s basic needs, including the failure to provide adequate health care, supervision, clothing, nutrition, housing as well as their physical, emotional, social, educational and safety needs. Society generally believes there are necessary behaviors a caregiver must provide in order for a child to develop physically, socially, and emotionally. Causes of neglect may result from several parenting problems including mental disorders, substance abuse, domestic violence, unemployment, unplanned pregnancy, and poverty.

Exam Probability: **Medium**

29. *Answer choices:*

(see index for correct answer)

- a. Reparenting
- b. Childhaven
- c. Mary Ellen Wilson
- d. Cybertip.ca

Guidance: level 1

:: Social theories ::

_________ is the basis of modern economic and social systems in industrialized, standardized mass production and mass consumption. The concept is named for Henry Ford. It is used in social, economic, and management theory about production, working conditions, consumption, and related phenomena, especially regarding the 20th century.

Exam Probability: **High**

30. *Answer choices:*

(see index for correct answer)

- a. Conservatives
- b. Neotribalism

- c. Somatocentrism
- d. Fordism

Guidance: level 1

:: Adoption, fostering, orphan care and displacement ::

_________ is a psychological model attempting to describe the dynamics of long-term and short-term interpersonal relationships between humans. " _________ is not formulated as a general theory of relationships; it addresses only a specific facet": how human beings respond in relationships when hurt, separated from loved ones, or perceiving a threat.

Exam Probability: **High**

31. *Answer choices:*

(see index for correct answer)

- a. Foundling hospital
- b. Attachment theory
- c. Maternal deprivation
- d. Runaway

Guidance: level 1

:: Humanistic psychology ::

_________ is a psychological perspective that rose to prominence in the mid-20th century in answer to the limitations of Sigmund Freud`s psychoanalytic theory and B. F. Skinner`s behaviorism. With its roots running from Socrates through the Renaissance, this approach emphasizes individuals` inherent drive towards self-actualization, the process of realizing and expressing one`s own capabilities and creativity.

Exam Probability: **Low**

32. *Answer choices:*

(see index for correct answer)

- a. The Journal of Humanistic Counseling
- b. Paul Rebillot
- c. Humanistic psychology
- d. Primal Integration

Guidance: level 1

:: Linguistic rights ::

In sociology, a _________ refers to a category of people who experience relative disadvantage as compared to members of a dominant social group. _________ membership is typically based on differences in observable characteristics or practices, such as: ethnicity , race , religion , sexual orientation , disability, or gender identity. Utilizing the framework of intersectionality, it is important to recognize that an individual may simultaneously hold membership in multiple _________ s . Likewise, individuals may also be part of a _________ in regard to some characteristics, but part of a dominant group in regard to others.

Exam Probability: **Low**

33. *Answer choices:*

(see index for correct answer)

- a. Minority group
- b. Collioure
- c. Regional language
- d. Moldovan schools in Transnistria

Guidance: level 1

:: Philosophy of life ::

Ethics or moral philosophy is a branch of philosophy that involves systematizing, defending, and recommending concepts of right and wrong conduct. The field of ethics, along with aesthetics, concerns matters of value, and thus comprises the branch of philosophy called axiology.

Exam Probability: **Low**

34. *Answer choices:*

(see index for correct answer)

- a. Ambivalence
- b. personal life

Guidance: level 1

:: Scientific method ::

In the social sciences and life sciences, a case study is a research method involving an up-close, in-depth, and detailed examination of a subject of study, as well as its related contextual conditions.

Exam Probability: **High**

35. *Answer choices:*

(see index for correct answer)

- a. Case study
- b. Case studies
- c. Isotope dilution
- d. Open-label trial

Guidance: level 1

:: Social work ::

The __________ is the worldwide association of schools of social work, other tertiary level social work educational programs, and social work educators. The IASSW promotes the development of social work education throughout the world, develops standards to enhance quality of social work education, encourages international exchange, provides forums for sharing social work research and scholarship, and promotes human rights and social development through policy and advocacy activities.

Exam Probability: **Low**

36. *Answer choices:*

(see index for correct answer)

- a. Social learning tools
- b. IRISS
- c. 6-15-99 Club
- d. International Association of Schools of Social Work

Guidance: level 1

:: United States Department of Health and Human Services agencies ::

The _________ is a branch of the U.S. Department of Health and Human Services. It is charged with improving the quality and availability of treatment and rehabilitative services in order to reduce illness, death, disability, and the cost to society resulting from substance abuse and mental illnesses. The Administrator of SAMHSA reports directly to the Secretary of the U.S. Department of Health and Human Services. SAMHSA`s headquarters building is located outside of Rockville, Maryland.

Exam Probability: **High**

37. *Answer choices:*

(see index for correct answer)

- a. Public Health Service
- b. Administration for Children and Families

Guidance: level 1

:: Criminology ::

_________ is the theory of how the self-identity and behavior of individuals may be determined or influenced by the terms used to describe or classify them. It is associated with the concepts of self-fulfilling prophecy and stereotyping. _________ holds that deviance is not inherent in an act, but instead focuses on the tendency of majorities to negatively label minorities or those seen as deviant from standard cultural norms. The theory was prominent during the 1960s and 1970s, and some modified versions of the theory have developed and are still currently popular. A stigma is defined as a powerfully negative label that changes a person`s self-concept and social identity.

Exam Probability: **High**

38. *Answer choices:*

(see index for correct answer)

- a. Acting out
- b. Computational criminology
- c. Labeling theory
- d. Left realism

Guidance: level 1

:: Methods in sociology ::

_________ is one type of data collection method typically used in qualitative research and ethnography. It is a widely used methodology in many disciplines, particularly cultural anthropology, European ethnology, sociology, communication studies, human geography and social psychology. Its aim is to gain a close and intimate familiarity with a given group of individuals and their practices through an intensive involvement with people in their cultural environment, usually over an extended period of time.

Exam Probability: **Medium**

39. *Answer choices:*

(see index for correct answer)

- a. KH Coder

- b. Field research
- c. Participant observation
- d. Content analysis

Guidance: level 1

:: Behavior ::

_________ refers to behavior-change procedures that were employed during the 1970s and early 1980s. Based on methodological behaviorism, overt behavior was modified with presumed consequences, including artificial positive and negative reinforcement contingencies to increase desirable behavior, or administering positive and negative punishment and/or extinction to reduce problematic behavior. For the treatment of phobias, habituation and punishment were the basic principles used in flooding, a subcategory of desensitization.

Exam Probability: **Medium**

40. *Answer choices:*

(see index for correct answer)

- a. Homology
- b. Theory of planned behavior
- c. social learning theory
- d. Challenging behaviour

Guidance: level 1

:: Punishments ::

_________ in criminal law is a period of supervision over an offender, ordered by the court instead of serving time in prison.

Exam Probability: **High**

41. *Answer choices:*

(see index for correct answer)

- a. Diesel therapy
- b. Riding the rail
- c. Postenpflicht
- d. Tarring and feathering

Guidance: level 1

:: Nonverbal communication ::

It includes the use of visual cues such as body language , distance and physical environments/appearance, of voice and of touch . It can also include the use of time and eye contact and the actions of looking while talking and listening, frequency of glances, patterns of fixation, pupil dilation, and blink rate .

Exam Probability: **Medium**

42. *Answer choices:*

(see index for correct answer)

- a. Sensory analysis
- b. Writing
- c. Nonverbal communication
- d. Discrimination testing

Guidance: level 1

:: Rape ::

__________ is a form of acquaintance rape. The two phrases are often used interchangeably, but __________ specifically refers to a rape in which there has been some sort of romantic or potentially sexual relationship between the two parties. Acquaintance rape also includes rapes in which the victim and perpetrator have been in a non-romantic, non-sexual relationship, for example as co-workers or neighbors. __________ is particularly prevalent on college campuses, where it frequently occurs in situations involving alcohol or other __________ drugs, which may facilitate the execution of drug-facilitated sexual assault .

Exam Probability: **Low**

43. *Answer choices:*

(see index for correct answer)

- a. Rape culture
- b. Miryang gang rape

- c. Drug facilitated sexual assault
- d. Date rape

Guidance: level 1

:: Affordable housing ::

_________ International , generally referred to as _________ or simply Habitat, is an international, non-governmental, and nonprofit organization, which was founded in 1976 by Linda and Millard Fuller. Habitat has been devoted to building "simple, decent, and affordable" housing, a self-described "Christian housing ministry," and has addressed the issues of poverty housing all over the world. The international operational headquarters are located in Americus, Georgia, United States, with the administrative headquarters located in Atlanta. There are five area offices located around the world: United States and Canada; Africa and the Middle East ; Asia-Pacific ; Europe and Central Asia ; and Latin America and the Caribbean .

Exam Probability: **Medium**

44. *Answer choices:*

(see index for correct answer)

- a. Homeless International
- b. War on Poverty
- c. Housing Benefit
- d. Habitat for Humanity

Guidance: level 1

:: Child abuse ::

_________ is a form of psychotherapy in which the therapist actively assumes the role of a new or surrogate parental figure for the client, in order to treat psychological disturbances caused by defective, even abusive, parenting. The underlying assumption is that all mental illness results principally from such parenting, even including schizophrenia and bipolar disorder. Despite the therapy`s association with the theory and practice of transactional analysis in the field of psychology, there is not much empirical evidence for its effectiveness, and some significant suggestions that it can in fact be harmful; most mental health practitioners do not regard _________ as a legitimate form of therapy.

Exam Probability: **Medium**

45. *Answer choices:*

(see index for correct answer)

- a. Auguste Ambroise Tardieu
- b. Reparenting
- c. Lloyd deMause
- d. Emotional dysregulation

Guidance: level 1

:: Social work ::

_________ is an interdisciplinary field with the objective of meeting human needs through an applied knowledge base, focusing on prevention as well as remediation of problems, and maintaining a commitment to improving the overall quality of life of service populations. The process involves the study of social technologies , service technologies , and scientific innovations that are designed to ameliorate problems and enhance the quality of life of individuals, families and communities to improve the delivery of service with better coordination, accessibility and accountability. The mission of _________ is to promote a practice that involves simultaneously working at all levels of society in the process of promoting the autonomy of individuals or groups, making informal or formal _________ systems more efficient and effective, and advocating for positive social change within society.

Exam Probability: **Low**

46. *Answer choices:*

(see index for correct answer)

- a. Life story work
- b. Richard Clarke Cabot
- c. Human services
- d. Psychoeducation

Guidance: level 1

:: Marriage ::

In family law and public policy, _________ is an ongoing, periodic payment made by a parent for the financial benefit of a child following the end of a marriage or other relationship. Child maintenance is paid directly or indirectly by an obligor to an obligee for the care and support of children of a relationship that has been terminated, or in some cases never existed. Often the obligor is a non-custodial parent. The obligee is typically a custodial parent, a caregiver, a guardian, or the state.

Exam Probability: **Medium**

47. *Answer choices:*

(see index for correct answer)

- a. Same-sex marriage
- b. Article 12 of the European Convention on Human Rights
- c. Child Support
- d. Eligible bachelor

Guidance: level 1

:: Ethnicity ::

_________ is the act of making something more native; transformation of some service, idea, etc. to suit a local culture, especially through the use of more indigenous people in administration, employment, etc.

Exam Probability: **Medium**

48. *Answer choices:*

(see index for correct answer)

- a. Polyethnicity
- b. The Race Question
- c. Indigenization
- d. Ethnic majority

Guidance: level 1

:: Counseling ::

_________ is an immediate and short-term psychological care aimed at assisting individuals in a crisis situation in order to restore equilibrium to their bio-psycho-social functioning and to minimize the potential of long-term psychological trauma.

Exam Probability: **Medium**

49. *Answer choices:*

(see index for correct answer)

- a. Crisis intervention
- b. Logotherapy
- c. Online counseling
- d. Genomic counseling

Guidance: level 1

:: Labor ::

The _________ generally describes an approach whereby labour unions aim to satisfy members` demands for resolving grievances and securing benefits through methods other than direct grassroots-oriented pressure on employers. It is often contrasted to the organising model, and to rank and file organization.

Exam Probability: **Medium**

50. *Answer choices:*

(see index for correct answer)

- a. Ethical Trading Initiative
- b. Senior Community Service Employment Program
- c. Factors of production
- d. Service model

Guidance: level 1

:: Affordable housing ::

__________ is housing which is deemed affordable to those with a median household income or below as rated by the national government or a local government by a recognized housing affordability index. Most of the literature on __________ refers to mortgages and number of forms that exist along a continuum – from emergency shelters, to transitional housing, to non-market rental , to formal and informal rental, indigenous housing, and ending with affordable home ownership.

Exam Probability: **Low**

51. *Answer choices:*

(see index for correct answer)

- a. Millennial Housing Commission
- b. Habitat for Humanity
- c. Housing Benefit
- d. Conrad Egan

Guidance: level 1

:: Law ::

A __________ is an introductory and expressionary statement in a document that explains the document`s purpose and underlying philosophy. When applied to the opening paragraphs of a statute, it may recite historical facts pertinent to the subject of the statute. It is distinct from the long title or enacting formula of a law.

Exam Probability: **High**

52. *Answer choices:*

(see index for correct answer)

- a. Unofficial law
- b. Preamble
- c. Secret law
- d. Oral law

Guidance: level 1

:: Sex crimes ::

_________ is the act of exposing in a public or semi-public context those parts of one`s body that are not normally exposed for example, the breasts, genitals or buttocks. The practice may arise from a desire or compulsion to expose themselves in such a manner to groups of friends or acquaintances, or to strangers for their amusement or sexual satisfaction or to shock the bystander. Exposing oneself only to an intimate partner is normally not regarded as _________ . In law, the act of _________ may be called indecent exposure, "exposing one`s person", or other expressions.

Exam Probability: **Medium**

53. *Answer choices:*

(see index for correct answer)

- a. Military sexual trauma
- b. Somaly Mam
- c. Exhibitionism
- d. Haiti Sex abuse

Guidance: level 1

:: Types of organization ::

A society is a group of individuals involved in persistent social interaction, or a large social group sharing the same geographical or social territory, typically subject to the same political authority and dominant cultural expectations. _________ are characterized by patterns of relationships between individuals who share a distinctive culture and institutions; a given society may be described as the sum total of such relationships among its constituent of members. In the social sciences, a larger society often exhibits stratification or dominance patterns in subgroups.

Exam Probability: **High**

54. *Answer choices:*

(see index for correct answer)

- a. Trade association
- b. Richard Watts Charities
- c. twelve-step
- d. Societies

Guidance: level 1

:: Family law ::

The _________ of 1988 provides federal funding to US states in support of prevention, assessment, investigation, prosecution, and treatment activities and provides grants to public agencies and nonprofit organizations for demonstration programs and projects. Additionally, it identifies the federal role in supporting research, evaluation, technical assistance, and data collection activities; establishes the Office on Child Abuse and Neglect; and mandates the National Clearinghouse on Child Abuse and Neglect Information. It also sets forth a minimum definition of child abuse and neglect.

Exam Probability: **High**

55. *Answer choices:*

(see index for correct answer)

- a. Child Abuse Prevention and Treatment Act
- b. Family Group Conference
- c. Filial responsibility laws
- d. Boll case

Guidance: level 1

:: Community-building organizations ::

__________ is the United States` first and largest organization uniting families and allies with people who are lesbian, gay, bisexual, transgender, or queer . __________ National is the national organization, which provides support to the __________ network of local chapters. __________ has nearly 400 chapters across the United States, with more than 200,000 members and supporters.

Exam Probability: **High**

56. *Answer choices:*

(see index for correct answer)

- a. VISTE
- b. PFLAG
- c. Toronto Public Space Committee
- d. Lifewater International

Guidance: level 1

:: Child safety ::

The term _________ is generally synonymous with international parental kidnapping, child snatching, and child stealing. However, the more precise legal usage of _________ originates in private international law and refers to the illegal removal of children from their home by an acquaintance or family member to a foreign country. In this context, "illegal" is normally taken to mean "in breach of custodial rights" and "home" is defined as the child`s habitual residence. As implied by the "breach of custodial rights," the phenomenon of _________ generally involves an illegal removal that creates a jurisdictional conflict of laws whereby multiple authorities and jurisdictions could conceivably arrive at seemingly reasonable and conflicting custodial decisions with geographically limited application. Such a result often strongly affects a child`s access and connection to half their family and may cause the loss of their former language, culture, name and nationality, it violates numerous children`s rights, and can cause severe psychological and emotional trauma to the child and family left behind.

Exam Probability: **High**

57. *Answer choices:*

(see index for correct answer)

- a. Child safety seat
- b. Childproofing
- c. Adam Walsh Child Protection and Safety Act
- d. Building Sites Bite

Guidance: level 1

:: Rape ::

A __________ is any drug that is an incapacitating agent which—when administered to another person—incapacitates the person and renders them vulnerable to a drug-facilitated sexual assault, including rape. One of the most common types of drug-facilitated sexual assaults are those in which a victim consumes a recreational drug such as alcohol that was administered surreptitiously. The other most common form of drug-facilitated sexual assault involves the non-surreptitiously-administered consumption of alcohol. Here, the victims in these cases are drinking voluntarily which then makes them unable to make informed decisions and/or give consent.

Exam Probability: **Medium**

58. *Answer choices:*

(see index for correct answer)

- a. Rape fantasy
- b. Date rape drug
- c. Rape crisis center
- d. Rape culture

Guidance: level 1

:: bad_topic ::

Elisabeth Lillian Wehner , better known as _________ , was an American author. She is best known for her 1943 novel A Tree Grows in Brooklyn, which became an immediate bestseller and is now considered one of the great American novels of the 20th century. The book was later adapted to the screen in the movie of the same title, directed by Elia Kazan, and starring Dorothy McGuire, Joan Blondell, James Dunn and Peggy Ann Garner.

Exam Probability: **High**

59. *Answer choices:*

(see index for correct answer)

- a. learned optimism
- b. James Farrer
- c. Betty Smith
- d. Masahiko Aoki

Guidance: level 1

Social institutions

Institutions are "stable, valued, recurring patterns of behavior", or mechanisms of social order, which govern the behaviour of a set of individuals within a given community. Institutions are identified with a social purpose, transcending individuals and intentions by mediating the rules that govern living behavior.

The term "institution" commonly applies to both informal institutions such as customs, or behavior patterns important to a society, and to particular formal institutions created by entities such as the government and public services. Primary or meta-institutions are institutions such as the family that are broad enough to encompass other institutions.

:: Free trade agreements ::

A _________ is a wide-ranging taxes, tariff and trade treaty that often includes investment guarantees. It exists when two or more countries agree on terms that helps them trade with each other. The most common _________ s are of the preferential and free trade types are concluded in order to reduce tariffs, quotas and other trade restrictions on items traded between the signatories.

Exam Probability: **Medium**

1. *Answer choices:*

(see index for correct answer)

- a. Commonwealth free trade
- b. Trade agreement
- c. Trade, Development and Cooperation Agreement
- d. Central Asian Union

Guidance: level 1

:: Sociological terminology ::

Living organisms including humans are _________ when they live collectively in interacting populations, whether they are aware of it, and whether the interaction is voluntary or involuntary.

Exam Probability: **Low**

2. *Answer choices:*

(see index for correct answer)

- a. Social
- b. Gemeinschaft
- c. Semi-periphery countries
- d. Social microcosm

Guidance: level 1

:: Family ::

Siblings play a unique role in one another's lives that simulates the companionship of parents as well as the influence and assistance of friends. Because siblings often grow up in the same household, they have a large amount of exposure to one another, like other members of the immediate family. However, though a _________ can have both hierarchical and reciprocal elements, this relationship tends to be more egalitarian and symmetrical than with family members of other generations. Furthermore, _________ s often reflect the overall condition of cohesiveness within a family.

Exam Probability: **High**

3. *Answer choices:*

(see index for correct answer)

- a. Extended family
- b. Charter of the Rights of the Family
- c. Grandchild
- d. Sibling relationship

Guidance: level 1

:: Price indices ::

A _________ measures changes in the price level of market basket of consumer goods and services purchased by households.

Exam Probability: **High**

4. *Answer choices:*

(see index for correct answer)

- a. Quality bias
- b. Lipstick index
- c. Internet Domain Name Index
- d. Christmas Price Index

Guidance: level 1

:: Philosophy of love ::

_________ is a Greco-Christian term referring to love, "the highest form of love, charity" and "the love of God for man and of man for God". The word is not to be confused with philia, brotherly love, or philautia, self-love, as it embraces a universal, unconditional love that transcends and persists regardless of circumstance. It goes beyond just the emotions to the extent of seeking the best for others. The noun form first occurs in the Septuagint, but the verb form goes as far back as Homer, translated literally as affection, as in "greet with affection" and "show affection for the dead". Other ancient authors have used forms of the word to denote love of a spouse or family, or affection for a particular activity, in contrast to eros .

Exam Probability: **High**

5. *Answer choices:*

(see index for correct answer)

- a. Courage
- b. Agape
- c. Philosophy of Love: A Partial Summing-Up
- d. Symposium

Guidance: level 1

:: Parent education program ::

_________ is a parent education program based on the Gordon Model by Thomas Gordon. Dr. Gordon taught the first P.E.T. course in 1962 and the courses proved to be so popular with parents that he began training instructors throughout the United States to teach it in their communities. Over the next several years, the course spread to all 50 states. In 1970, Dr. Gordon wrote the _________ book which gave many more parents access to this new parenting philosophy. As a result, people in many parts of the world became interested in making the course available in their countries. The book became a best-seller and was updated in 2000 revised book.

Exam Probability: **High**

6. *Answer choices:*

(see index for correct answer)

- a. Systematic Training for Effective Parenting
- b. Parent education program

Guidance: level 1

:: Sexual revolution ::

_________ is an American men`s lifestyle and entertainment magazine. It was founded in Chicago in 1953, by Hugh Hefner and his associates, and funded in part by a $1,000 loan from Hefner`s mother. Notable for its centerfolds of nude and semi-nude models , _________ played an important role in the sexual revolution and remains one of the world`s best-known brands, having grown into _________ Enterprises, Inc. , with a presence in nearly every medium. In addition to the flagship magazine in the United States, special nation-specific versions of _________ are published worldwide.

Exam Probability: **Medium**

7. *Answer choices:*

(see index for correct answer)

- a. Sexual revolution
- b. Peace movement
- c. Hippie
- d. Playboy

Guidance: level 1

:: Theories of history ::

Systems theory is the interdisciplinary study of systems. A system is a cohesive conglomeration of interrelated and interdependent parts that is either natural or man-made. Every system is delineated by its spatial and temporal boundaries, surrounded and influenced by its environment, described by its structure and purpose or nature and expressed in its functioning. In terms of its effects, a system can be more than the sum of its parts if it expresses synergy or emergent behavior. Changing one part of the system usually affects other parts and the whole system, with predictable patterns of behavior. For systems that are self-learning and self-adapting, the positive growth and adaptation depend upon how well the system is adjusted with its environment. Some systems function mainly to support other systems by aiding in the maintenance of the other system to prevent failure. The goal of systems theory is systematically discovering a system`s dynamics, constraints, conditions and elucidating principles that can be discerned and applied to systems at every level of nesting, and in every field for achieving optimized equifinality.

Exam Probability: **High**

8. *Answer choices:*

(see index for correct answer)

- a. apocalyptic
- b. Marxist
- c. System theory

Guidance: level 1

:: Market research ::

Demography is the statistical study of populations, especially human beings. As a very general science, it can analyze any kind of dynamic living population, i.e., one that changes over time or space . Demography encompasses the study of the size, structure, and distribution of these populations, and spatial or temporal changes in them in response to birth, migration, aging, and death. Based on the demographic research of the earth, earth`s population up to the year 2050 and 2100 can be estimated by demographers. _________ are quantifiable characteristics of a given population.

Exam Probability: **High**

9. *Answer choices:*

(see index for correct answer)

- a. Telescopic observations strategic framework

- b. Call to action
- c. Marketing research
- d. Demographics

Guidance: level 1

:: Teratogens ::

Fetal alcohol spectrum disorders are a group of conditions that can occur in a person whose mother drank alcohol during pregnancy. Problems may include an abnormal appearance, short height, low body weight, small head size, poor coordination, low intelligence, behavior problems, and problems with hearing or seeing. Those affected are more likely to have trouble in school, legal problems, participate in high-risk behaviors, and have trouble with alcohol or other drugs. The most severe form of the condition is known as _________ . Other types include partial _________ , alcohol-related neurodevelopmental disorder and alcohol-related birth defects . Some accept only FAS as a diagnosis, seeing the evidence as inconclusive with respect to other types.

Exam Probability: **Medium**

10. *Answer choices:*

(see index for correct answer)

- a. Fetal alcohol syndrome
- b. Fetal alcohol spectrum disorder
- c. Jervine
- d. Valproic acid

Guidance: level 1

:: Early childhood education in the United States ::

Head Start is a program of the United States Department of Health and Human Services that provides comprehensive early childhood education, health, nutrition, and parent involvement services to low-income children and their families. The program`s services and resources are designed to foster stable family relationships, enhance children`s physical and emotional well-being, and establish an environment to develop strong cognitive skills. The transition from preschool to elementary school imposes diverse developmental challenges that include requiring the children to engage successfully with their peers outside the family network, adjust to the space of a classroom, and meet the expectations the school setting provides.

Exam Probability: **Low**

11. *Answer choices:*

(see index for correct answer)

- a. Kent Place School
- b. Kindermusik
- c. Head Start Program
- d. The Mother Company

Guidance: level 1

:: Misogyny ::

The _________ or Taleban, who refer to themselves as the Islamic Emirate of Afghanistan , are a Sunni Islamic fundamentalist political movement and military organization in Afghanistan currently waging war within that country. Since 2016, the _________ `s leader is Mawlawi Hibatullah Akhundzada. The leadership is based in Quetta, Pakistan.

Exam Probability: **High**

12. *Answer choices:*

(see index for correct answer)

- a. Finkbeiner test
- b. Make me a sandwich
- c. Ernie Awards
- d. The First Blast of the Trumpet Against the Monstruous Regiment of Women

Guidance: level 1

:: Demography ::

_________ is the period of development towards an intimate relationship wherein a couple get to know each other and decide if there will be an engagement. A _________ may be an informal and private matter between two people or may be a public affair, or a formal arrangement with family approval. Traditionally, in the case of a formal engagement, it has been perceived that it is the role of a male to actively "court" or "woo" a female, thus encouraging her to understand him and her receptiveness to a proposal of marriage.

Exam Probability: **High**

13. *Answer choices:*

(see index for correct answer)

- a. Courtship
- b. Maternal death
- c. Infant mortality
- d. Preston curve

Guidance: level 1

:: Child care ::

_________ , otherwise known as day care, is the care and supervision of a child or multiple children at a time, whose ages range from six weeks to thirteen years. _________ is the action or skill of looking after children by a day-care center, nannies, babysitter, teachers or other providers. _________ is a broad topic that covers a wide spectrum of professionals, institutions, contexts, activities, and social and cultural conventions. Early _________ is an equally important and often overlooked component of child development. _________ providers can be children`s first teachers, and therefore play an integral role in systems of early childhood education. Quality care from a young age can have a substantial impact on the future successes of children. The main focus of childcare is on the development of the child, whether that be mental, social, or psychological.

Exam Probability: **High**

14. *Answer choices:*

(see index for correct answer)

- a. Primary caregiver
- b. Family child care
- c. Child care
- d. Childcare voucher scheme

Guidance: level 1

:: Social philosophy ::

The term " _________ " was coined by the American sociologist C. Wright Mills in his 1959 book The _________ to describe the type of insight offered by the discipline of sociology. The term is used in introductory textbooks in sociology to explain the nature of sociology and its relevance in daily life.

Exam Probability: **Low**

15. *Answer choices:*

(see index for correct answer)

- a. Avant-garde
- b. Sociological imagination
- c. Philosophy of social science
- d. Biological determination

Guidance: level 1

:: Economic sociology ::

_________ is the perception and actuality that one is cared for, has assistance available from other people, and most popularly, that one is part of a supportive social network. These supportive resources can be emotional , informational , or companionship ; tangible or intangible .

Exam Probability: **High**

16. *Answer choices:*

(see index for correct answer)

- a. Precariat
- b. Office humor
- c. Social support
- d. Economic sociology

Guidance: level 1

:: Sexual revolution ::

The __________ , also known as a time of sexual liberation, was a social movement that challenged traditional codes of behavior related to sexuality and interpersonal relationships throughout the United States and subsequently, the wider world, from the 1960s to the 1980s. Sexual liberation included increased acceptance of sex outside of traditional heterosexual, monogamous relationships . The normalization of contraception and the pill, public nudity, pornography, premarital sex, homosexuality, masturbation, alternative forms of sexuality, and the legalization of abortion all followed.

Exam Probability: **Medium**

17. *Answer choices:*

(see index for correct answer)

- a. Free love
- b. The personal is political
- c. Combined oral contraceptive pill

- d. Hair

Guidance: level 1

:: Pediatrics ::

__________ , also known as cot death or crib death, is the sudden unexplained death of a child of less than one year of age. Diagnosis requires that the death remains unexplained even after a thorough autopsy and detailed death scene investigation. SIDS usually occurs during sleep. Typically death occurs between the hours of 00:00 and 09:00. There is usually no noise or evidence of struggle.

Exam Probability: **High**

18. *Answer choices:*

(see index for correct answer)

- a. Child life specialist
- b. Pediatric stroke
- c. Sudden infant death syndrome
- d. Apgar score

Guidance: level 1

:: Rites of passage ::

A _________ is a ceremony or ritual of the passage which occurs when an individual leaves one group to enter another. It involves a significant change of status in society. In cultural anthropology the term is the Anglicisation of rite de passage, a French term innovated by the ethnographer Arnold van Gennep in his work Les rites de passage, "The Rites of Passage". The term is now fully adopted into anthropology as well as into the literature and popular cultures of many modern languages.

Exam Probability: **Medium**

19. *Answer choices:*

(see index for correct answer)

- a. Queima das Fitas
- b. Praxe
- c. Walkabout
- d. Rite of passage

Guidance: level 1

:: Obstetrics ::

The __________ , commonly called the bag of waters, sometimes the membranes, is the sac in which the fetus develops in amniotes. It is a thin but tough transparent pair of membranes that hold a developing embryo until shortly before birth. The inner of these fetal membranes, the amnion, encloses the amniotic cavity, containing the amniotic fluid and the fetus. The outer membrane, the chorion, contains the amnion and is part of the placenta. On the outer side, the __________ is connected to the yolk sac, the allantois and, via the umbilical cord, to the placenta.

Exam Probability: **Low**

20. *Answer choices:*

(see index for correct answer)

- a. Oocyte selection
- b. Vivette Glover
- c. Cervical mucus plug
- d. Amniotic sac

Guidance: level 1

:: Family ::

A _________ , blended family, or bonus family, is a family where at least one parent has children that are not genetically related to the other spouse or partner. Either parent, or both, may have children from previous relationships. Children in a _________ may live with one biological parent, or they may live with each biological parent for a period of time. In addition, visitation rights mean that children in stepfamilies often have contact with both biological parents, even if they permanently live with only one.

Exam Probability: **Medium**

21. *Answer choices:*

(see index for correct answer)

- a. Stepfather-in-law
- b. DINKY
- c. Stepfamily
- d. Family in advertising

Guidance: level 1

:: Family ::

_________ s are the parents of a person`s father or mother – paternal or maternal. Every sexually-reproducing living organism who is not a genetic chimera has a maximum of four genetic _________ s, eight genetic great-_________ s, sixteen genetic great-great- _________ s, thirty-two genetic great-great-great- _________ s, etc. In the history of modern humanity, around 30,000 years ago, the number of modern humans who lived to be a _________ increased. It is not known for certain what spurred this increase in longevity but largely results in the improved medical technology and living standard, but it is generally believed that a key consequence of three generations being alive together was the preservation of information which could otherwise have been lost; an example of this important information might have been where to find water in times of drought.

Exam Probability: **Medium**

22. *Answer choices:*

(see index for correct answer)

- a. Sole Survivor Policy
- b. Parents, Families and Friends of Lesbians and Gays
- c. Family caregivers
- d. Family cookbooks

Guidance: level 1

:: Derivatives (finance) ::

_________ is any bodily activity that enhances or maintains physical fitness and overall health and wellness. It is performed for various reasons, to aid growth and improve strength, preventing aging, developing muscles and the cardiovascular system, honing athletic skills, weight loss or maintenance, improving health and also for enjoyment. Many individuals choose to _________ outdoors where they can congregate in groups, socialize, and enhance well-being.

Exam Probability: **High**

23. *Answer choices:*

(see index for correct answer)

- a. Swap ratio
- b. Basis trading
- c. Exercise
- d. Accumulator

Guidance: level 1

:: Human ecology ::

_________ is an interdisciplinary and transdisciplinary study of the relationship between humans and their natural, social, and built environments. The philosophy and study of _________ has a diffuse history with advancements in ecology, geography, sociology, psychology, anthropology, zoology, epidemiology, public health, and home economics, among others.

Exam Probability: **Medium**

24. *Answer choices:*

(see index for correct answer)

- a. Human ecology
- b. Pollution haven hypothesis
- c. New Suburbanism
- d. Ecology and Society

Guidance: level 1

:: Divorce ::

__________ , also known as dissolution of marriage, is the process of terminating a marriage or marital union. __________ usually entails the canceling or reorganizing of the legal duties and responsibilities of marriage, thus dissolving the bonds of matrimony between a married couple under the rule of law of the particular country or state. __________ laws vary considerably around the world, but in most countries __________ requires the sanction of a court or other authority in a legal process, which may involve issues of distribution of property, child custody, alimony , child visitation / access, parenting time, child support, and division of debt. In most countries, monogamy is required by law, so __________ allows each former partner to marry another person; where polygyny is legal but polyandry is not, __________ allows the woman to marry another person.

Exam Probability: **Low**

25. *Answer choices:*

(see index for correct answer)

- a. Irreconcilable differences
- b. Dear John letter
- c. Divorce
- d. Divorce party

Guidance: level 1

:: Statistical terminology ::

The __________ is the value separating the higher half from the lower half of a data sample . For a data set, it may be thought of as the "middle" value. For example, in the data set , the __________ is 6, the fourth largest, and also the fourth smallest, number in the sample. For a continuous probability distribution, the __________ is the value such that a number is equally likely to fall above or below it.

Exam Probability: **High**

26. *Answer choices:*

(see index for correct answer)

- a. Median
- b. confounding
- c. Univariate
- d. Control group

Guidance: level 1

:: Marriage ::

A _________ is a document issued, either by a church or state authority, authorizing a couple to marry. The procedure for obtaining a license varies between countries and has changed over time. _________ s began to be issued in the Middle Ages, to permit a marriage which would otherwise be illegal .

Exam Probability: **Low**

27. *Answer choices:*

(see index for correct answer)

- a. Catherinette
- b. Marriage license
- c. Third-party custody
- d. Parenting plan

Guidance: level 1

:: Ethically disputed judicial practices ::

_________ or physical punishment is a punishment intended to cause physical pain on a person. It is most often practised on minors, especially in home and school settings. Common methods include spanking or paddling. It has also historically been used on adults, particularly on prisoners and enslaved people. Other common methods include flagellation and caning.

Exam Probability: **Medium**

28. *Answer choices:*

(see index for correct answer)

- a. Frisking
- b. Corporal punishment
- c. Bill of attainder
- d. death penalty

Guidance: level 1

:: Types of marriage ::

_________ is the practice of marrying within a specific social group, caste or ethnic group, rejecting those from others as unsuitable for marriage or other close personal relationships.

Exam Probability: **High**

29. *Answer choices:*

(see index for correct answer)

- a. Jumping the broom
- b. Open marriage
- c. Endogamy
- d. Transnational marriage

Guidance: level 1

:: Utilitarianism ::

_________ is a school of thought that argues that the pursuit of pleasure and intrinsic goods are the primary or most important goals of human life. A hedonist strives to maximize net pleasure . However upon finally gaining said pleasure, happiness may remain stationary.

Exam Probability: **Low**

30. *Answer choices:*

(see index for correct answer)

- a. The Methods of Ethics
- b. Hedonism
- c. Act utilitarianism
- d. The Theory of Good and Evil

Guidance: level 1

:: Marriage ::

A __________ , antenuptial agreement, or premarital agreement , commonly abbreviated as prenup, is a contract entered into prior to marriage, civil union, or any agreement prior to the main agreement by the people intending to marry or contract with each other. The content of a __________ can vary widely, but commonly includes provisions for division of property and spousal support in the event of divorce or breakup of marriage. They may also include terms for the forfeiture of assets as a result of divorce on the grounds of adultery; further conditions of guardianship may be included as well. It should not be confused with the historic marriage settlement which was concerned not primarily with the effects of divorce but with the establishment and maintenance of dynastic families.

Exam Probability: **High**

31. *Answer choices:*

(see index for correct answer)

- a. Mating
- b. Chattel marriage
- c. Prenuptial agreement
- d. Matrimonial website

Guidance: level 1

:: Social institutions ::

_________ is a marriage that takes place after a previous marital union has ended, as through divorce or widowhood.Some individuals are more likely to remarry than others; the likelihood can differ based on previous relationship status , level of interest in establishing a new romantic relationship, gender, race, and age among other factors. Those who choose not to remarry may prefer alternative arrangements like cohabitation or living apart together. _________ also provides mental and physical health benefits. However, although remarried individuals tend to have better health than individuals who do not repartner, they still generally have worse health than individuals who have remained continuously married. Someone who repeatedly remarries is referred to as a serial wedder.

Exam Probability: **Medium**

32. *Answer choices:*

(see index for correct answer)

- a. Remarriage
- b. Institutional abuse
- c. Institutional analysis
- d. Washington County Closed-Circuit Educational Television Project

Guidance: level 1

:: Marriage ::

__________ occurs when a married person engages in sexual activity with someone other than his or her spouse. From a different perspective, it also applies to a single person having sex with a married person. From a religious perspective, it could also have a third interpretation as referring to sex between people who are not in a conjugal relationship.

Exam Probability: **Low**

33. *Answer choices:*

(see index for correct answer)

- a. Widow
- b. Nubile
- c. Bridal theology
- d. Extramarital sex

Guidance: level 1

:: Interpersonal relationships ::

An intimate relationship is an interpersonal relationship that involves physical or emotional __________ . Although an intimate relationship is commonly a sexual relationship, it may also be a non-sexual relationship involving family, friends, or acquaintances.

Exam Probability: **Low**

34. *Answer choices:*

(see index for correct answer)

- a. social ties
- b. sexual relationship
- c. interpersonal relationship

Guidance: level 1

:: Types of marriage ::

_________ is a formal marriage or an informal union entered into by an individual before reaching a certain age, specified by several global organizations such as UNICEF as minors under the age of 18. The legally prescribed marriageable age in some jurisdictions is below 18 years, especially in the case of girls; and even when the age is set at 18 years, many jurisdictions permit earlier marriage with parental consent or in special circumstances, such as teenage pregnancy. In certain countries, even when the legal marriage age is 18, cultural traditions take priority over legislative law. _________ violates the rights of children; it affects both boys and girls, but it is more common among girls. _________ has widespread and long term consequences for child brides and grooms. According to several UN agencies, comprehensive sexuality education can prevent such a phenomenon.

Exam Probability: **Low**

35. *Answer choices:*

(see index for correct answer)

- a. Sororate marriage

- b. Civil marriage
- c. Transnational marriage
- d. Common-law marriage

Guidance: level 1

:: Sexualization ::

__________ is a person`s perception of the aesthetics or sexual attractiveness of their own body. It involves how a person sees themselves, compared to the standards that have been set by society. The Austrian neurologist and psychoanalyst Paul Schilder coined the phrase body-image in his book The Image and Appearance of the Human Body .Human society has at all times placed great value on beauty of the human body, but a person`s perception of their own body may not correspond to society`s standards.

Exam Probability: **High**

36. *Answer choices:*

(see index for correct answer)

- a. Body image
- b. Exploitation of women in mass media
- c. Sexuality in music videos
- d. Sexual objectification

Guidance: level 1

:: Philosophy of education ::

_________ is the process of facilitating learning, or the acquisition of knowledge, skills, values, beliefs, and habits. _________ al methods include storytelling, discussion, teaching, training, and directed research. _________ frequently takes place under the guidance of educators and also learners may also educate themselves. _________ can take place in formal or informal settings and any experience that has a formative effect on the way one thinks, feels, or acts may be considered _________ al. The methodology of teaching is called pedagogy.

Exam Probability: **High**

37. *Answer choices:*

(see index for correct answer)

- a. Percy Hughes
- b. Education
- c. Faculty of Arts
- d. Waldorf education

Guidance: level 1

:: Anti-aging substances ::

_________ is a hormone produced by the placenta after implantation. The presence of hCG is detected in some pregnancy tests . Some cancerous tumors produce this hormone; therefore, elevated levels measured when the patient is not pregnant may lead to a cancer diagnosis and, if high enough, paraneoplastic syndromes, however, it is not known whether this production is a contributing cause, or an effect of carcinogenesis. The pituitary analog of hCG, known as luteinizing hormone , is produced in the pituitary gland of males and females of all ages.

Exam Probability: **High**

38. *Answer choices:*

(see index for correct answer)

- a. Growth hormone
- b. Resveratrol
- c. Human chorionic gonadotropin
- d. FGF21

Guidance: level 1

:: Child abuse ::

_________ includes the physical, psychological, or sexual abuse of one sibling by another. A fourth category that brought attention from researchers during the first decade of the 21st century is sibling relational abuse, which derives from relational aggression.

Exam Probability: **Low**

39. *Answer choices:*

(see index for correct answer)

- a. Child neglect
- b. Psychosocial short stature
- c. Ed Block Courage Award
- d. Jersey child abuse investigation 2008

Guidance: level 1

:: Adoption ::

__________ is a process whereby a person assumes the parenting of another, usually a child, from that person`s biological or legal parent or parents. Legal __________ s permanently transfers all rights and responsibilities, along with filiation, from the biological parent or parents.

Exam Probability: **Medium**

40. *Answer choices:*

(see index for correct answer)

- a. Adoption
- b. Adoption law
- c. International adoption

- d. Adoption Covenant

Guidance: level 1

:: Bullying ::

__________ is a type of harassment technique that relates to a sexual nature and the unwelcome or inappropriate promise of rewards in exchange for sexual favors. __________ includes a range of actions from mild transgressions to sexual abuse or assault. Harassment can occur in many different social settings such as the workplace, the home, school, churches, etc. Harassers or victims may be of any gender.

Exam Probability: **High**

41. *Answer choices:*

(see index for correct answer)

- a. Shut up
- b. Relational aggression
- c. Sexual harassment
- d. Envy

Guidance: level 1

:: Matchmaking ::

_________ is a formalized matchmaking process which has the purpose of encouraging eligible singles to meet large numbers of new potential partners in a very short period of time. It was created in 1998 by Antony Beilinsohn after his rabbi, Yaacov Deyo, encouraged him to conceive of innovative ways for eligible Jewish singles to meet and marry.

Exam Probability: **High**

42. *Answer choices:*

(see index for correct answer)

- a. Dating agency
- b. Speed dating
- c. Robert Stewart Sparks
- d. Giggem

Guidance: level 1

:: Polyamory ::

_________ is the practice of, or desire for, intimate relationships with more than one partner, with the consent of all partners involved. It has been described as "consensual, ethical, and responsible non-monogamy". People who identify as polyamorous believe in an open relationship with a conscious management of jealousy; they reject the view that sexual and relational exclusivity are necessary for deep, committed, long-term loving relationships.

Exam Probability: **High**

43. *Answer choices:*

(see index for correct answer)

- a. Polyfidelity
- b. Paint Your Wagon
- c. Polyamory
- d. Primary and secondary

Guidance: level 1

:: Egalitarianism ::

__________ , or equalitarianism, is a school of thought within political philosophy that prioritizes equality for all people. Egalitarian doctrines are generally characterized by the idea that all human persons are equal in fundamental worth or moral status. According to the Merriam-Webster Dictionary, the term has two distinct definitions in modern English, namely either as a political doctrine that all people should be treated as equals and have the same political, economic, social and civil rights, or as a social philosophy advocating the removal of economic inequalities among people, economic __________ , or the decentralization of power. Some sources define __________ as the point of view that equality reflects the natural state of humanity.

Exam Probability: **High**

44. *Answer choices:*

(see index for correct answer)

- a. Neohumanism

- b. Equal opportunity
- c. Social pension
- d. Egalitarianism

Guidance: level 1

:: Social philosophy ::

__________ is a family of consequentialist ethical theories that promotes actions that maximize happiness and well-being for the majority of a population. Although different varieties of __________ admit different characterizations, the basic idea behind all of them is to in some sense maximize utility, which is often defined in terms of well-being or related concepts. For instance, Jeremy Bentham, the founder of __________ , described utility as

Exam Probability: **Low**

45. *Answer choices:*

(see index for correct answer)

- a. Golden Rule
- b. Subjective expected utility
- c. Utilitarianism
- d. Agency

Guidance: level 1

:: Marriage ::

_________ is the property of being from the same kinship as another person. In that aspect, _________ is the quality of being descended from the same ancestor as another person.

Exam Probability: **High**

46. *Answer choices:*

(see index for correct answer)

- a. Eternity ring
- b. Wedlease
- c. Consanguinity
- d. The History of Human Marriage

Guidance: level 1

:: Conceptions of self ::

The term "the looking glass self" was created by American sociologist Charles Horton Cooley in 1902, and introduced into his own work "Human Nature and the Social Order". It is described as, our reflection of how we think we appear to others. To further explain would be how oneself imagines how other view him/her. An example would be one`s mother would view their child as flawless, while another person would think differently. Cooley takes into account three steps when using "the looking glass self". Step one is how one imagines one looks to other people. Step two is how one imagines the judgment of others based on how one thinks they view them. Step three is how one thinks of how the person views them based on their previous judgments.

Exam Probability: **Medium**

47. *Answer choices:*

(see index for correct answer)

- a. Looking-glass self
- b. self-concept

Guidance: level 1

:: Demography ::

__________ services are defined as "educational, comprehensive medical or social activities which enable individuals, including minors, to determine freely the number and spacing of their children and to select the means by which this may be achieved". __________ may involve consideration of the number of children a woman wishes to have, including the choice to have no children, as well as the age at which she wishes to have them. These matters are influenced by external factors such as marital situation, career considerations, financial position, any disabilities that may affect their ability to have children and raise them, besides many other considerations. If sexually active, __________ may involve the use of contraception and other techniques to control the timing of reproduction. Other techniques commonly used include sexuality education, prevention and management of sexually transmitted infections, pre-conception counseling and management, and infertility management. __________ as defined by the United Nations and the World Health Organization encompasses services leading up to conception and does not promote abortion as a __________ method, although levels of contraceptive use reduces the need for abortion.

Exam Probability: **Medium**

48. *Answer choices:*

(see index for correct answer)

- a. Social determinants of health
- b. Maximum life span
- c. Family planning
- d. Human population control

Guidance: level 1

:: Bullying ::

__________ is the practice of singling out a person or group for unmerited blame and consequent negative treatment. __________ may be conducted by individuals against individuals , individuals against groups , groups against individuals , and groups against groups.

Exam Probability: **High**

49. *Answer choices:*

(see index for correct answer)

- a. SuEllen Fried
- b. Scapegoating
- c. Culture of fear
- d. Personal boundaries

Guidance: level 1

:: Family ::

Domestic violence is violence or other abuse by one person against another in a domestic setting, such as in marriage or cohabitation. It may be termed intimate partner violence when committed by a spouse or partner in an intimate relationship against the other spouse or partner, and can take place in heterosexual or same-sex relationships, or between former spouses or partners. Domestic violence can also involve violence against children, parents, or the elderly. It takes a number of forms, including physical, verbal, emotional, economic, religious, reproductive, and sexual abuse, which can range from subtle, coercive forms to marital rape and to violent physical abuse such as choking, beating, female genital mutilation, and acid throwing that results in disfigurement or death. Domestic murders include stoning, bride burning, honor killings, and dowry deaths.

Exam Probability: **Medium**

50. *Answer choices:*

(see index for correct answer)

- a. Family reunion
- b. Family secret
- c. Family Violence
- d. Family Equality Council

Guidance: level 1

:: Legal research ::

_________ is an abstract concept of management of complex systems according to a set of rules and trends. In systems theory, these types of rules exist in various fields of biology and society, but the term has slightly different meanings according to context. For example.

Exam Probability: **Low**

51. *Answer choices:*

(see index for correct answer)

- a. Computer-assisted legal research
- b. Regulation

Guidance: level 1

:: Social status ::

_________ theory is a sociological and psychological theory that studies the social behavior in the interaction of two parties that implement a cost-benefit analysis to determine risks and benefits. Also, the theory involves economic relationships, it occurs when each party have goods that the other parties value. _________ theory suggests that these calculations occur on romantic relationships, friendships, professional relationships and ephemeral relationships as simple as exchanging words with a customer at the cash register. _________ theory says that if the costs of the relationship are higher than the rewards, such as a lot of effort or money put into a relationship and not reciprocated, this could lead to issues.

Exam Probability: **Low**

52. *Answer choices:*

(see index for correct answer)

- a. Social mobility
- b. Social exchange
- c. Sponsored mobility
- d. caste system

Guidance: level 1

:: Cohort study methods ::

A _________ is a research design that involves repeated observations of the same variables over short or long periods of time . It is often a type of observational study, although they can also be structured as longitudinal randomized experiments.

Exam Probability: **Low**

53. *Answer choices:*

(see index for correct answer)

- a. Longitudinal study
- b. Retrospective cohort study
- c. Cohort study
- d. Prospective cohort study

Guidance: level 1

:: Casual sex ::

_________ is sexual activity that takes places outside a romantic relationship and implies an absence of commitment, emotional attachment, or familiarity between sexual partners. Examples are sexual activity while casually dating, one-night stands, extramarital sex, prostitution, or swinging.

Exam Probability: **High**

54. *Answer choices:*

(see index for correct answer)

- a. Dark room
- b. Sexual addiction
- c. Glory hole
- d. Casual sex

Guidance: level 1

:: Contraception for males ::

__________ is a surgical procedure for male sterilization or permanent contraception. During the procedure, the male vas deferens are cut and tied or sealed so as to prevent sperm from entering into the urethra and thereby prevent fertilization of a female through sexual intercourse. Vasectomies are usually performed in a physician`s office, medical clinic, or, when performed on an animal, in a veterinary clinic—hospitalization is not normally required as the procedure is not complicated, the incisions are small, and the necessary equipment routine.

Exam Probability: **Medium**

55. *Answer choices:*

(see index for correct answer)

- a. Vas-occlusive contraception
- b. Gamendazole
- c. Vasectomy
- d. Reversible inhibition of sperm under guidance

Guidance: level 1

:: Futurology ::

Systems theory is the interdisciplinary study of systems. A system is a cohesive conglomeration of interrelated and interdependent parts that is either natural or man-made. Every system is delineated by its spatial and temporal boundaries, surrounded and influenced by its environment, described by its structure and purpose or nature and expressed in its functioning. In terms of its effects, a system can be more than the sum of its parts if it expresses synergy or emergent behavior. Changing one part of the system usually affects other parts and the whole system, with predictable patterns of behavior. For systems that are self-learning and self-adapting, the positive growth and adaptation depend upon how well the system is adjusted with its environment. Some systems function mainly to support other systems by aiding in the maintenance of the other system to prevent failure. The goal of systems theory is systematically discovering a system`s dynamics, constraints, conditions and elucidating principles that can be discerned and applied to systems at every level of nesting, and in every field for achieving optimized equifinality.

Exam Probability: **High**

56. *Answer choices:*

(see index for correct answer)

- a. World Future Society
- b. Global brain
- c. Future car technologies
- d. Future Character of Conflict

Guidance: level 1

:: Criminology ::

In criminology, _________ proposes that exploiting the process of socialization and social learning builds self-control and reduces the inclination to indulge in behavior recognized as antisocial. It derives from functionalist theories of crime and was developed by Ivan Nye , who proposed that there were three types of control.

Exam Probability: **High**

57. *Answer choices:*

(see index for correct answer)

- a. Immigration and crime
- b. Aging offender
- c. Sociobiological theories of rape
- d. Sexual violence by intimate partners

Guidance: level 1

:: Family ::

A _________ is an occasion when many members of an extended family congregate. Sometimes reunions are held regularly, for example on the same date of every year.

Exam Probability: **High**

58. *Answer choices:*

(see index for correct answer)

- a. Foster parent
- b. Devrani
- c. Sibling-in-law
- d. Family reunion

Guidance: level 1

:: Semantics ::

__________ is a type of meaning in which a phrase, statement or resolution is not explicitly defined, making several interpretations plausible. A common aspect of __________ is uncertainty. It is thus an attribute of any idea or statement whose intended meaning cannot be definitively resolved according to a rule or process with a finite number of steps.

Exam Probability: **Medium**

59. *Answer choices:*

(see index for correct answer)

- a. Applied semantics
- b. Concept
- c. Ambiguity
- d. Q-based narrowing

Guidance: level 1

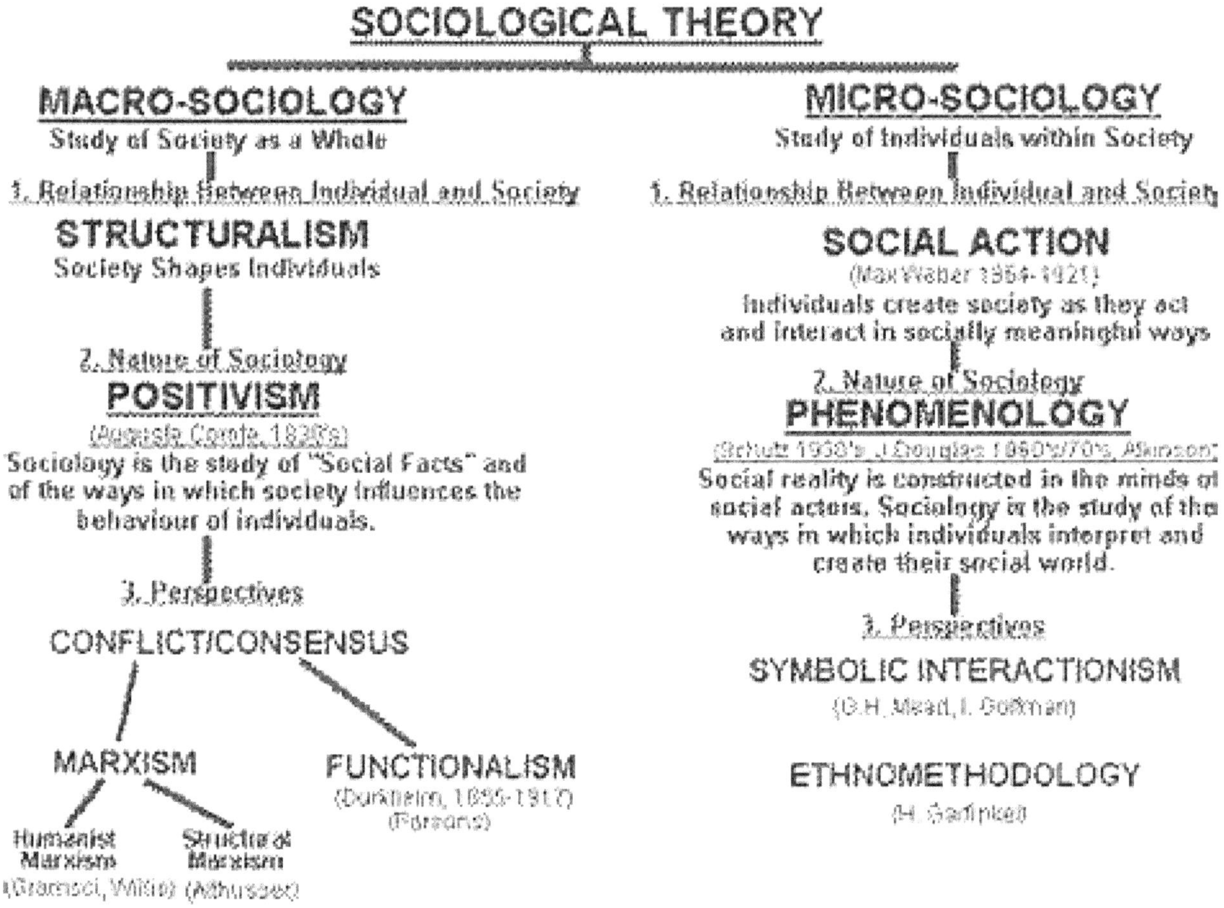

Social theories

Social theories are analytical frameworks, or paradigms, that are used to study and interpret social phenomena. A tool used by social scientists, social theories relate to historical debates over the validity and reliability of different methodologies (e.g. positivism and antipositivism), the primacy of either structure or agency, as well as the relationship between contingency and necessity. Social theory in an informal nature, or authorship based outside of academic social and political science, may be referred to as "social criticism" or "social commentary", or "cultural criticism" and may be associated both with formal cultural and literary scholarship, as well as other non-academic or journalistic forms of writing.

:: Individualism ::

__________ or personal freedoms are personal guarantees and freedoms that the government cannot abridge, either by law or by judicial interpretation, without due process. Though the scope of the term differs between countries, __________ may include the freedom of conscience, freedom of press, freedom of religion, freedom of expression, freedom of assembly, the right to security and liberty, freedom of speech, the right to privacy, the right to equal treatment under the law and due process, the right to a fair trial, and the right to life. Other __________ include the right to own property, the right to defend oneself, and the right to bodily integrity. Within the distinctions between __________ and other types of liberty, distinctions exist between positive liberty/positive rights and negative liberty/negative rights.

Exam Probability: **Low**

1. *Answer choices:*

(see index for correct answer)

- a. Viveza criolla
- b. Individualist anarchism in France
- c. Individualism Index
- d. Autonomy

Guidance: level 1

:: Critical theory ::

__________ is the reflective assessment and critique of society and culture by applying knowledge from the social sciences and the humanities. As a term, __________ has two meanings with different origins and histories: the first originated in sociology and the second originated in literary criticism, whereby it is used and applied as an umbrella term that can describe a theory founded upon critique; thus, the theorist Max Horkheimer described a theory as critical insofar as it seeks "to liberate human beings from the circumstances that enslave them."

Exam Probability: **Medium**

2. *Answer choices:*

(see index for correct answer)

- a. Non-simultaneity
- b. Critical theory
- c. Science wars
- d. Symbolic annihilation

Guidance: level 1

:: Sociobiology ::

__________ is a social system in which men hold primary power and predominate in roles of political leadership, moral authority, social privilege and control of property. Some patriarchal societies are also patrilineal, meaning that property and title are inherited by the male lineage.

Exam Probability: **Low**

3. *Answer choices:*

(see index for correct answer)

- a. The Origins of Virtue
- b. Patriarchy
- c. Eusociality
- d. Gyne

Guidance: level 1

:: History of international trade ::

__________ is a national economic policy that is designed to maximize the exports of a nation. __________ was dominant in modernized parts of Europe from the 16th to the 18th centuries before falling into decline, although some commentators argue that it is still practiced in the economies of industrializing countries in the form of economic interventionism.

Exam Probability: **Medium**

4. *Answer choices:*

(see index for correct answer)

- a. Trade coin
- b. North American fur trade

- c. The Imperialism of Free Trade
- d. Ancient Egyptian trade

Guidance: level 1

:: Social philosophy ::

The __________ is a school of social theory and critical philosophy associated with the Institute for Social Research, at Goethe University Frankfurt. Founded in the Weimar Republic , during the European interwar period , the __________ comprised intellectuals, academics, and political dissidents who were ill-fitted to the contemporary socio-economic systems of the 1930s. The Frankfurt theoreticians proposed that social theory was inadequate for explaining the turbulent political factionalism and reactionary politics occurring in ostensibly liberal capitalist societies in the 20th century. Critical of capitalism and of Marxism–Leninism as philosophically inflexible systems of social organisation, the School's critical theory research indicated alternative paths to realising the social development of a society and a nation.

Exam Probability: **Medium**

5. *Answer choices:*

(see index for correct answer)

- a. Frankfurt School
- b. Threat of force
- c. Marxist philosophy
- d. Ernest R. House

Guidance: level 1

:: Political science ::

_________ is the belief or attitude that individuals who form an elite—a select group of people with a certain ancestry, intrinsic quality, high intellect, wealth, special skills, or experience—are more likely to be constructive to society as a whole, and therefore deserve influence or authority greater than that of others.

Exam Probability: **Medium**

6. *Answer choices:*

(see index for correct answer)

- a. Ethnosymbolism
- b. Elitism
- c. Low politics
- d. Political history

Guidance: level 1

:: Critical theory ::

_________ is a field of critical theory that emerged in the early 1990s out of the fields of queer studies and women`s studies. _________ is often cast in the dual role of method and method`s foil. _________ includes both queer readings of texts and the theorization of `queerness` itself. Heavily influenced by the work of Lauren Berlant, Leo Bersani, Judith Butler, Lee Edelman, Jack Halberstam, and Eve Kosofsky Sedgwick, _________ builds both upon feminist challenges to the idea that gender is part of the essential self and upon gay/lesbian studies` close examination of the socially constructed nature of sexual acts and identities. Whereas gay/lesbian studies focused its inquiries into natural and unnatural behavior with respect to homosexual behavior, _________ expands its focus to encompass any kind of sexual activity or identity that falls into normative and deviant categories. Italian feminist and film theorist Teresa de Lauretis coined the term _________ for a conference she organized at the University of California, Santa Cruz in 1990 and a special issue of Differences: A Journal of Feminist Cultural Studies she edited based on that conference.

Exam Probability: **Medium**

7. *Answer choices:*

(see index for correct answer)

- a. Freudo-Marxism
- b. Subversion and containment
- c. Aesthetic Journalism
- d. Critical technical practice

Guidance: level 1

:: Welfare reform ::

_________ s are changes in the operation of a given welfare system, with the goals of reducing the number of individuals dependent on government assistance, keeping the welfare systems affordable, and assisting recipients become self-sufficient. Classical liberals, libertarians, and conservatives generally argue that welfare and other tax-funded services reduce incentives to work, exacerbate the free-rider problem, and intensify poverty. Socialists, on the other hand, generally criticize _________ because it usually minimizes the public safety net, and strengthens the capitalist economic system. _________ is constantly debated because of the varying opinions on the government's determined balance of providing guaranteed welfare benefits, and promoting self-sufficiency.

Exam Probability: **High**

8. *Answer choices:*

(see index for correct answer)

- a. New Deal
- b. Workfare in the United Kingdom
- c. Get Britain Working
- d. Welfare reform

Guidance: level 1

:: Socialism ::

A _________ is the lowest remuneration that employers can legally pay their workers—the price floor below which workers may not sell their labor. Most countries had introduced _________ legislation by the end of the 20th century.

Exam Probability: **Low**

9. *Answer choices:*

(see index for correct answer)

- a. To the Finland Station
- b. Social dividend
- c. Radical Society
- d. Labour voucher

Guidance: level 1

:: Political terminology ::

_________ is the transition to a more democratic political regime, including substantive political changes moving in a democratic direction. It may be the transition from an authoritarian regime to a full democracy, a transition from an authoritarian political system to a semi-democracy or transition from a semi-authoritarian political system to a democratic political system.

Exam Probability: **High**

10. *Answer choices:*

(see index for correct answer)

- a. Dirty tricks
- b. Oropolitics
- c. Political constitution
- d. Party line

Guidance: level 1

:: Liberalism ::

_________ is any effort to procure economic and social rights, political rights or equality, often for a specifically disenfranchised group, or more generally, in discussion of such matters. _________ stems from ex manus capere . Among others, Karl Marx discussed political _________ in his 1844 essay "On the Jewish Question", although often in addition to the term human _________ . Marx`s views of political _________ in this work were summarized by one writer as entailing "equal status of individual citizens in relation to the state, equality before the law, regardless of religion, property, or other `private` characteristics of individual people."

Exam Probability: **High**

11. *Answer choices:*

(see index for correct answer)

- a. John Stuart Mill Institute
- b. Commercial liberalism

- c. Habeas corpus
- d. Emancipation

Guidance: level 1

:: Queer studies ::

_________ : Feminism and the Subversion of Identity is a book by the philosopher Judith Butler, in which the author argues that gender is a kind of improvised performance. The work is influential in feminism, women`s studies, and lesbian and gay studies, and has also enjoyed widespread popularity outside of traditional academic circles. Butler`s ideas about gender came to be seen as foundational to queer theory and the advancing of dissident sexual practices during the 1990s.

Exam Probability: **Low**

12. *Answer choices:*

(see index for correct answer)

- a. Epistemology of the Closet
- b. Gender Trouble
- c. Chicago Gay and Lesbian Hall of Fame
- d. Undoing Gender

Guidance: level 1

:: Ethnic cleansing ::

_________ is the systematic forced removal of ethnic, racial and/or religious groups from a given territory by a more powerful ethnic group, often with the intent of making it ethnically homogeneous. The forces applied may be various forms of forced migration , intimidation, as well as genocide and genocidal rape.

Exam Probability: **Low**

13. *Answer choices:*

(see index for correct answer)

- a. Wola massacre
- b. Captivity of Kodavas at Seringapatam
- c. Cham issue
- d. Ethnic cleansing

Guidance: level 1

:: Prejudice and discrimination ::

__________ , also known as reservation in India and Nepal, positive discrimination / action in the United Kingdom, and employment equity in Canada and South Africa, is the policy of promoting the education and employment of members of groups that are known to have previously suffered from discrimination. Historically and internationally, support for __________ has sought to achieve goals such as bridging inequalities in employment and pay, increasing access to education, promoting diversity, and redressing apparent past wrongs, harms, or hindrances.

Exam Probability: **Medium**

14. *Answer choices:*

(see index for correct answer)

- a. Affirmative action
- b. The True-Born Englishman
- c. White supremacy
- d. Accents

Guidance: level 1

:: Structuralism ::

The __________ is a concept in the psychoanalytic theory of Jacques Lacan. The __________ is based on the belief that infants recognize themselves in a mirror or other symbolic contraption which induces apperception from the age of about six months.

Exam Probability: **Medium**

15. *Answer choices:*

(see index for correct answer)

- a. Aldo van Eyck
- b. Herman Hertzberger
- c. Mirror stage
- d. Alf Sommerfelt

Guidance: level 1

:: Theories of history ::

_________ is a theory and method of working-class self-emancipation. As a theory, it relies on a method of socioeconomic analysis that views class relations and social conflict using a materialist interpretation of historical development and takes a dialectical view of social transformation. It originates from the works of 19th-century German philosophers Karl Marx and Friedrich Engels.

Exam Probability: **High**

16. *Answer choices:*

(see index for correct answer)

- a. apocalyptic
- b. Marxist

- c. system theory

Guidance: level 1

:: Welfare reform ::

The _________ was a series of programs, public work projects, financial reforms, and regulations enacted by President Franklin D. Roosevelt in the United States between 1933 and 1936. It responded to needs for relief, reform, and recovery from the Great Depression. Major federal programs included the Civilian Conservation Corps , the Civil Works Administration , the Farm Security Administration , the National Industrial Recovery Act of 1933 and the Social Security Administration . They provided support for farmers, the unemployed, youth and the elderly. The _________ included new constraints and safeguards on the banking industry and efforts to re-inflate the economy after prices had fallen sharply. _________ programs included both laws passed by Congress as well as presidential executive orders during the first term of the presidency of Franklin D. Roosevelt.

Exam Probability: **Low**

17. *Answer choices:*

(see index for correct answer)

- a. Workfare
- b. Get Britain Working
- c. New Deal
- d. Welfare reform

Guidance: level 1

:: Frankfurt School ::

_________ : Reflections From Damaged Life is a 1951 book by the philosopher Theodor W. Adorno and a seminal text in critical theory. Adorno started writing it during World War II, in 1944, while he lived as an exile in America, and completed it in 1949. It was originally written for the fiftieth birthday of his friend and collaborator Max Horkheimer, who had co-authored the earlier book Dialectic of Enlightenment with Adorno.

Exam Probability: **Low**

18. *Answer choices:*

(see index for correct answer)

- a. One-Dimensional Man
- b. Arcades Project
- c. Minima Moralia
- d. Konrad Ott

Guidance: level 1

:: Consensus reality ::

__________ , in semiotics and postmodernism, is an inability of consciousness to distinguish reality from a simulation of reality, especially in technologically advanced postmodern societies. __________ is seen as a condition in which what is real and what is fiction are seamlessly blended together so that there is no clear distinction between where one ends and the other begins. It allows the co-mingling of physical reality with virtual reality and human intelligence with artificial intelligence . Individuals may find themselves, for different reasons, more in tune or involved with the hyperreal world and less with the physical real world. Some famous theorists of __________ /hyperrealism include Jean Baudrillard, Albert Borgmann, Daniel J. Boorstin, Neil Postman and Umberto Eco.

Exam Probability: **Medium**

19. *Answer choices:*

(see index for correct answer)

- a. Argumentum ad populum
- b. Perspectivism
- c. Common sense
- d. Hyperreality

Guidance: level 1

:: Liberal democracies ::

_________ , officially the Hellenic Republic , also known as Hellas , is a country located in Southern and Southeast Europe, with a population of approximately 11 million as of 2016. Athens is the nation`s capital and largest city, followed by Thessaloniki.

Exam Probability: **High**

20. *Answer choices:*

(see index for correct answer)

- a. South Africa
- b. Guyana
- c. Barbados
- d. Greece

Guidance: level 1

:: Critical theory ::

An _________ is a collection of normative beliefs and values that an individual or group holds for other than purely epistemic reasons. In other words, these rely on basic assumptions about reality that may or may not have any factual basis. The term is especially used to describe systems of ideas and ideals which form the basis of economic or political theories and resultant policies. In these there are tenuous causal links between policies and outcomes owing to the large numbers of variables available, so that many key assumptions have to be made. In political science the term is used in a descriptive sense to refer to political belief systems

Exam Probability: **Low**

21. *Answer choices:*

(see index for correct answer)

- a. Ideology
- b. Post-Critical
- c. Eugenio Donato
- d. Critical realism

Guidance: level 1

:: Critical theory ::

The _________ is an area in social life where individuals can come together to freely discuss and identify societal problems, and through that discussion influence political action. Such a discussion is called public debate and is defined as the expression of views on matters that are of concern to the public—often, but not always, with opposing or diverging views being expressed by participants in the discussion. Public debate takes place mostly through the mass media, but also at meetings or through social media, academic publications and government policy documents. The term was originally coined by German philosopher Jürgen Habermas who defined "the _________ as a virtual or imaginary community which does not necessarily exist in any identifiable space". Communication scholar Gerard A. Hauser defines it as "a discursive space in which individuals and groups associate to discuss matters of mutual interest and, where possible, to reach a common judgment about them". The _________ can be seen as "a theater in modern societies in which political participation is enacted through the medium of talk" and "a realm of social life in which public opinion can be formed".

Exam Probability: **Medium**

22. *Answer choices:*

(see index for correct answer)

- a. Ctheory
- b. Public sphere
- c. Eugenio Donato
- d. Neo-Gramscianism

Guidance: level 1

:: Structuralism ::

__________ is an approach to linguistics originating from the work of Swiss linguist Ferdinand de Saussure and is part of the overall approach of structuralism. __________ involves collecting a corpus of utterances and then attempting to classify all of the elements of the corpus at their different linguistic levels: the phonemes, morphemes, lexical categories, noun phrases, verb phrases, and sentence types.

Exam Probability: **Low**

23. *Answer choices:*

(see index for correct answer)

- a. Shadrach Woods
- b. Objet petit a

- c. Mythologies
- d. Structural linguistics

Guidance: level 1

:: International relations theory ::

_________ is the policy of a nation seeking to extend or retain its authority over other people or territories, generally with the aim of opening trade opportunities. The colonising country seeks to benefit from the colonised country or land mass. In the process, colonisers imposed their religion, economics, and medicinal practices on the natives. Some argue this was a positive move toward modernisation, while other scholars counter that this is an intrinsically Eurocentric rationalisation, given that modernisation is itself a concept introduced by Europeans. _________ is largely regarded as a relationship of domination of an indigenous majority by a minority of foreign invaders where the latter rule in pursuit of its interests.

Exam Probability: **High**

24. *Answer choices:*

(see index for correct answer)

- a. Foundationalism
- b. Hegemonic stability theory
- c. Politics Among Nations
- d. Colonialism

Guidance: level 1

:: Post-structuralism ::

_________ was either a continuation or a rejection of the intellectual project that preceded it—structuralism. Structuralism proposes that one may understand human culture by means of a structure—modeled on language —that differs from concrete reality and from abstract ideas—a "third order" that mediates between the two. Post-structuralist authors all present different critiques of structuralism, but common themes include the rejection of the self-sufficiency of structuralism and an interrogation of the binary oppositions that constitute its structures. Writers whose works are often characterised as post-structuralist include: Roland Barthes, Jacques Derrida, Michel Foucault, Gilles Deleuze, Judith Butler, Jean Baudrillard, Julia Kristeva, and Jürgen Habermas, although many theorists who have been called "post-structuralist" have rejected the label.

Exam Probability: **Low**

25. *Answer choices:*

(see index for correct answer)

- a. Post-structuralism
- b. Post-anarchism
- c. Abjection
- d. Fashionable Nonsense

Guidance: level 1

:: Structuralism ::

_________ is a compendium-like text by French semiotician Roland Barthes, originally published under the title of "Éléments de Sémiologie" in the French review Communications . The English translation by Annette Lavers and Colin Smith has been published independently as a short book.

Exam Probability: **High**

26. *Answer choices:*

(see index for correct answer)

- a. Alf Sommerfelt
- b. Giancarlo De Carlo
- c. Halen Estate
- d. Elements of Semiology

Guidance: level 1

:: Liberal democracies ::

_________ , formal name: the Kingdom of _________ , is a Scandinavian Nordic country in Northern Europe. It borders Norway to the west and north and Finland to the east, and is connected to Denmark in the southwest by a bridge-tunnel across the Öresund, a strait at the Swedish-Danish border. At 450,295 square kilometres , _________ is the largest country in Northern Europe, the third-largest country in the European Union and the fifth largest country in Europe by area. _________ has a total population of 10.2 million of which 2.5 million have a foreign background. It has a low population density of 22 inhabitants per square kilometre . The highest concentration is in the southern half of the country.

Exam Probability: **High**

27. *Answer choices:*

(see index for correct answer)

- a. Grenada
- b. Serbia
- c. Trinidad and Tobago
- d. Namibia

Guidance: level 1

:: Political theories ::

_________ or discursive democracy is a form of democracy in which deliberation is central to decision-making. It adopts elements of both consensus decision-making and majority rule. _________ differs from traditional democratic theory in that authentic deliberation, not mere voting, is the primary source of legitimacy for the law.

Exam Probability: **High**

28. *Answer choices:*

(see index for correct answer)

- a. Neo-Nazism
- b. Centrism

- c. Third International Theory
- d. Deliberative democracy

Guidance: level 1

:: Sociological terminology ::

__________ , a topic in the humanities and social sciences, is both a historical period , as well as the ensemble of particular socio-cultural norms, attitudes and practices that arose in the wake of the Renaissance—in the "Age of Reason" of 17th-century thought and the 18th-century "Enlightenment". Some commentators consider the era of __________ to have ended by 1930, with World War II in 1945, or the 1980s or 1990s; the following era is called post __________ . The term "contemporary history" is also used to refer to the post-1945 timeframe, without assigning it to either the modern or postmodern era.

Exam Probability: **Low**

29. *Answer choices:*

(see index for correct answer)

- a. Peer group
- b. Socialite
- c. Clown society
- d. Third culture kid

Guidance: level 1

:: Communism-based civil wars ::

The _________ , also known as the Second Indochina War, and in Vietnam as the Resistance War Against America or simply the American War, was an undeclared war in Vietnam, Laos, and Cambodia from 1 November 1955 to the fall of Saigon on 30 April 1975. It was the second of the Indochina Wars and was officially fought between North Vietnam and South Vietnam. North Vietnam was supported by the Soviet Union, China, and other communist allies; South Vietnam was supported by the United States, South Korea, the Philippines, Australia, Thailand and other anti-communist allies. The war is considered a Cold War-era proxy war from some US perspectives. It lasted some 19 years with direct U.S. involvement ending in 1973 following the Paris Peace Accords, and included the Laotian Civil War and the Cambodian Civil War, resulting in all three countries becoming communist states in 1975.

Exam Probability: **Medium**

30. *Answer choices:*

(see index for correct answer)

- a. Communist insurgency in the Philippines
- b. Insurgency of the Communist Party of the Philippines
- c. Cambodian Civil War

Guidance: level 1

:: Frankfurt School ::

_________ : Studies in the Ideology of Advanced Industrial Society is a 1964 book by the philosopher Herbert Marcuse, in which the author offers a wide-ranging critique of both contemporary capitalism and the Communist society of the Soviet Union, documenting the parallel rise of new forms of social repression in both these societies, as well as the decline of revolutionary potential in the West. He argues that "advanced industrial society" created false needs, which integrated individuals into the existing system of production and consumption via mass media, advertising, industrial management, and contemporary modes of thought.

Exam Probability: **High**

31. *Answer choices:*

(see index for correct answer)

- a. Felix Weil
- b. Richard Lehun
- c. One-Dimensional Man
- d. Alfred Schmidt

Guidance: level 1

:: Civil disobedience ::

A __________ is an act of voluntary and intentional abstention from using, buying, or dealing with a person, organization, or country as an expression of protest, usually for moral, social, political, or environmental reasons. The purpose of a __________ is to inflict some economic loss on the target, or to indicate a moral outrage, to try to compel the target to alter an objectionable behavior.

Exam Probability: **Low**

32. *Answer choices:*

(see index for correct answer)

- a. Boycott
- b. Gherao
- c. Henry David Thoreau
- d. Nikki Craft

Guidance: level 1

:: Sociological theories ::

__________ is the application of the evolutionary concept of natural selection to human society. The term itself emerged in the 1880s, and it gained widespread currency when used after 1944 by opponents of these ways of thinking. The majority of those who have been categorized as social Darwinists did not identify themselves by such a label.

Exam Probability: **High**

33. *Answer choices:*

(see index for correct answer)

- a. Political Parties
- b. Generationism
- c. Field theory
- d. Social Darwinism

Guidance: level 1

:: Consensus reality ::

In science and philosophy, a __________ is a distinct set of concepts or thought patterns, including theories, research methods, postulates, and standards for what constitutes legitimate contributions to a field.

Exam Probability: **High**

34. *Answer choices:*

(see index for correct answer)

- a. Common sense
- b. Paradigm
- c. World view
- d. Contextualism

Guidance: level 1

:: Devolution in the United Kingdom ::

_________ refers to a grant-in-aid of a specified amount from the Federal government of the United States to individual states and local governments to help support various broad purpose programs, such as law enforcement, social services, public health, and community development. _________ s have less oversight from the federal government and provide flexibility to each state in terms of designing and implementing programs. _________ s, categorical grants, and general revenue sharing are three types of federal government grants-in-aid programs.

Exam Probability: **High**

35. *Answer choices:*

(see index for correct answer)

- a. Scottish Government
- b. Commission on Scottish Devolution
- c. Barnett formula
- d. Wales Act 1978

Guidance: level 1

:: Social philosophy ::

In anthropology, _________ refers to the weakening of ties between a location and the cultural entities such as people, objects, languages, or traditions associated with that location.

Exam Probability: **Medium**

36. *Answer choices:*

(see index for correct answer)

- a. Deterritorialization
- b. Liberal eugenics
- c. Phenomenology
- d. Social democracy

Guidance: level 1

:: Divorce ::

_________ , also known as dissolution of marriage, is the process of terminating a marriage or marital union. _________ usually entails the canceling or reorganizing of the legal duties and responsibilities of marriage, thus dissolving the bonds of matrimony between a married couple under the rule of law of the particular country or state. _________ laws vary considerably around the world, but in most countries _________ requires the sanction of a court or other authority in a legal process, which may involve issues of distribution of property, child custody, alimony , child visitation / access, parenting time, child support, and division of debt. In most countries, monogamy is required by law, so _________ allows each former partner to marry another person; where polygyny is legal but polyandry is not, _________ allows the woman to marry another person.

Exam Probability: **High**

37. *Answer choices:*

(see index for correct answer)

- a. Divorce party
- b. Wakaresaseya
- c. Dear John letter
- d. Divorce

Guidance: level 1

:: Narcissism ::

_________ is the pursuit of gratification from vanity or egotistic admiration of one`s idealised self image and attributes. The term originated from Greek mythology, where the young Narcissus fell in love with his own image reflected in a pool of water. _________ is a concept in psychoanalytic theory, which was popularly introduced in Sigmund Freud`s essay On _________ . The American Psychiatric Association has listed the classification narcissistic personality disorder in its Diagnostic and Statistical Manual of Mental Disorders since 1968, drawing on the historical concept of megalomania.

Exam Probability: **Medium**

38. *Answer choices:*

(see index for correct answer)

- a. Self-righteousness
- b. Narcissism of small differences
- c. Narcissistic Personality Inventory
- d. Boasting

Guidance: level 1

:: Globalization-related theories ::

_________ is a broad movement that developed in the mid- to late 20th century across philosophy, the arts, architecture, and criticism and that marked a departure from modernism. The term has also more generally been applied to the historical era following modernity and the tendencies of this era.

Exam Probability: **Low**

39. *Answer choices:*

(see index for correct answer)

- a. Sociocultural
- b. Postmodernism
- c. Primitive accumulation of capital
- d. Economic development

Guidance: level 1

:: Marxist theory ::

_________ is a term coined by continental Europe socialists in the late 1930s that has come to refer to modern capitalism from World War 2 onward. Since 2016, the term has been used in the United States to refer to absurdities, crises, injustices, and inequality created by modern business development.

Exam Probability: **Medium**

40. *Answer choices:*

(see index for correct answer)

- a. Universal class
- b. Late capitalism

- c. Cultural Marxism
- d. Base and superstructure

Guidance: level 1

:: Freudo-Marxism ::

_________ : A Philosophical Inquiry into Freud is a book by the German philosopher and social critic Herbert Marcuse, in which the author proposes a non-repressive society, attempts a synthesis of the theories of Karl Marx and Sigmund Freud, and explores the potential of collective memory to be a source of disobedience and revolt and point the way to an alternative future. Its title alludes to Freud`s Civilization and Its Discontents . The 1966 edition has an added "political preface".

Exam Probability: **Low**

41. *Answer choices:*

(see index for correct answer)

- a. Eros and Civilization
- b. Shulamith Firestone
- c. Reimut Reiche

Guidance: level 1

:: Welfare state ::

The __________ is a form of government in which the state protects and promotes the economic and social well-being of the citizens, based upon the principles of equal opportunity, equitable distribution of wealth, and public responsibility for citizens unable to avail themselves of the minimal provisions for a good life. Sociologist T. H. Marshall described the modern __________ as a distinctive combination of democracy, welfare, and capitalism.

Exam Probability: **Low**

42. *Answer choices:*

(see index for correct answer)

- a. Welfare state
- b. Big government
- c. Cash transfers
- d. European social model

Guidance: level 1

:: Fair division ::

An __________ is a provision made in accordance with a legal framework of a society. Typically, __________ s are based on concepts of principle which are themselves based in concepts of social equality or enfranchisement. In recent decades the meaning of the word has been extended to encompass informal expectations of social relationships, social conventions and social norms which are considered unreasonable or unduly prescriptive upon others.

Exam Probability: **High**

43. *Answer choices:*

(see index for correct answer)

- a. Entitlement
- b. Spite
- c. Proportional
- d. Stromquist moving-knife procedure

Guidance: level 1

:: Sociological terminology ::

__________ is an awareness of shared interests, objectives, standards, and sympathies cause creating a psychological sense of unity of groups or classes. It refers to the ties in a society that bind people together as one. The term is generally employed in sociology, and the other social sciences as well as in philosophy, bioethics, and Catholic social teaching. It is also a significant concept in Catholic social teaching therefore it is a core concept in Christian democratic political ideology.

Exam Probability: **Medium**

44. *Answer choices:*

(see index for correct answer)

- a. Solidarity

- b. Emotive
- c. Famous for being famous
- d. Cultural baggage

Guidance: level 1

:: Islamic terrorism ::

_________ is a militant Sunni Islamist multi-national organization founded in 1988 by Osama bin Laden, Abdullah Azzam, and several other Arab volunteers during the Soviet invasion of Afghanistan.

Exam Probability: **Medium**

45. *Answer choices:*

(see index for correct answer)

- a. Ansar al-Jihad al-Alami
- b. Al-Qaeda
- c. Islamic Jihad for the Liberation of Palestine
- d. 2011 Abuja police headquarters bombing

Guidance: level 1

:: Casual sex ::

_________ involves the transfer of goods or services from one person or entity to another, often in exchange for money. A system or network that allows _________ is called a market.

Exam Probability: **Low**

46. *Answer choices:*

(see index for correct answer)

- a. Cruising for sex
- b. Trade
- c. One-night stand
- d. Anonymous sex

Guidance: level 1

:: Demography ::

_________ refers to the population shift from rural areas to urban areas, the gradual increase in the proportion of people living in urban areas, and the ways in which each society adapts to this change. It is predominantly the process by which towns and cities are formed and become larger as more people begin living and working in central areas. Although the two concepts are sometimes used interchangeably, _________ should be distinguished from urban growth: _________ is "the proportion of the total national population living in areas classed as urban", while urban growth refers to "the absolute number of people living in areas classed as urban". The United Nations projected that half of the world`s population would live in urban areas at the end of 2008. It is predicted that by 2050 about 64% of the developing world and 86% of the developed world will be urbanized. That is equivalent to approximately 3 billion urbanites by 2050, much of which will occur in Africa and Asia. Notably, the United Nations has also recently projected that nearly all global population growth from 2017 to 2030 will be by cities, about 1.1 billion new urbanites over the next 13 years.

Exam Probability: **Low**

47. *Answer choices:*

(see index for correct answer)

- a. Birth dearth
- b. Data Sharing for Demographic Research
- c. Urbanization
- d. Population ageing

Guidance: level 1

:: Sociological theories ::

_________ is a "condition in which society provides little moral guidance to individuals". This evolves from conflict of belief systems and causes breakdown of social bonds between an individual and the community . In a person this can progress into a dysfunctional ability to integrate within normative situations of their social world e.g., an unruly personal scenario that results in fragmentation of social identity and rejection of values.

Exam Probability: **Medium**

48. *Answer choices:*

(see index for correct answer)

- a. expert power
- b. Anomie
- c. power-control theory
- d. Preference theory

Guidance: level 1

:: Recycling ::

Originated by the philosopher Jacques Derrida, _________ is an approach to understanding the relationship between text and meaning. Derrida`s approach consisted of conducting readings of texts looking for things that run counter to the intended meaning or structural unity of a particular text. The purpose of _________ is to show that the usage of language in a given text, and language as a whole, are irreducibly complex, unstable, or impossible. Throughout his readings, Derrida hoped to show _________ at work.

Exam Probability: **High**

49. *Answer choices:*

(see index for correct answer)

- a. Deconstruction
- b. Urban lumberjacking
- c. Doy Bags
- d. Composition leather

Guidance: level 1

:: Gender studies ::

__________ is romantic attraction, sexual attraction, or sexual behavior toward both males and females, or to more than one sex or gender. It may also be defined as romantic or sexual attraction to people of any sex or gender identity, which is also known as pansexuality.

Exam Probability: **High**

50. *Answer choices:*

(see index for correct answer)

- a. Gender Studies and Human Rights Documentation Centre
- b. Bisexuality
- c. Bem Sex-Role Inventory

- d. Sally Gary

Guidance: level 1

:: Marxist theory ::

The _________ comprises those engaged in waged or salaried labour, especially in manual-labour occupations and industrial work. Working-class occupations include blue-collar jobs, some white-collar jobs, and most pink-collar jobs. Members of the _________ rely for their income exclusively upon their earnings from wage labour; thus, according to the more inclusive definitions, the category can include almost all of the working population of industrialized economies, as well as those employed in the urban areas of non-industrialized economies or in the rural workforce.

Exam Probability: **High**

51. *Answer choices:*

(see index for correct answer)

- a. Communist society
- b. Fictitious capital
- c. Commodity production
- d. Working class

Guidance: level 1

:: Globalization-related theories ::

__________ is the process in which a nation is being improved in the sector of the economic, political, and social well-being of its people. The term has been used frequently by economists, politicians, and others in the 20th and 21st centuries. The concept, however, has been in existence in the West for centuries. "Modernization, "westernization", and especially "industrialization" are other terms often used while discussing __________ . __________ has a direct relationship with the environment and environmental issues. __________ is very often confused with industrial development, even in some academic sources.

Exam Probability: **Medium**

52. *Answer choices:*

(see index for correct answer)

- a. post-industrial
- b. Industrialisation
- c. Sociocultural
- d. Economic development

Guidance: level 1

:: Political terminology ::

The term __________ in common usage refers to a tendency of people sharing a particular racial, religious, ethnic, social, or cultural identity to form exclusive political alliances, instead of engaging in traditional broad-based party politics, or promote their particular interests without regard for interests of a larger political group. In academic usage, the term has been used to refer to a wide range of political activities and theoretical analysis rooted in experiences of injustice shared by different social groups. In this usage, __________ typically aim to reclaim greater self-determination and political freedom for marginalized groups through understanding their distinctive nature and challenging externally imposed characterizations, instead of organizing solely around belief systems or party affiliations. Identity is used as a tool to articulate political claims, promote political ideologies, and guide political action with the aim of asserting group distinctiveness and gaining power and recognition in the context of perceived inequality or injustice.

Exam Probability: **Low**

53. *Answer choices:*

(see index for correct answer)

- a. Goulash Communism
- b. Salami tactics
- c. Identity politics
- d. Dirty tricks

Guidance: level 1

:: Authoritarianism ::

__________ is a form of radical, right-wing, authoritarian ultranationalism, characterized by dictatorial power, forcible suppression of opposition, and strong regimentation of society and of the economy, which came to prominence in early 20th-century Europe. The first fascist movements emerged in Italy during World War I before it spread to other European countries. Opposed to liberalism, Marxism, and anarchism, __________ is placed on the far-right within the traditional left–right spectrum.

Exam Probability: **High**

54. *Answer choices:*

(see index for correct answer)

- a. Fascism
- b. Anti-authoritarianism
- c. Totalitarianism
- d. Ingroup

Guidance: level 1

:: Sociology of culture ::

The term _________ has a range of meanings within the contexts of sociology, of political philosophy, and of colloquial use. In sociology and in everyday usage, it is a synonym for "ethnic pluralism", with the two terms often used interchangeably, for example, a cultural pluralism in which various ethnic groups collaborate and enter into a dialogue with one another without having to sacrifice their particular identities. It can describe a mixed ethnic community area where multiple cultural traditions exist or a single country within which they do . Groups associated with an aboriginal or autochthonous ethnic group and foreigner ethnic groups are often the focus.

Exam Probability: **Medium**

55. *Answer choices:*

(see index for correct answer)

- a. Symbolic boundaries
- b. Monoculturalism
- c. Sanankuya
- d. Collectivism

Guidance: level 1

:: Critical theory ::

_________ may be conceived as a universe of what is self-evident or given, a world that subjects may experience together. For Edmund Husserl, the _________ is the fundamental for all epistemological enquiries. The concept has its origin in biology and cultural Protestantism.

Exam Probability: **Low**

56. *Answer choices:*

(see index for correct answer)

- a. Lifeworld
- b. Arthur Kroker
- c. Epic and Novel
- d. Thing theory

Guidance: level 1

:: Social concepts ::

__________ are legal, social, or ethical principles of freedom or entitlement; that is, __________ are the fundamental normative rules about what is allowed of people or owed to people, according to some legal system, social convention, or ethical theory. __________ are of essential importance in such disciplines as law and ethics, especially theories of justice and deontology.

Exam Probability: **High**

57. *Answer choices:*

(see index for correct answer)

- a. Social dilemma
- b. Class consciousness

- c. Rights
- d. State of affairs

Guidance: level 1

:: Economic problems ::

A __________ is a metaphor used to represent an invisible barrier that keeps a given demographic from rising beyond a certain level in a hierarchy.

Exam Probability: **Low**

58. *Answer choices:*

(see index for correct answer)

- a. Glass ceiling
- b. Expenditure cascades
- c. Economic stagnation
- d. Affluenza

Guidance: level 1

:: Postmodern theory ::

__________ is the economic or cultural state or condition of society which is said to exist after modernity . Some schools of thought hold that modernity ended in the late 20th century – in the 1980s or early 1990s – and that it was replaced by __________ , while others would extend modernity to cover the developments denoted by __________ , while some believe that modernity ended after World War II. The idea of the post-modern condition is sometimes characterised as a culture stripped of its capacity to function in any linear or autonomous state like regressive isolationism, as opposed to the progressive mindstate of modernism.

Exam Probability: **Low**

59. *Answer choices:*

(see index for correct answer)

- a. Post-processual archaeology
- b. Christine Buci-Glucksmann
- c. Postcolonial literature
- d. Michael Joyce

Guidance: level 1

INDEX: Correct Answers

Foundations of Sociology

1. : Public policy

2. : Authority

3. : Self-esteem

4. a: India

5. b: Community

6. : Social isolation

7. d: Mother

8. c: Urbanization

9. b: Role

10. c: Perception

11. a: Vietnam War

12. d: Parent

13. b: Law enforcement

14. a: Organization

15. : Prison

16. c: Solidarity

17. a: Double standard

18. c: Soviet Union

19. c: Differential association

20. c: Ethics

21. : Taboo

22. d: Caste

23. c: Body image

24. b: Marxist

25. b: Analysis

26. : Observation

27. b: Mores

28. a: Mortality rate

29. c: Principle

30. : Goal

31. d: Medicalization

32. b: Childbirth

33. b: France

34. b: Social

35. c: Economic inequality

36. a: Poverty line

37. a: Patriarchy

38. d: Sexual violence

39. b: Afghanistan

40. a: Social work

41. a: Life chances

42. c: Endogamy

43. b: Resocialization

44. a: Socialism

45. b: Reinforcement

46. b: Medicaid

47. d: Torture

48. : Capitalism

49. a: Polygamy

50. a: Probation

51. : Christianity

52. b: Consciousness

53. c: Recidivism

54. : Achieved status

55. c: Daniel

56. a: High school

57. b: Homelessness

58. b: Ethnic group

59. a: Accountability

Globalization

1. c: Minimum wage

2. : World War II

3. d: Child care

4. d: Emancipation

5. d: Social services

6. : Ku Klux Klan

7. c: Social justice

8. : Economic development

9. b: Consumerism

10. b: Social change

11. d: Buddleja davidii

12. d: Equal opportunity

13. d: Deindustrialization

14. : Medicaid

15. a: Reconstruction Era

16. c: Unemployment

17. d: Immigration

18. b: Sociology

19. b: Novel ecosystem

20. a: Conservatives

21. : Desegregation

22. b: Interracial marriage

23. c: Taxes

24. c: Social construction

25. c: Resource distribution

26. a: Social structure

27. a: Merchant

28. d: Subsidy

29. d: Independence

30. : Melting pot

31. a: Affirmative action

32. a: Civil Rights

33. d: Middle class

34. a: New Deal

35. c: Nationalism

36. b: Property

37. b: Wage

38. c: Gender inequality

39. a: Hypothetico-deductive model

40. a: Colonialism

41. a: Hurricane Katrina

42. d: Discrimination

43. c: Peace

44. b: Industrialization

45. b: Conservation movement

46. c: World population

47. c: Ideologies

48. b: Biogeochemistry

49. : Resource

50. : Racial segregation

51. d: Racism

52. d: Indentured

53. d: Americanization

54. b: Islam

55. c: Multiculturalism

56. : Genocide

57. d: Trade

58. a: Adoption

59. : Self-determination

Ideologies

1. b: Italy

2. : Institute for Social Research

3. c: George Herbert Mead

4. c: Free trade

5. d: Social justice

6. b: Friendship

7. a: Afghanistan

8. : Castell

9. c: Sexual violence

10. b: Relations of production

11. b: Priest

12. c: Women only space

13. d: Public sphere

14. a: Impression management

15. c: On Religion

16. c: Ruling class

17. b: Interdependence

18. a: Religion

19. a: Contradiction

20. b: Theodor W. Adorno

21. b: Mexico

22. : Accountability

23. c: Civil religion

24. b: Islam

25. b: Shopping

26. b: Gender role

27. c: Working Class

28. c: China

29. c: India

30. : Solidarity

31. : Japan

32. : Communism

33. : Sex-positive feminism

34. b: Racism

35. : Economic mobility

36. : Consciousness

37. a: Diaspora

38. c: Ableism

39. b: Radical cheerleading

40. c: Actor

41. b: Positivism

42. : Development theory

43. b: Economic globalization

44. b: Neo-Marxism

45. c: Heterosexism

46. a: Hegemony

47. : Socialism

48. : Nazism

49. c: Feudalism

50. b: Social inequality

51. b: Mode of production

52. d: Immanuel

53. c: Traditional society

54. b: Queer theory

55. b: Veil

56. c: Divorce

57. d: Collective behavior

58. c: Gender inequality

59. a: Counter-Reformation

Social Issues

1. a: Homelessness

2. d: Debt crisis

3. d: Prostitution

4. a: Sexual desire

5. b: Social construction

6. b: Self-disclosure

7. a: Eating disorder

8. c: Demographics

9. b: National Crime Victimization Survey

10. : Aggression

11. c: Brown

12. : Amniocentesis

13. : Racial profiling

14. a: Homosexuality

15. : Incest

16. : Crisis

17. c: Fetal Alcohol Syndrome

18. a: Ethnic group

19. a: Threat

20. b: Restorative justice

21. c: Tobacco

22. d: Disease

23. b: Child development

24. b: Interdependence

25. a: Verbal abuse

26. : Stressor

27. c: Upper class

28. b: Income Tax

29. d: Dictatorship

30. a: Ascribed status

31. c: Columbine High School

32. a: Wealth

33. b: Student debt

34. : Greenpeace

35. d: Wage

36. c: Grameen Bank

37. d: Islamophobia

38. b: Criminology

39. b: Underclass

40. : Casual sex

41. : Adverse effect

42. c: Double standard

43. d: Androgyny

44. d: Systems theory

45. c: Industrialization

46. a: Drug Abuse

47. c: Same-sex marriage

48. a: Median

49. a: Prenatal

50. c: Sandwich generation

51. b: Stereotype

52. d: Feminization of poverty

53. a: Trade

54. : Sex education

55. : Activism

56. b: Castration

57. a: World War II

58. : Drug courts

59. a: Autonomy

Social change

1. b: JUSTICE

2. a: American Indian Movement

3. c: Ethnocentrism

4. : Corporate welfare

5. b: Child abuse

6. b: Statutory rape

7. b: CARE

8. d: Organized crime

9. : Breeder

10. : Critical race theory

11. d: Torture

12. b: Population transfer

13. a: Internment

14. : Zero population growth

15. c: Hobo

16. a: Leadership

17. : Sexual harassment

18. d: Centenarian

19. d: Hate speech

20. d: Crime prevention

21. : Resource mobilization

22. : Court

23. c: Social integration

24. c: Social justice

25. : Aryan Nations

26. b: Arab Spring

27. d: AIDES

28. d: Same-sex marriage

29. c: Imagination

30. a: Birth control

31. d: Public housing

32. a: Harassment

33. : Radical feminism

34. b: Vietnam War

35. : Recession

36. a: Non-governmental organization

37. b: Female infanticide

38. b: Corporate crime

39. c: Criminology

40. a: Femininity

41. b: Resource depletion

42. : Embezzlement

43. d: Economic globalization

44. b: Structural adjustment

45. b: Slavery

46. d: Racism

47. : Occupy Wall Street

48. a: Disobedience

49. b: Sexual abuse

50. c: Pornography

51. d: Biological warfare

52. : Ageism

53. d: Demographic transition

54. a: Violence against women

55. c: Migrant worker

56. b: Affirmative action

57. c: Wealth

58. : Circumcision

59. d: Fetal alcohol syndrome

Criminology

1. c: Legalization

2. b: CSI effect

3. c: Social structure

4. d: Smuggling

5. d: Mental disorder

6. a: Private prison

7. d: Duress

8. b: Prostitution

9. b: Assault

10. c: Terrorism

11. b: Defendant

12. : Rational choice

13. c: Patriarchy

14. : Public policy

15. c: Ideology

16. d: National Crime Victimization Survey

17. b: Corruption

18. : Jurisdiction

19. b: Child pornography

20. a: Homeland

21. d: Socioeconomics

22. b: Capitalism

23. a: Probation

24. d: Motion

25. c: Auburn system

26. c: Grand jury

27. d: Embezzlement

28. c: Workplace violence

29. d: Criminal justice

30. b: Phishing

31. d: Violent Crime Control and Law Enforcement Act

32. d: Social learning theory

33. : Reintegrative shaming

34. a: Civil liberties

35. b: Diversion program

36. d: Antisocial personality disorder

37. : Total institution

38. d: Voir dire

39. : Corpus delicti

40. : School violence

41. d: Adjudication

42. : Forgery

43. a: Rights

44. : Domestic violence

45. a: Postmodernism

46. d: Prosecutor

47. a: Security

48. d: Al-Qaeda

49. b: Genocide

50. d: Gun Control

51. b: Organized Crime Control Act

52. c: Status offense

53. a: Principle

54. d: Acquittal

55. a: Uniform

56. d: Due process

57. b: Robbery

58. a: Parole board

59. : Mass murder

Social work

1. d: Child abuse

2. c: Diagnostic and Statistical Manual of Mental Disorders

3. : Field research

4. b: Risk assessment

5. a: Harassment

6. d: Family Support Act

7. c: C. Henry Kempe

8. b: Descriptive statistics

9. d: Sectarian

10. c: Least restrictive environment

11. c: Ethics

12. d: Social class

13. a: Emotional intelligence

14. : Informed consent

15. a: Perception

16. b: Self-concept

17. c: Methamphetamine

18. a: Modernity

19. c: Sampling bias

20. c: Social constructionism

21. b: Facilitator

22. d: Organization

23. : Federalism

24. a: Black Church

25. : Tobacco

26. : New Deal

27. d: Ethical dilemma

28. d: Vigil

29. : Child neglect

30. d: Fordism

31. b: Attachment theory

32. c: Humanistic psychology

33. a: Minority group

34. c: Ethical theory

35. b: Case studies

36. d: International Association of Schools of Social Work

37. c: Substance Abuse and Mental Health Services Administration

38. c: Labeling theory

39. c: Participant observation

40. : Behavior modification

41. : Probation

42. c: Nonverbal communication

43. d: Date rape

44. d: Habitat for Humanity

45. b: Reparenting

46. c: Human services

47. c: Child Support

48. c: Indigenization

49. a: Crisis intervention

50. d: Service model

51. : Affordable housing

52. b: Preamble

53. c: Exhibitionism

54. d: Societies

55. a: Child Abuse Prevention and Treatment Act

56. b: PFLAG

57. : International child abduction

58. b: Date rape drug

59. c: Betty Smith

Social institutions

1. b: Trade agreement

2. a: Social

3. d: Sibling relationship

4. : Consumer price index

5. b: Agape

6. c: Parent Effectiveness Training

7. d: Playboy

8. c: System theory

9. d: Demographics

10. a: Fetal alcohol syndrome

11. c: Head Start Program

12. : Taliban

13. a: Courtship

14. c: Child care

15. b: Sociological imagination

16. c: Social support

17. : Sexual revolution

18. c: Sudden infant death syndrome

19. d: Rite of passage

20. d: Amniotic sac

21. c: Stepfamily

22. : Grandparent

23. c: Exercise

24. a: Human ecology

25. c: Divorce

26. a: Median

27. b: Marriage license

28. b: Corporal punishment

29. c: Endogamy

30. b: Hedonism

31. c: Prenuptial agreement

32. a: Remarriage

33. d: Extramarital sex

34. d: Intimacy

35. : Child marriage

36. a: Body image

37. b: Education

38. c: Human chorionic gonadotropin

39. : Sibling abuse

40. a: Adoption

41. c: Sexual harassment

42. b: Speed dating

43. c: Polyamory

44. d: Egalitarianism

45. c: Utilitarianism

46. c: Consanguinity

47. a: Looking-glass self

48. c: Family planning

49. b: Scapegoating

50. c: Family Violence

51. b: Regulation

52. b: Social exchange

53. a: Longitudinal study

54. d: Casual sex

55. c: Vasectomy

56. : Interdependence

57. : Social control theory

58. d: Family reunion

59. c: Ambiguity

Social theories

1. : Civil liberties

2. b: Critical theory

3. b: Patriarchy

4. : Mercantilism

5. a: Frankfurt School

6. b: Elitism

7. : Queer theory

8. d: Welfare reform

9. : Minimum wage

10. : Democratization

11. d: Emancipation

12. b: Gender Trouble

13. d: Ethnic cleansing

14. a: Affirmative action

15. c: Mirror stage

16. d: Marxism

17. c: New Deal

18. c: Minima Moralia

19. d: Hyperreality

20. d: Greece

21. a: Ideology

22. b: Public sphere

23. d: Structural linguistics

24. d: Colonialism

25. a: Post-structuralism

26. d: Elements of Semiology

27. : Sweden

28. d: Deliberative democracy

29. : Modernity

30. d: Vietnam War

31. c: One-Dimensional Man

32. a: Boycott

33. d: Social Darwinism

34. b: Paradigm

35. : Block grant

36. a: Deterritorialization

37. d: Divorce

38. : Narcissism

39. b: Postmodernism

40. b: Late capitalism

41. a: Eros and Civilization

42. a: Welfare state

43. a: Entitlement

44. a: Solidarity

45. b: Al-Qaeda

46. b: Trade

47. c: Urbanization

48. b: Anomie

49. a: Deconstruction

50. b: Bisexuality

51. d: Working class

52. d: Economic development

53. c: Identity politics

54. a: Fascism

55. : Multiculturalism

56. a: Lifeworld

57. c: Rights

58. a: Glass ceiling

59. : Postmodernity

CPSIA information can be obtained
at www.ICGtesting.com
Printed in the USA
LVHW011538301019
635718LV00004B/524/P

9 781538 851098